Anaerobic Parasitic Protozoa

Genomics and Molecular Biology

Edited by

C. Graham Clark

Department of Infectious and Tropical Diseases
London School of Hygiene and Tropical Medicine
London
UK

Patricia J. Johnson

Department of Microbiology, Immunology and Molecular Genetics
University of California
Los Angeles
USA

Rodney D. Adam

Departments of Medicine and Immunobiology
University of Arizona College of Medicine
Tucson, AZ
USA

Caister Academic Press

Caister Academic Press
Norfolk, UK

www.caister.com

British Library Cataloguing-in-Publication Data
A catalogue record for this book is available from the British Library

ISBN: 978-1-904455-61-5

Cover image adapted from Figure 2.2.

Printed and bound in Great Britain

Contents

	List of Contributors	v
	Preface	ix
1	*Entamoeba histolytica*: Genome Status and Web Resources Elisabet Caler and Hernan Lorenzi	1
2	The Genome of *Giardia* and other Diplomonads Jan O. Andersson, Jon Jerlström-Hultqvist and Staffan G. Svärd	23
3	The Genome of *Trichomonas vaginalis* Jane M. Carlton, Shehre-Banoo Malik, Steven A. Sullivan, Thomas Sicheritz-Ponten, Petrus Tang and Robert P. Hirt	45
4	Dissecting the Actin Cytoskeleton of *Entamoeba histolytica* from a Genomic Perspective Chung Chau Hon, Kumiko Nakada-Tsukui, Tomoyoshi Nozaki and Nancy Guillén	81
5	New Insights into the Composition and Function of the Cytoskeleton in *Giardia lamblia* and *Trichomonas vaginalis* Heidi G. Elmendorf, Richard D. Hayes, Shweta Srivastava and Patricia J. Johnson	119
6	*Entamoeba histolytica* Genomic Analyses Upinder Singh and Gretchen M. Ehrenkaufer	157
7	Manipulation of Gene Expression in *Entamoeba histolytica* Rivka Bracha, Nancy Guillén and David Mirelman	175
8	*Giardia*: Nuclear and Chromosomal Structure and Replication Rodney D. Adam and Staffan G. Svärd	193
9	The *Blastocystis* Mitochondrion-like Organelles Anastasios D. Tsaousis, Alexandra Stechmann, Karleigh A. Hamblin, Mark van der Giezen, Vicente Pérez-Brocal and C. Graham Clark	205
	Index	221

Other Books of Interest

Salmonella: From Genome to Function	2011
Insect Virology	2010
Environmental Microbiology: Current Technology and Water Applications	2010
Sensory Mechanisms in Bacteria: Molecular Aspects of Signal Recognition	2010
Bifidobacteria: Genomics and Molecular Aspects	2010
Molecular Phylogeny of Microorganisms	2010
Nanotechnology in Water Treatment Applications	2010
Iron Uptake and Homeostasis in Microorganisms	2010
Caliciviruses: Molecular and Cellular Virology	2010
Epstein-Barr Virus: Latency and Transformation	2010
Lentiviruses and Macrophages: Molecular and Cellular Interactions	2010
RNA Interference and Viruses: Current Innovations and Future Trends	2010
Retroviruses: Molecular Biology, Genomics and Pathogenesis	2010
Frontiers in Dengue Virus Research	2010
Metagenomics: Theory, Methods and Applications	2010
Microbial Population Genetics	2010
Borrelia: Molecular Biology, Host Interaction and Pathogenesis	2010
Aspergillus: Molecular Biology and Genomics	2010
Environmental Molecular Microbiology	2010
Neisseria: Molecular Mechanisms of Pathogenesis	2010
Lab-on-a-Chip Technology: Biomolecular Separation and Analysis	2009
Lab-on-a-Chip Technology: Fabrication and Microfluidics	2009
Real-Time PCR: Current Technology and Applications	2009
ABC Transporters in Microorganisms	2009
Pili and Flagella: Current Research and Future Trends	2009
Bacterial Polysaccharides: Current Innovations and Future Trends	2009
Microbial Toxins: Current Research and Future Trends	2009
Acanthamoeba: Biology and Pathogenesis	2009
Bacterial Secreted Proteins: Secretory Mechanisms and Role in Pathogenesis	2009
Lactobacillus Molecular Biology: From Genomics to Probiotics	2009
Mycobacterium: Genomics and Molecular Biology	2009
Clostridia: Molecular Biology in the Post-genomic Era	2009
Plant Pathogenic Bacteria: Genomics and Molecular Biology	2009
Microbial Production of Biopolymers and Polymer Precursors	2009
Epigenetics	2008
RNA and the Regulation of Gene Expression	2008

Contributors

Rodney D. Adam
Departments of Medicine and Immunobiology
University of Arizona College of Medicine
Tucson, AZ
USA

adamr@u.arizona.edu

Jan O. Andersson
Department of Evolution, Genomics and Systematics
EBC
Uppsala University
Uppsala
Sweden

jan.andersson@ebc.uu.se

Rivka Bracha
Weizmann Institute of Science
Department of Biological Chemistry
Rehovot
Israel

rivka.bracha@weizmann.ac.il

Elisabet Caler
Informatics Department
J. Craig Venter Institute
9704 Medical Center Drive
Rockville, MD
USA

ecaler@jcvi.org

Jane M. Carlton
Department of Medical Parasitology
New York University Langone Medical Center
New York, NY
USA

jane.carlton@nyumc.org

C. Graham Clark
Department of Infections and Tropical Diseases
London School of Hygiene and Tropical Medicine
London
UK

graham.clark@lshtm.ac.uk

Gretchen M. Ehrenkaufer
Division of Infectious Diseases
Department of Internal Medicine
Stanford University
Stanford, CA
USA

gehrenk@stanford.edu

Heidi G. Elmendorf
Department of Biology
Georgetown University
Washington, DC
USA

hge@georgetown.edu

Mark van der Giezen
Centre for Eukaryotic Evolutionary Microbiology
School of Biosciences
University of Exeter
Exeter
UK

m.vandergiezen@exeter.ac.uk

Nancy Guillén
Unité de Biologie Cellulaire du Parasitisme
INSERM U786
Institut Pasteur
Paris
France

nguillen@pasteur.fr

Karleigh A. Hamblin
Current address:
DSTL
Porton Down
Salisbury
Wiltshire
UK

kahamblin@dstl.gov.uk

Richard D. Hayes
Molecular Biology Institute
UCLA
Los Angeles, CA
USA

rdhayes@ucla.edu

Robert P. Hirt
Institute for Cell and Molecular Biosciences
Newcastle University
Newcastle upon Tyne
UK

r.p.hirt@ncl.ac.uk

Chung Chau Hon
Unité de Biologie Cellulaire du Parasitisme
INSERM U786
Institut Pasteur
Paris
France

bigchung@pasteur.fr

Jon Jerlström-Hultqvist
Department of Cell and Molecular Biology
BMC
Uppsala University
Uppsala
Sweden

jon.jerlstromhultqvist@icm.uu.se

Patricia J. Johnson
Department of Microbiology, Immunology and Molecular Genetics
UCLA
Los Angeles, CA
USA

johnsonp@ucla.edu

Hernan Lorenzi
Informatics Department
J. Craig Venter Institute
Rockville, MD
USA

hlorenzi@jcvi.org

Shehre-Banoo Malik
Department of Medical Parasitology
New York University Langone Medical Center
New York, NY
USA

shehre-banoo.malik@nyumc.org

David Mirelman
Weizmann Institute of Science
Department of Biological Chemistry
Rehovot
76100
Israel

david.mirelman@weizmann.ac.il

Kumiko Nakada-Tsukui
National Institute of Infectious Diseases
Department of Parasitology
Shinjuku-ku
Tokyo
Japan

kumiko@nih.go.jp

Tomoyoshi Nozaki
National Institute of Infectious Diseases
Department of Parasitology
Shinjuku-ku
Tokyo
Japan

nozaki@nih.go.jp

Vicente Pérez-Brocal
Department of Infections and Tropical Diseases
London School of Hygiene and Tropical Medicine
London
UK

perez_vicbro@gva.es

Thomas Sicheritz-Ponten
Center for Biological Sequence Analysis
Department of Systems Biology
Technical University of Denmark
Lyngby
DK

thomas@cbs.dtu.dk

Upinder Singh
Division of Infectious Diseases
Department of Internal Medicine
Stanford University
Stanford, CA
USA

usingh@stanford.edu

Shweta Srivastava
Department of Biology
Georgetown University
Washington, DC
USA

ss767@georgetown.edu

Alexandra Stechmann
Centre for Comparative Genomics and Evolutionary Bioinformatics
Department of Biochemistry and Molecular Biology
Dalhousie University
Halifax
Canada

astechmann@gmail.com

Steven A. Sullivan
Department of Medical Parasitology
New York University Langone Medical Center
New York, NY
USA

steven.sullivan@nyumc.org

Staffan G. Svärd
Department of Cell and Molecular Biology
BMC
Uppsala University
Uppsala
Sweden

staffan.svard@icm.uu.se

Petrus Tang
Bioinformatics Center
Chang Gung University
Taoyuan
Taiwan

petang@mail.cgu.edu.tw

Anastasios D. Tsaousis
Centre for Comparative Genomics and Evolutionary Bioinformatics
Department of Biochemistry and Molecular Biology
Dalhousie University
Halifax
Canada

anastasios.tsaousis@dal.ca

Preface

From the outside, the inclusion of *Entamoeba, Giardia* and *Trichomonas* in the same volume may seem odd as the organisms are very different from each other in many ways. Two are flagellates, one an amoeba, and they variously live in the urogenital tract, the small bowel and the large bowel. However, they all share the characteristic of being anaerobic, or at best microaerotolerant, unicellular eukaryotes and parasitic in humans. They also grow in axenic culture in very similar media and, historically, there has been a lot of cross-fertilization between the research communities that study the three organisms. In many ways this is due to the influence of one man, Louis S. Diamond, who sadly died while this volume was in production. He was largely responsible for developing the media in which these organisms grow. In addition, a number of the authors in the current volume have benefited from his expertise and generosity either by working with him directly or by being trained by the many 'offspring' of his small laboratory at the National Institute of Health in Bethesda, Maryland, USA, where he spent almost all of his professional career. Without the ability to grow these organisms in axenic culture almost none of the work presented in these nine chapters would have been possible, and we dedicate this volume to his memory.

It is fortuitous that the *Entamoeba, Giardia* and *Trichomonas* genome projects all came to fruition within a relatively short space of time. While the fine-tuning of the genome sequences and the sequence analyses will continue for some time, this is clearly a good time to take stock of what we have learned to date. The first three chapters summarize the genome projects and many of the features of the data they generated. The first chapter by Caler and Lorenzi gives an overview of the *Entamoeba histolytica* genome project as well as comparative analyses that have been performed with the genomes of two additional species, *E. dispar* and *E. invadens.* Andersson *et al.* describe the *Giardia* genome project in Chapter 2, bringing the story up to date with preliminary data from a second *Giardia* genome and from the related organism *Spironucleus.* In Chapter 3, Carlton *et al.* describe the trials that faced those trying to sequence the *Trichomonas* genome, which turned out to be 10-fold larger than anticipated. The chapters also describe the various databases that contain the genome information and how the data can be accessed.

All three organisms are highly motile, but use dramatically different mechanisms of locomotion. The next two chapters describe insights that have been obtained from the genome sequences by those with a particular interest in the cytoskeleton of these organisms. *Entamoeba* has an actin-based cytoskeleton, and in Chapter 4 Hon *et al.* investigate its organization and function through mining the genome sequence data, and their comparative analyses reveal both similarities and significant differences between the *Entamoeba* cytoskeleton and those studied in other eukaryotic cells. Microtubule-based cytoskeletons characterize both *Giardia* and *Trichomonas,* and in Chapter 5 Elmendorf *et al.* describe the cytoskeletal components identified in the two genomes as well as comparing the organisms with each other.

Post-genomic analyses take many forms, some of which are described in the next three chapters. In Chapter 6, Singh and Ehrenkaufer review the use to date of two types of genome-dependent methodologies to study *Entamoeba* – microarrays and proteomic analyses. These are being employed in a number of ways but the initial focus, not surprisingly, is on linking genes, gene expression and protein expression to differences in virulence between strains of *E. histolytica* and related species. Whole-genome analyses need to be complemented by single-gene analysis techniques, and these are reviewed in Chapter 7 by Bracha *et al.* Unlike many eukaryotes, including *Trichomonas, E. histolytica* has been reluctant to allow genes to be 'knocked out' by homologous recombination. As a result, a range of alternative techniques have had to be employed including a unique method of epigenetic gene silencing, which is among the techniques described here.

Ultimately, the genome sequences need to lead back to the biology of the whole cell. Some such studies are discussed in the previous chapters, but in Chapter 8 Adam and Svärd describe aspects of the unique nuclear and chromosomal organisation of *Giardia.* Having two nuclei, the relationship and interplay between the two has intrigued researchers for many years and our current understanding is reviewed here, along with a discussion of whether *Giardia* is clonal or sexual.

The book concludes with a chapter that at first sight seems to be completely unrelated to the rest. In Chapter 9, Tsaousis *et al.* describe our current understanding of the mitochondrion-like organelles of *Blastocystis.* While *Blastocystis* is unrelated to the organisms considered in the preceding chapters and does not yet have a nuclear genome sequence available for study, it does share some important characteristics with *Entamoeba, Giardia* and *Trichomonas* – it is an anaerobic parasitic unicellular eukaryote. The other three organisms all contain highly derived mitochondria that appear to have arisen independently, known as mitosomes in *Entamoeba* and *Giardia* and hydrogenosomes in *Trichomonas.* The mitochondrion-like organelles of *Blastocystis* are much less derived and give us an insight into what an intermediate evolutionary stage may have looked like.

The published genome sequences of *Entamoeba, Giardia* and *Trichomonas* represent the end of the beginning, and we are now entering the second phase of genome analyses. With recent developments in next-generation sequencing we can expect a rapid accumulation of genome sequences from additional strains and new species of these organisms over the next few years; indeed, this is already under way. Translating the huge amount of new data into a greater understanding of the biology and evolution of these organisms and the diseases they cause will be a challenge, and we hope that the overview of genome studies in these organisms provided by this book will inspire others to join us in this quest!

Entamoeba histolytica: Genome Status and Web Resources

1

Elisabet Caler and Hernan Lorenzi

Abstract

Entamoeba histolytica is the causative agent of amoebic dysentery and liver abscess that affect millions of people annually worldwide resulting in tens of thousands deaths. Loftus *et al.* published an annotation of the draft genome assembly in 2005. However, subsequent identification of assembly artefacts and the availability of additional sequence and functional data made necessary a thorough revision of its entire sequence and annotation. The genome of *E. histolytica* has now been re-assembled and re-annotated, incorporating significant structural and functional modifications to existing gene models. The new 20 million basepair genome assembly contains 8201 predicted genes; known and novel transposable elements have been mapped and characterized, functional assignments have been revised and updated, and additional information has been incorporated, including metabolic pathways, Gene Ontology assignments, curation of transporters, and generation of gene families. All information is available through GenBank and Pathema, a web resource that unifies and displays up to date genome information. Here we present the work done and the current Web Resources available to the community for research and data mining.

Introduction

The genus *Entamoeba* belongs to the super-group Amoebozoa and is composed of numerous species some of which are pathogens of animals, including humans. In particular, *E. histolytica* is the aetiological agent responsible for amoebic dysentery and amoebic liver abscess. These diseases affect approximately 50 million people every year, resulting in nearly 100,000 deaths, and amoebiasis represents the fourth leading cause of death worldwide due to a protozoan infection, being surpassed only by malaria, African trypanosomiasis and leishmaniasis (Ali *et al.*, 2008; Haque *et al.*, 2003; Petri *et al.*, 2000; WHO, 1998). *E. histolytica* is a small, nucleated cell that lacks mitochondria, and has a simple life cycle, consisting of a uninucleate trophozoite that feeds mainly on bacteria and divides by binary fission, and a cyst with four nuclei that is the transmission stage. *E. dispar* and *E. coli* are among the related species in humans that are morphologically similar to *E. histolytica* but with no apparent pathogenic properties (Ali *et al.*, 2008). Most *Entamoeba* species are found in the digestive tract of their hosts, the exceptions being *E. gingivalis* and *E. moshkovskii*, which are primarily found in the mouth and in fresh water anoxic sediments, respectively (Ali *et al.*, 2008). However, increasing numbers of reports from Bangladesh, Australia and India, among other locations, have demonstrated that *E. moshkovskii* also can establish asymptomatic infections in humans with high prevalence (Ali *et al.*, 2003; Fotedar *et al.*, 2008; Parija and Khairnar, 2005).

The combined prevalence of *E. histolytica* and *E. dispar* infection in the US is approximately 4%; however, certain groups have a much higher incidence of infection and disease, in particular institutionalized populations, patients infected with HIV, and travelers to or immigrants from endemic areas (Ali *et al.*, 2008; Freedman *et al.*, 2006;

Petri, 1996). Worldwide, these *Entamoeba* species infect approximately 10% of the population, but the prevalence is much higher in certain developing countries from Central and South America, Africa, and Asia and depends largely on cultural habits, age, socio-economic status, sanitation conditions and population density (Ali *et al.*, 2008).

To date, the pathogenicity of *E. histolytica* is not completely understood, and the number of cases may be underestimated. From the clinical point of view, variation in virulence can be attributable in part to parasite genetic diversity (Ali *et al.*, 2007). For example, different populations with similar rates of infection can have noticeably different serological titres. These also correlate with results of animal studies using representative genotypes (Nozaki *et al.*, 2006). One isolate from a seronegative population did not give rise to liver abscesses, even after inoculating a hamster with 4 times the usual number of trophozoites. In addition, this innocuous isolate was unable to destroy monolayers of HeLa or CHO cells, an *in vitro* marker of virulence. This avirulent phenotype was associated with a specific genotype, but how the specific markers of this genotype are related to pathogenicity is not known (Nozaki *et al.*, 2006).

Sequence and assembly of *Entamoeba* species genomes: *E. histolytica*, *E. dispar* and *E. invadens*

Re-assembly of the *E. histolytica* genome and assembly of related species

The purpose of the *E. histolytica* genome sequencing, as for most other pathogen genomes sequenced to date, was to generate a tool for both research and medical use. As more is learned from sequencing and about the role that genome architecture and composition play in infection and disease progression, genomic sequence information is becoming increasingly important for biomedical research. Genome information allows for better understanding of pathogenic processes and consequently helps improve the prevention, diagnosis, and treatment of the disease.

The genome sequence initiative was a joint effort between The Sanger Institute and the J. Craig Venter Institute (JCVI; formerly The Institute for Genomic Research, TIGR). The initial genome sequence, assembly and automated annotation of *E. histolytica* strain HM-1:IMSS was published in 2005 (Loftus *et al.*, 2005). This initial assembly presented multiple problems. Indeed, sequence analysis using intra-scaffold dot-plots revealed multiple artifactual tandem duplications located at the boundaries between neighbouring contigs (a scaffold is an arrangement of ordered contigs and a contig refers to an contiguous assembled sequence). In addition, gene predictions on these artifactual tandem duplications and on unidentified or unmasked repetitive regions caused an overprediction of genes by approximately 18%. In an attempt to improve the quality of the previously published genome draft, the original sequence reads from JCVI used to perform the original assembly were combined with additional reads from The Sanger Institute, not previously incorporated. The resulting assembly was deposited at the National Centre for Biotechnology Information (NCBI) with the accession number AAFB02000000. It represents the last and best of the re-assembly efforts for this genome. A comparative description of the features of the two *E. histolytica* assemblies is presented in Table 1.1.

Among the challenges for assembling the genomic sequence of this organism were the low GC content and the fact that nearly 19% of the genome is composed of repetitive elements. In general, the capacity for sequence assemblers to produce a single contig is limited by repeats. Almost identical reads from different copies of the same repeat cannot be distinguished by the assembler and might lead to genome misassemblies. To avoid these artefacts, Celera Assembler (Huson *et al.*, 2001) first removes potential repetitive regions by identifying any group of nearly identical reads whose number exceeds the average genome coverage. These groups of reads are then separated from the rest and categorized as 'surrogates'. In consequence, genomes rich in long transposable elements tend to produce shorter contigs that are interrupted at either repetitive sequences or sequencing gaps. To minimize the number of reads classified as surrogates, the per cent identity threshold above which reads are considered to be identical was increased for *E. histolytica*.

Table 1.1 Comparative analysis between the old and current *E. histolytica* genome assemblies

	New *E. histolytica* assembly	Old *E. histolytica* assembly
Genome		
Size (bp)	20,800,560	23,361,983
GC Content (%)	24.2	24.1
Number of genes*	8201	9985
Mean gene length (bp)	1260.9	1170.7
Number of genes/kbp	0.39	0.43
Longest gene (bp)	15,210	15,210
Shortest gene (bp)	147	96
Per cent coding	49.5	50.0
Per cent genes with introns	24.4	24.9
Exons		
Number	10,696	13,176
Mean number per gene	1.3	1.3
GC content (%)	28.0	28.1
Mean length (bp)	962.0	886.1
Longest exon (bp)	15,210	15,208
Shortest exon (bp)	3	1
Total length (bp)	10,289,260	11,675,669
Introns		
Number	2536	3191
GC content (%)	19.3	21.7
Mean length (bp)	74.3	100.0
Longest intron (bp)	869	3296
Shortest intron (bp)	11	8
Total length (bp)	188,331	319,223
Intergenic regions		
GC content (%)	20.5	20.0
Mean length (bp)	708.7	823.5
Longest intergenic region (bp)	16,340	13,773
Shortest intergenic region (bp)	2	1

*Number of genes includes pseudogenes.

Repetitive regions can also affect the final length of scaffolds. Indeed, after building genomic contigs, Celera Assembler tries to concatenate contigs into scaffolds using pair-mate read information from medium- and large-insert genomic libraries. Pair-mates are reads from two ends of the same clone insert, and are therefore known to be linked in the genome. In large insert libraries these reads may prove to be located in different contigs, therefore allowing them to be concatenated. In highly repetitive genomes, such as that of *E. histolytica,* the likelihood of a pair-mate read being part of a surrogate and therefore not informative for scaffold building is relatively high. Moreover, since in *E. histolytica* transposable elements tend to form large repeat clusters (see below), the likelihood of genomic clones bridging such clusters decreases dramatically. As a result, scaffolds in this parasite are relatively small and composed of just a few contigs. Similar assembly strategies were utilized to obtain the assemblies for *E. dispar* and *E. invadens*. Their draft genomes were sequenced, assembled and annotated at the JCVI, and the genome assemblies and annotations are deposited in the GenBank database at NCBI with the following accession numbers: AANV02000000 for *E. dispar* and AANW02000000 for *E. invadens*.

Genome features

The new *E. histolytica* assembled draft genome consists of 20 million basepairs (Mb) of sequence assembled into 2173 contigs organized in 1498 scaffolds. Because the precise karyotype for this organism is not known it is not possible to assign scaffolds to chromosomes (mapping). The genome has a GC content of 24.2% and contains 8201 predicted genes of which 3820 code for hypothetical proteins (see Table 1.1). tRNA genes are organized into tandem arrays ranging in size from 490 to 1775 basepairs (bp) and containing up to 5 tRNA genes in each unit. The total estimated number of tRNA genes that are present in these arrays is approximately 4500 (Clark *et al.*, 2006). One remarkable aspect of this genome is the highly repetitive nature of its sequence (see below). We were able to determine that 19% of the genome is composed of repeats of various origins, some of them unique to *E. histolytica* (Lorenzi *et al.*, 2008).

Genome architecture: repeat content

Transposable elements in E. histolytica *and related species*

Transposable elements (TEs) are classified into two broad classes depending on their transposition mechanisms. Class I includes two distinct types of TEs, long terminal repeat (LTR) and non-LTR retroelements, both requiring reverse transcription from a RNA intermediate (retrotransposons). LTR retroelements are reverse transcribed, replicated, and then transposed as double-stranded DNA. Non-LTR retroelements consist of short or long interspersed nuclear elements (SINEs and LINEs, respectively) and are transposed by reverse transcription of a mRNA intermediary directly into the site of integration (Nishihara and Okada, 2008; Wicker *et al.*, 2007). Class II TEs consist of elements that transpose DNA directly (transposons). Both classes of TEs are found scattered along the *E. histolytica* genome, and there is also a third group composed of unclassified TEs whose replication mechanisms are unknown (Lorenzi *et al.*, 2008).

Class I elements

Three subfamilies of LINE and SINE elements have been identified in *E. histolytica* (Bakre *et al.*, 2005b; Pritham *et al.*, 2005; Sharma *et al.*, 2001). The three LINEs are of similar sizes ranging from 4645 bp to 4845 bp. Consensus Eh_LINE1 and Eh_LINE2 sequences contain two open reading frames (ORFs) spanning most of their lengths. The 5′ ORF (ORF-1) codes for a polypeptide with some similarity to proteins containing coiled-coil domains (Bakre *et al.*, 2005a), while the protein encoded by the second ORF (ORF-2) contains a central reverse transcriptase (RT) domain followed by a putative CCHC-type DNA binding motif and an endonuclease domain similar to site-specific Type-II restriction endonucleases (Bakre *et al.*, 2005b; Sharma *et al.*, 2001; Van Dellen *et al.*, 2002). On the other hand, Eh_LINE3 contains only an ORF-2 with its corresponding RT and nuclease domains. Van Dellen *et al.* have proposed the existence of some Eh_LINE1 and Eh_LINE2 copies harbouring a single long ORF spanning both ORF-1 and ORF-2, based on the reconstruction of consensus sequences for those retroelements (Van Dellen *et al.*, 2002). These long ORFs seem to be interrupted by a conserved AAGCA duplication at position 1442 in 80% of Eh_LINE1s or a CG deletion at position 1249 in 92% of Eh_LINE2s, generating two different predicted ORFs (Bakre *et al.*, 2005a). However, no Eh_LINE has been identified containing such a long ORF in the new *E. histolytica* assembly suggesting these elements probably code for two separate polypeptides (Lorenzi *et al.*, 2008).

Even though the *E. histolytica* LINEs have the potential to encode an RT and nuclease domain related to site-specific restriction endonucleases, and in contrast to other non-LTR retroelements from the R2 group in other organisms, none of the three *E. histolytica* LINEs display any integration site preference. However, they seem to have the tendency to insert in AT-rich intergenic regions, frequently near other repetitive sequences (Bakre *et al.*, 2005b; Lorenzi *et al.*, 2008). These integration events generate direct target site duplications of up to 20 bp flanking the element (Van Dellen *et al.*, 2002). Eh_LINE1 is the most abundant of the three subfamilies with approximately 90 complete and 650 fragmented copies of the element in the genome followed by Eh_LINE2 (70 complete and 440 truncated copies) and Eh_LINE3 (10 complete and 80 fragmented) (Lorenzi *et al.*, 2008).

It is not clear whether these LINEs are still able to transpose in *E. histolytica*. Although no element has been identified containing both ORFs intact, there are at least some Eh_LINE1 and Eh_LINE2 elements harbouring an intact ORF-1 or an ORF-2 with conserved RT and endonuclease domains (Bakre *et al.*, 2005b; Lorenzi *et al.*, 2008). It has been reported that LINEs containing a non-functional mutation of the reverse transcriptase gene are still able to transpose using the transposition machinery encoded by other functional LINE elements (Biessmann *et al.*, 1994; Jensen *et al.*, 1994). Therefore, it seems likely that LINEs in *E. histolytica* are able to transpose using the required replication machinery *in trans*. Moreover, RT-PCR studies and the identification of several expressed sequence tags (ESTs) that show more than 98% identity to each of the three Eh_LINE subfamilies suggest that these elements are transcribed and might still be active in this parasite (Bakre *et al.*, 2005b; Lorenzi *et al.*, 2008).

Three LINE subfamilies homologous to Eh_LINEs 1–3 have also been identified in *E. dispar* (Lorenzi *et al.*, 2008; Sharma *et al.*, 2001). Consensus sequences for Ed_LINE1 and Ed_LINE2 harbour two ORFs equivalent to those found in Eh_LINE1 and Eh_LINE2, respectively. Likewise, Ed_LINE3 contains a single ORF similar to the putative RT encoded by Eh_LINE3. Relative abundance among LINE subfamilies resembles that in *E. histolytica*, with Ed_LINE1 being the most common element with 573 copies, followed by Ed_LINE2 (287 copies) and Ed_LINE3 (44 copies). No element has been identified containing both predicted ORFs intact (Lorenzi *et al.*, 2008). A single family of LINEs related to the three subfamilies from *E. histolytica* and *E. dispar* has been reported in both *E. invadens* and *E. moshkovskii* (Pritham *et al.*, 2005). However, LINE elements have been poorly amplified in these species with only 69 and 10 examples having been detected in *E. invadens* and *E. moshkovskii*, respectively (Lorenzi *et al.*, 2008; Pritham *et al.*, 2005).

Phylogenetic analyses based on either the conserved RT or endonuclease sequences demonstrate that the LINE subfamilies found in the four *Entamoeba* species have a monophyletic origin from an ancestral element related to the R4 clade within the R2 group of non-LTR retroelements (Lorenzi *et al.*, 2008; Pritham *et al.*, 2005; Van Dellen *et al.*, 2002). The same studies also show that in the common ancestor of *E. histolytica* and *E. dispar*, after diverging from the lineage that led to *E. invadens*, the LINE population expanded and split into two separate subfamilies giving rise to Eh/Ed_LINE1 and Eh/Ed_LINE2. Subsequently, in the same ancestor, a second LINE subfamily diverged from Eh/Ed_LINE1 giving rise to Eh/Ed_LINE3 (Lorenzi *et al.*, 2008; Van Dellen *et al.*, 2002).

In addition to LINEs, three different families of short non-autonomous SINE retroelements have been described in *E. histolytica* ranging from 538 bp to 656 bp in size (Bakre *et al.*, 2005b; Lorenzi *et al.*, 2008; Shire and Ackers, 2007). Eh_SINE1 is the most abundant with 264 complete and 181 truncated copies mapped to the new *E. histolytica* assembly, followed by Eh_SINE2 (94 complete and 162 fragmented repeats) and Eh_SINE3 (9 intact and 40 truncated copies). All *E. histolytica* LINEs and SINEs have the same 5′ sequence AGAT. The fact that Eh_SINE1 and Eh_LINE1 also share their last 70 bp suggests that they use similar transposition mechanisms (Bakre *et al.*, 2005b; Van Dellen *et al.*, 2002). An analogous situation also exists with the pair Eh_SINE2/Eh_LINE2, where the last 160 bases are approximately 70% identical (Clark *et al.*,

2007). Phylogenetic analysis shows a completely different landscape for Eh_SINE3 where the 5′ end is related to the first 240 bp from Eh_SINE2 but the 3′ end resembles that of Eh_SINE1. This result suggests that Eh_SINE3 originated as the result of recombination between Eh_SINE1 and Eh_SINE2 giving rise to a chimeric element, Eh_SINE3 (Lorenzi *et al.*, 2008). The chimeric nature of Eh_SINE3 and/or a putative 7 bp deletion with respect to the Eh_SINE1 consensus sequence at its 3′ end could be possible explanations for the low abundance of this element in *E. histolytica* (Shire and Ackers, 2007). If that is the case, then it does not seem to have limited the transposition of Ed_SINE1, an element homologous to Eh_SINE3 in *E. dispar* (Shire and Ackers, 2007). Indeed, Ed_SINE1 is the most abundant SINE retrotransposon in this parasite, where 425 copies have been mapped. Two additional SINE families are present in *E. dispar*, Ed_SINE2 and Ed_SINE3, which are homologous to Eh_SINE2 and Eh_SINE1 respectively. While Ed_SINE2 is found inserted 190 times, just a few copies of Ed_SINE3 are present in this genome (Lorenzi *et al.*, 2008).

Although it is not known whether SINEs are still active, cDNA evidence from the GenBank EST database supports the hypothesis that Eh_SINE1 and Eh_SINE2 are transcribed and probably able to transpose. However, no ESTs matching Eh_SINE3 or any of the SINEs from *E. dispar* were identified (Lorenzi *et al.*, 2008).

Class II elements

In contrast to the retroelements, DNA transposons have not been amplified successfully in either *E. histolytica* or *E. dispar* and just a few vestigial copies remain as evidence of their existence in the common ancestor of these two parasites and *E. moshkovskii* (Lorenzi *et al.*, 2008; Pritham *et al.*, 2005). Less than 10 representatives from the *Mutator* and *mariner* superfamilies of DNA transposons have been identified in *E. histolytica* and *E. dispar*: some truncated copies of an element coding for a putative transposase related to the *Mutator* transposon *EMULE* and one predicted ORF coding for a 335 amino acid (aa) protein similar to a transposase from the *mariner* DNA transposon *Hydargos*, both families also being present in *E. invadens* and *E. moshkovskii*. Notably, the opposite situation is observed in the genomes of *E. invadens* and *E. moshkovskii*, where class II elements are widespread but the amplification of retrotransposons seems to have been almost non-existent (Lorenzi *et al.*, 2008; Pritham *et al.*, 2005). Indeed, many families of DNA transposons belonging to superfamilies Tc1/*mariner*, *hAT*, *Mutator* and *piggyBac* are present in these two genomes. A recent quantification of repeats in the *E. invadens* assembly shows that DNA transposons represent more than 9% of the genome while LINEs only make up 0.1% (Lorenzi *et al.*, 2008; Pritham *et al.*, 2005).

It has been proposed that such a disparity in repeat type composition among the four *Entamoeba* species could be influenced by their effective population size, host range and the fact that *E. histolytica* and *E. dispar* seem to have gone through recent population bottlenecks (Ghosh *et al.*, 2000; Pritham *et al.*, 2005). Lynch and Conery have shown that class I and class II transposable elements have different estimated critical effective population sizes above which they could not proliferate in their host genome (Lynch and Conery, 2003). The same study also shows that proliferation of class I and class II repeats are affected differently by genome size. In this regard, DNA transposons seem to have a higher genome size threshold of around 30 Mb below which they cannot be fixed in the population, while that threshold would be close to 10 Mb for non-LTR retrotransposons (Lynch and Conery, 2003). Therefore, it seems reasonable that genome size differences between *E. histolytica* and *E. dispar* (~20 Mb) versus *E. invadens* (~40 Mb) and/or variations in their effective population sizes may have contributed to skew their repeat composition (Lynch and Conery, 2003; Pritham *et al.*, 2005).

Unclassified transposable elements

A third group of transposable elements without similarity to any known class I or class II repeat has been identified recently in *E. histolytica* (Lorenzi *et al.*, 2008). Eh_ERE1 is a 7160 bp TE composed of a 2719 bp core region flanked by two 2221 bp inverted repeats. The core sequence contains a predicted ORF coding for a putative 369 aa polypeptide with some similarity to proteins from two NCBI clusters of orthologous proteins (COGs) that contain chromosome segregation ATPases (COG1196) and DNA repair ATPases (COG419). A total of 777 examples of Eh_ERE1 have been identified covering almost 5% of the *E. histolytica* genome. Although all Eh_ERE1 integrations are truncated, there are about 100 potentially complete copies where the truncation coincides with the end of a scaffold. The fact that the single predicted ORF detects some *E. histolytica* entries in the GenBank EST database suggests that the ORF is expressed. Moreover, analysis of non-synonymous to synonymous changes at the third codon position (dN/dS analysis) shows that the ORF is under purifying selection (Lorenzi *et al.*, 2008). Altogether, the evidence seems to indicate Eh_ERE1 is still active in this parasite. The core region of Eh_ERE1 was also identified in *E. dispar*, where it is frequently associated with Ed_SINE1 and Ed_LINE1, and in *E. invadens*, indicating that this element was present in the common ancestor of these parasites (Lorenzi *et al.*, 2008). In *E. dispar*, Ed_ERE1 represent about 2% of the genome while in *E. invadens* it is not very well conserved and only spans 0.4% of the genomic sequence.

A second transposable element that falls within this category is a 1936-bp repeat called Eh_ERE2 (Lorenzi *et al.*, 2008). Unlike Eh_ERE1 elements, Eh_ERE2 is specific to *E. histolytica*. It contains a 1892 bp core region flanked by two 22-bp imperfect inverted repeats and harbouring a single predicted ORF. Insertion of Eh_ERE2 into the genome generates 3–14-bp direct target site duplications flanking the element. The single ORF codes for a 173-aa polypeptide with no homology to any known protein and no EST spanning the single Eh_ERE2 ORF has been identified. However, dN/dS analysis suggests that this ORF is also under purifying selection and therefore might still be functional. There are approximately 70 complete and 730 truncated copies of Eh_ERE2, representing a total of 3.5% of the *E. histolytica* genome (Lorenzi *et al.*, 2008).

Neither Eh_ERE1 nor Eh_ERE2 seem to recognize any specific sequence when transposed into the genome. However, they are frequently integrated into AT-rich intergenic regions and no instances were found where the insertion interrupted a protein-coding gene.

Genomic distribution of transposable elements in E. histolytica

Various studies have indicated that transposable elements are not uniformly distributed along the *E. histolytica* genome. Bakre *et al.* (Bakre *et al.*, 2005a) analysed the composition of the adjacent 2-kilobasepair (Kbp) regions flanking LINEs and SINEs, concluding that these elements do not seem to be located in telomeric regions but instead are frequently found near protein-coding genes. When a second retrotransposon was identified in the flanking regions, 50% of the time it was within 0.1 Kbp of the first, suggesting that LINEs and SINEs are prone to integrate next to each other forming clusters. Similarly, Lorenzi *et al.* (Lorenzi *et al.*, 2008) computed the distribution of inter-repeat distances for all transposable elements identified in this genome showing that, although the mean distance is approximately 3 Kbp, 50% of the repeats are closer than 390 bp indicating that they form clusters. Eh_ERE1 is the element most frequently found in clusters (467 copies) followed by Eh_LINE1 (363 copies) and Eh_ERE2 (336 copies). Most repeats within clusters are long transposable elements (LINEs and EREs) except for Eh_ERE1, which also frequently clusters with Eh_SINE1 and Eh_SINE2.

Repeats are not homogeneously distributed along the *E. histolytica* genome. About 90% of the genomic sequence has a repeat density below its mean expected value of 4 repeats every 10 Kbp while a small fraction of the genome is highly enriched in repeat clusters. Comparative sequence analysis between the genomes of *E. dispar* and *E. histolytica* has shown that repeat clusters are frequently found at break points in synteny between these two parasites; this strongly supports the idea that repeat-rich regions could contribute to genome instability and therefore to genetic variability and speciation in these parasites (Lorenzi *et al.*, 2008).

The consensus sequences of the novel *Entamoeba* repeat elements have been deposited in Repbase, one of the most commonly used databases of repetitive elements (Jurka *et al.*, 2005).

Genome architecture: gene statistics and gene organization

The *E. histolytica* genome contains 8201 annotated genes. On average, genes are 1261 bp long and contain 1.3 exons, with a mean exon size of 962 bp and a mean intron size of 74 bp. Intergenic regions are 709 bp on average. However, due to the fact that 19% of the genome is occupied by repetitive elements, most genes are compactly distributed with 50% of them being separated by less than 282 bp. Moreover, we have determined that approximately 4140 of these genes are arranged into 948 highly compacted cluster-like structures of up to 20 consecutive genes, each separated from the next by less than 282 bp. Interestingly, 350 of these clusters are composed of genes transcribed from the same DNA strand, resembling bacterial operons; however, no evidence has been found for polycistronic transcription based on current EST data and genome alignments.

To date, 1088 (13%) of the predicted genes in the genome are supported by EST and/or cDNA data. Of the remaining 87% of protein encoding genes, 59% have significant similarity at the amino acid level to known proteins in other species (BLASTP hits with a per cent identity of > 30%, coverage of > 70% and an e-value of < 1×10^{-5}) or conserved protein domains (Hidden Markov Model (HMM) hits above a trusted cutoff and an e-value of < 1×10^{-3}). This cutoff is a good e-value that represents the number of hits that would be expected to have a score equal to or better than this by chance alone, generally 1×10^{-5}, and an empirical value represented by the 'Pfam (paralogous family) trusted cutoff', which is the lowest score for a sequence included in the full seed alignment of a given domain. 41% of the predicted proteome is composed of proteins with no homology to any other known protein (i.e. hypothetical proteins with no evidence).

The proteome of *E. histolytica* includes 897 protein families, clustered based on their domain composition. The proteome was searched against the Pfam (http://pfam.sanger.ac.uk/) and TIGRfam (http://www.tigr.org/TIGRFAMs/) HMM profiles using HMMER2 (Sonnhammer *et al.*, 1998). In order to identify new domains, all regions of proteins with known domain hits above a trusted domain cutoff are removed and the remaining peptide sequences are subject to 'all-vs-all' BLASP searches and subsequent clustering. Clustering of similar peptide sequences is done by linking any two peptide sequences having an identity above 30% over a span of at least 50 aa and an expected value of < 1×10^{-10}. The Jaccard coefficient of community was calculated for each linked pair of peptide sequences *a* and *b* as follows:

$$J_{a,b} = \frac{\text{number of distinct accessions matching } a \text{ and } b \text{ including } (a,b)}{\text{number of distinct accessions matching either } a \text{ or } b}$$

where $J_{a,b}$ is the Jaccard coefficient of community or link score between proteins *a* and *b* and represents a similarity measure between the two proteins. The associations between peptides with a link score above 0.6 were used to generate individual linkage clusters, and aligned using ClustalW (Thompson *et al.*, 1994) to construct multiple alignments of novel conserved protein domains not present in the Pfam and TIGRfam databases. Any *E. histolytica*-specific clusters containing three or more members were considered true domains for the purpose of building families. The peptides in alignments were searched back by BLASTP against the *E. histolytica* proteome to find additional members that might have been excluded during earlier stages due to the parameters employed. Full length protein sequences were then grouped on the basis of the presence of Pfam, TIGRfam and potential novel domains. Thus, proteins with exactly the same domain composition are grouped into the same protein families.

The 897 families generated by the Pfam algorithm are made up of 4620 proteins (56.6% of the proteome); 615 families are composed of 3 or more members (4060 proteins), and the remainder of the proteome corresponds to families of two or singletons (43.3%). The largest family is composed of 142 members, corresponding to Rab/Ras GTPase family of proteins, followed by the leucine-rich repeat/BspA family and a family of 105 proteins that contains protein kinase domains. The method also allowed the clustering of 92 families with 5 or more members, representing 12% of the proteome (1005 proteins), that harbour a total of 109 potentially novel *E. histolytica* domains.

To date, 48% of the proteins in the protein families and 55% of the proteins with some functional assignment have been manually curated. A total of 1093 proteins have been assigned EC numbers and 3474 genes contain Gene Ontology (GO) terms in their annotation. In some cases, GO terms have been assigned automatically through Pfam2GO, a tool that allows mapping between Pfam domains and GO terms and links proteins to GO terms directly. However, for all those families curated manually, GO assignments have been re-evaluated by expert curators. Additional evidence of the overall improvement in the quality of the annotation compared to the first annotation draft released in 2005 includes: the deletion of about 1700 genes mostly located within either transposable elements or artifactual tandem duplications; a 50% increase in the number of different Pfam or TIGRfam domains; and twice as many proteins containing either Pfam or TIGRfam domains.

Some protein families are frequently linked to transposable elements (Table 1.2). It is tempting to speculate that the pronounced expansion of repetitive elements that occurred in the *E. histolytica* genome has contributed to the appearance and amplification of protein families in this organism through repeated transposition and/or recombination events mediated by transposable elements. Although the biological significance of such amplification events is not clear, it is likely that the expansion of some gene families has increased the fitness and/or survival capacity of the parasite within the host. Accordingly, among the repeat-associated genes there are three gene families, related to a group of antibacterial genes from plants named AIG1, a family of leucine-rich/BspA-like proteins, and proteins belonging to the heat shock protein 70 family. In addition, there are four families of hypothetical proteins with at least seven members that have more than 70% of their genes linked to repeats. One of these families codes for a protein containing a single transmembrane domain close to its N-terminus. Although these families are also conserved in *E. dispar*, two of them are specifically amplified in *E. histolytica*.

Genome architecture: HAPPY mapping

The new *E. histolytica* assembly is composed of 1498 scaffolds, mostly due to the highly repetitive nature of this genome. In an attempt to determine the chromosome structure and generate pseudomolecules that contain an increased number of contigs incorporated into scaffolds, Dr. Justin Pachebat, University of Cambridge, UK, carried out a HAPPY mapping approach using 1618 genomic probes. In the context of the *E. histolytica* genome, HAPPY mapping (mapping based on the analysis of approximately haploid DNA samples using the polymerase chain reaction) can be used to determine the synteny and orientation of the current scaffolds across the genome. Since the initial experimental design was based on the first *E. histolytica* assembly, probes had to be remapped onto the newly assembled sequence using BLASTN. Only 1258 out of the

Table 1.2 *E. histolytica* protein families associated with transposable elements

Family ID	Family size	Repeat-associated genes	% of family	Description
Eh_2	114	41	36.0%	Leucine-rich/BspA-1ike protein
Eh_29	17	14	82.4%	AIG1-related family protein
Eh_42	15	11	73.3%	Hypothetical protein
Eh_56	13	12	92.3%	Hypothetical protein
Eh_64	12	10	83.3%	Hypothetical protein
Eh_87	9	8	88.9%	AIG1-related family protein
Eh_133	7	7	100.0%	Hypothetical protein

initial 1618 genomic probes mapped with 100% identity to a single locus and results had to be reinterpreted on the context of the new sequence. HAPPY mapping predicted a total of 59 linkage groups containing at least three probes and spanning 306 different scaffolds (http://pathema.jcvi.org/tigr-scripts/Entamoeba/PathemaHomePage.cgi). Some of these linkage groups could be further confirmed by comparative analysis among the new and old *E. histolytica* and the *E. dispar* assemblies (Fig. 1.1). Although not all linkage group have been able to be confirmed in the new or in the old assembly, some of them are compatible with the ordered contigs in the present scaffolds, indicating that even though more probes may be required to obtain a more accurate map, the technique offers a tremendous potential to elucidate the chromosome structure of *E. histolytica*.

Comparative genomics of *Entamoeba* spp.

Sybil comparative tool

Genome sequencing, re-assembly and re-annotation of *E. histolytica* has provided us with very useful information about the genome architecture, gene complement and overall composition of the organism. However, knowledge of the gene complement of a single species may not be sufficient to fully understand all aspects of the parasite's biology, such as pathogenicity, invasive capacity, tissue tropism, etc. In an attempt to obtain a deeper understanding of *E. histolytica* biology, the relationship between the genome structure and function across three *Entamoeba* species was investigated by the use of comparative genomics. For this purpose, the genomes of *E. dispar*, a closely related non-pathogenic amoeba, and *E.*

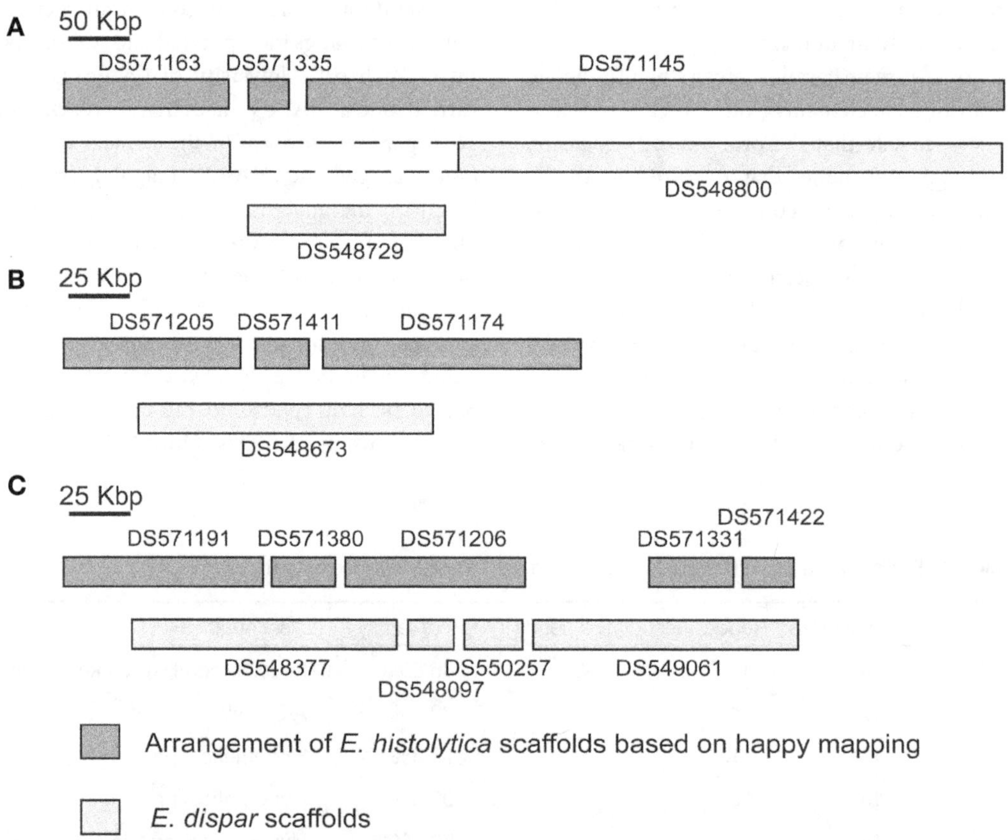

Figure 1.1 A-C: Examples of three different *E. histolytica* scaffold arrangements predicted by HAPPY mapping analysis (dark grey boxes) that agree with their syntenic counterparts in *E. dispar* (light grey boxes). GenBank accession numbers are indicated above or below boxes on the corresponding scaffolds.

invadens, a pathogen of reptiles, were sequenced to low coverage levels (5×) (Lorenzi *et al.*, 2008).

Comparative studies are fundamental not only for identifying conserved and unique regions and genes among species but also to understand the evolutionary changes that led to the acquisition of unique characteristics in each species. Three types of selection can be expected to help explain the divergence and similarities between species: *stabilizing selection* will preserve all traits responsible for similarities between species, *positive selection* will be acting on all divergent traits, while *neutral selection* will act on non-essential features. Comparative genomics provides an avenue to distinguish these and to analyse genomic regions and genes where selection forces have left their signature. Although tracing the events that led to speciation in eukaryotic genomes can be particularly challenging due to their complexity, *Entamoeba* species offer a good, manageable model to study the evolution of genomes. Comparative analysis between pathogenic and non-pathogenic amoebas will hopefully reveal evolutionary changes responsible for the acquisition of pathogenicity in *E. histolytica.*

The computational tool used for analysis and visualization of *Entamoeba* genome comparison data is called Sybil (http://sourceforge.net/projects/sybil), which has been implemented for an assortment of both prokaryotes and eukaryotes. Recently an integrated version of Sybil has been released to the public for comparison of *Entamoeba* genomes (http://pathema.jcvi.org/entamoeba). *Pathema-Entamoeba-Sybil* allows users to search for individual genes or clusters of orthologous genes of interest (defined as two or more homologous genes found in different species) and visualize their genomic context. A graphical comparative view is given of all the sequence regions that can be linked to a selected reference region by one or more gene or protein clusters. The reference sequence region can be specified in several ways, for example using the sequence name and start and end coordinates, or by using the identifier of a specific gene. The tool allows one to select which non-reference genomes

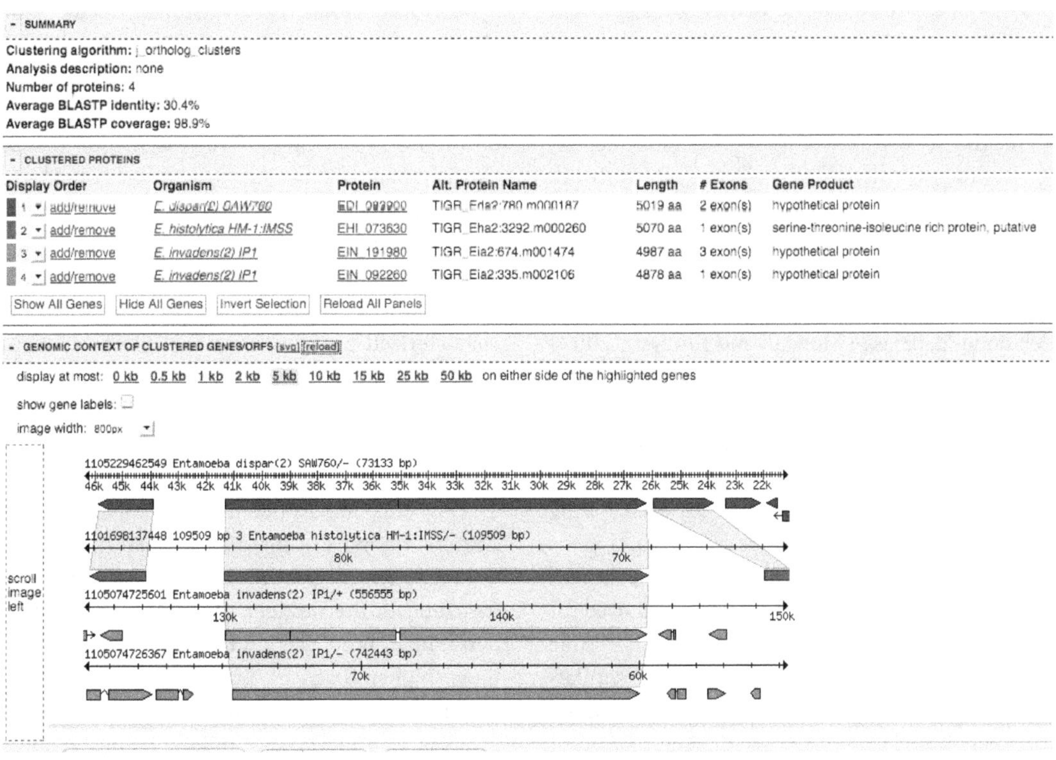

Figure 1.2 Screenshot of the Pathema-*Entamoeba*-Sybil graphical comparative view.

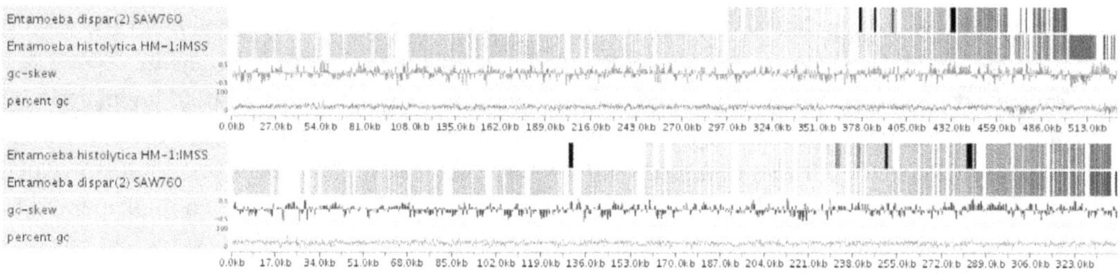

Figure 1.3 Pathema-*Entamoeba*-Sybil Gradient View. The graph represents a reference genome (bottom) and the interrogated genome (top). The colour gradient – yellow to blue (light grey to dark grey in this image) – allows gene locations and conserved gene order as well as rearrangements to be viewed.

to include in the comparison and to specify how many clustered genes a non-reference sequence region must have in common with the reference in order to be displayed. The resulting graphical display is interactive, allowing one to examine the genes and gene clusters in the view and to link the output to other relevant report pages (Fig. 1.2).

Pathema-Entamoeba-Sybil also presents a Synteny Gradient View that shows the conservation (or not) of synteny between the genomes with respect to a chosen reference (Fig. 1.3). The genes in the reference are coloured from yellow (light grey in Fig. 1.3) to blue (dark grey in Fig. 1.3) from left to right. The orthologous genes in the interrogated genomes are placed on top of those of the reference genome and are coloured based on their location in their respective genomes making it easy to identify regions of conserved gene order (synteny) as well as synteny break points, immediately highlighting rearrangement events between different strains or species in the selected regions.

Sybil extracts the information from a Chado relational database (Mungall and Emmert, 2007) that is populated with genome annotations and comparisons for the three *Entamoeba* genomes. Sybil relies on pre-computed comparisons that include 'all-vs-all' BLASTP protein searches for all proteomes being compared, and a clustering of orthologous and paralogous genes. Paralogous gene clusters are generated based on a Jaccard similarity coefficient, a statistical method used for comparing the similarity and diversity of sample sets (Haas *et al.*, 2005), using BLASTP similarity scores. Orthologous gene clusters are defined by a single linkage clustering of reciprocal best BLASTP hits between genes or paralogous clusters of genes from all pairs of genomes compared.

Additionally, Sybil also allows the mapping and visualization of repeat content, transposable element and retrotransposon locations (see section on repeats above).

Comparative genomics features

The genome characteristics of *E. histolytica* HM-1:IMSS in comparison with *E. dispar* SAW760 and *E. invadens* IP-1 is shown in Table 1.3.

All predicted proteins for the three amoeba genomes were compared as described above to identify species-specific genes and common core genes. Comparison of the three proteomes using Sybil revealed that these organisms share 5006 clusters of proteins (Jaccard clusters, as defined above) representing 71%, 68% and 47% of the protein coding genes of *E. histolytica*, *E. dispar* and *E. invadens*, respectively. Proteins within Jaccard clusters shared by all three species are considered to constitute the core proteome. Close inspection of the core proteome of *E. histolytica* shows that 42% correspond to proteins with putative functions assigned or that are similar to functionally characterized proteins, 8% are proteins containing conserved domains and 7% are hypothetical proteins conserved in other genera, while the remaining 43% are hypothetical proteins conserved among *Entamoeba* spp. only.

Fig. 1.4 shows the distribution of Jaccard clusters among the three species and it reflects the phylogenetic relationships among the amoebas. Indeed, the number of clusters only shared between *E. histolytica* and *E. dispar* is about 10 times higher than the number shared between either of these species and *E. invadens*. Only 619 genes in *E. histolytica* are unique to this genome while *E. dispar* and *E. invadens* contain 1185 and 5737 species-specific singletons, respectively.

Table 1.3 Genome summary for the three *Entamoeba* species, showing the individual genome statistics for genes, exons, introns and intergenic regions.

Genome	*E. histolytica*	*E. dispar*	*E. invadens*
Size (bp)	20,800,560	22,955,291	40,888,805
GC content (%)	24.2	23.5	29.9
Number of genes*	8201	8745	11,538
Mean gene length (bp)	1260.9	1229.2	1359.3
Number of genes/kbp	0.39	0.38	0.28
Longest gene (bp)	15,210	15,093	15,112
Shortest gene (bp)	147	234	189
Per cent coding	49.5	46.8	38.4
Per cent genes with introns	24.4	28.8	34.0
Exons			
Number	10,697	12,108	17,327
Mean number per gene	1.3	1.4	1.5
GC content (%)	28.0	27.7	38.5
Mean length (bp)	962.0	887.8	905.1
Longest exon (bp)	15,210	12,435	15,015
Shortest exon (bp)	3	3	2
Total length (bp)	10,289,260	10,749,589	15,683,489
Introns			
Number	2536	3359	5789
GC content (%)	19.3	18.0	23.9
Mean length (bp)	74.3	81.5	104.8
Longest intron (bp)	869	932	1091
Shortest intron (bp)	11	18	12
Total length (bp)	188,331	273,890	606,844
Intergenic regions			
GC content (%)	20.5	19.8	24.6
Mean length (bp)	708.7	683.3	1705.6
Longest intergenic region (bp)	16,340	18,422	52,231
Shortest intergenic region (bp)	2	2	4

*Number of genes includes pseudogenes.

Conclusion from comparative genomics

The comparative genome analysis of *E. histolytica, E. dispar* and *E. invadens* provides the core genetic outline of the species. Although significant differences are obvious in genome structure and gene content between the genomes of *E. histolytica* and *E. dispar* with respect to *E. invadens,* further analyses of *E. histolytica* and *E. dispar* are necessary to understand the origins

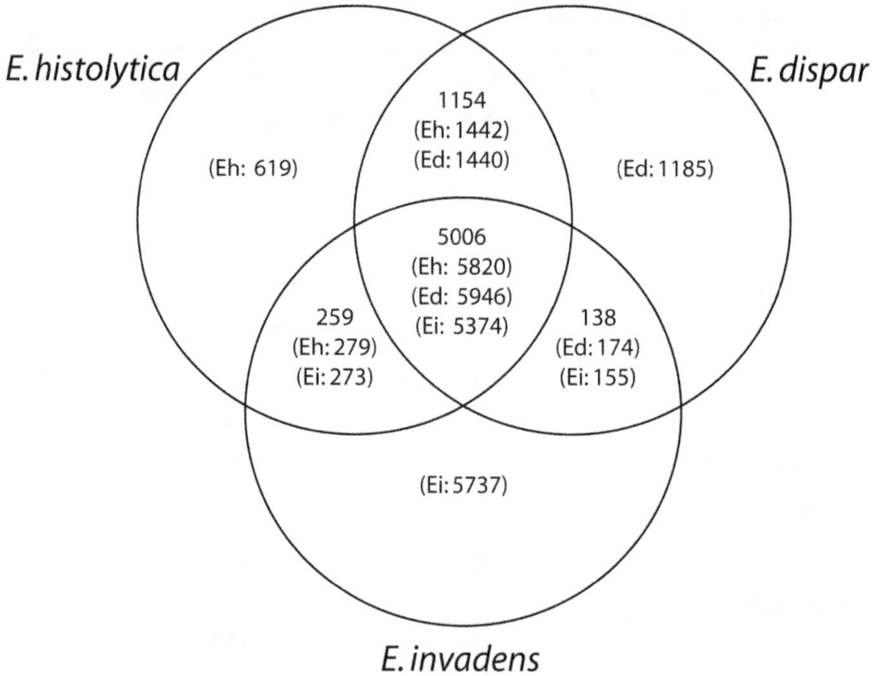

Figure 1.4 Venn diagram showing the number of Jaccard clusters shared by *E. histolytica* (Eh), *E. dispar* (Ed) and *E. invadens* (Ei). The number of genes corresponding to each group of Jaccard clusters is depicted within parentheses.

of the pathogenic traits that distinguish these two species. An obvious difference between (*E. histolytica–E. dispar*) and *E. invadens* is the lack of synteny; however, regions of microsynteny can be identified among the species.

A surprising observation is that the size of the *E. invadens* genome is almost twice that of the other two *Entamoeba* species. However, the increased genome size does not correlate with a proportional increase in gene count. Despite the *E. invadens* genome being 96% larger, there are only 40% more predicted genes. Larger genome size can be explained by the larger intergenic regions present in *E. invadens* that result in a less compacted genome (1700 bp vs. 700 bp; Table 1.3).

It is reasonable to speculate that the findings derived from comparative genomic analysis of the *Entamoeba* genomes may provide the basis for the development of target systems to identify strains and their phylogenetic relationships and epidemiology as well as alternative treatments for controlling amoebiasis in humans.

E. histolytica web resources

There are a number of Web Resources that contain information on *E. histolytica* and other *Entamoeba* species. These resources range from the highly analytical, containing a variety of analysis tools for genome and comparative genome analysis, to simple basic information and links to various other sites. A description of the most relevant sites follows.

Pathema

Pathema mission

Genome information for *E. histolytica, E. dispar* and *E. invadens* can be explored and visualized via the Pathema web site (http:/pathema.jcvi.org/entamoeba). Pathema is the web resource for the J. Craig Venter Institute's NIAID-funded Bioinformatics Resource Centre, one of eight centres with the mission of supporting bio-defence and infectious disease research. Pathema currently provides detailed curation of six target pathogens: *Bacillus anthracis, Clostridium botulinum,*

Burkholderia mallei, Burkholderia pseudomallei, Clostridium perfringens and *E. histolytica*, the only eukaryotic organism in the group. The central goal of Pathema is to provide an *Entamoeba* information hub that will help scientific progress and the understanding, detection, diagnosis and treatment of amoebiasis and other pathogens involved in new and re-emerging infectious diseases.

Pathema-*Entamoeba* provides a suite of bioinformatic tools for analysis specifically geared towards the analysis of genome data, including comparative genomics, protein motif searches, virulence factor and epitope listings, to facilitate the identification of potential targets for vaccine development, therapeutics, and diagnostics. Pathema-*Entamoeba* also contains an outreach component that includes training, documentation and organism-specific workshops.

Pathema-*Entamoeba* resources and tools

The home page of the Pathema-*Entamoeba* web site contains resources that include general information, functional curation of genes, pathogenicity-related genes, a comprehensive protein catalogue, comparative analysis tools (e.g. Sybil, see above) and functional genomics tools (Fig. 1.5).

Information on the differences in gene content between the original annotation and the current re-annotation of the *E. histolytica* genome is provided in a file mapping the old annotation to new annotation, together with a list of genes in the new annotation not present in the old annotation as well as a list of genes in the old annotation not present in the current. The differences between the annotations derive principally from the problems with the assembly generated in the initial draft genome as well as the methods of gene finding and curation employed. The first annotation relied mainly on automated methods to generate gene predictions, did not include EST information, and involved minimal or no manual scrutiny of gene structures and functional assignments. The second assembly gene-set was produced by a battery of up-to-date gene finders that included *ab initio* gene finders, homology-based gene finders, and EST data incorporation. A combination algorithm developed at JCVI, EvidenceModeler (Haas *et al.*, 2008), was used to generate the final gene-set that was released to Genbank in 2008.

The *Functional Curation* section provides details and documentation on the protocols and standard procedures utilized for the genome annotation of the three species, naming

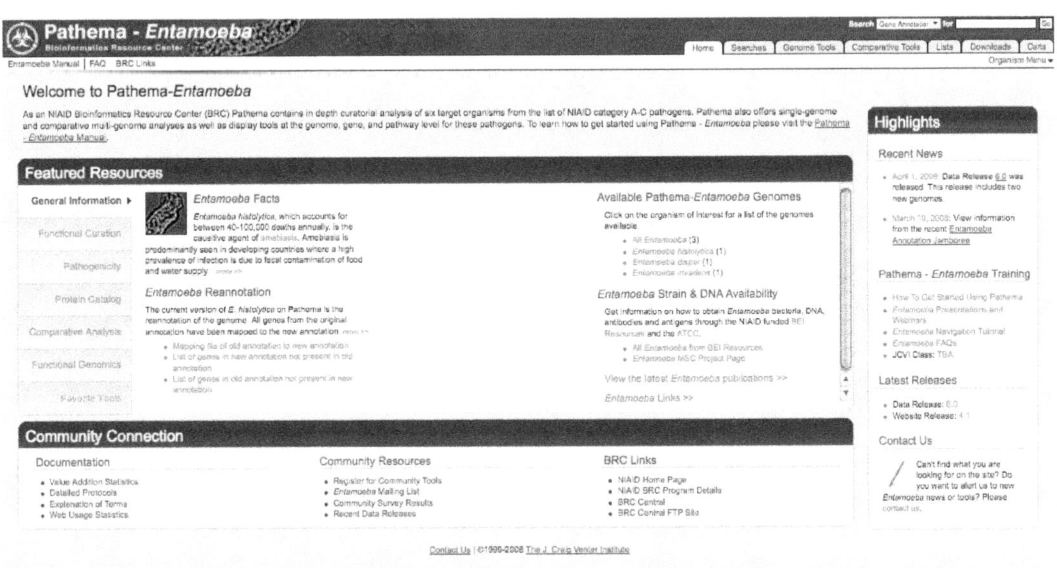

Figure 1.5 The Pathema-*Entamoeba* Web Resource Home Page, displaying the 'General Information' page. This includes '*Entamoeba* Facts', details about the re-annotation of the *E. histolytica* genome, available genomes, information about *Entamoeba* strains and DNA, and other *Entamoeba*-related information.

conventions followed to assign names to the gene-sets, and an explanation of Pathema terms. All documents are made available to the community. Community annotation is encouraged and is made possible upon registration on the Pathema site. Community annotation submissions are reviewed by expert curators and incorporated into the database after revision. This section allows the user to search for genes of interest by different features such as gene name, locus, gene symbol or gene annotation, and it allows for exact or inexact matches upon provision of a keyword. A multiplicity of gene tools and lists can be accessed from this page, such as *BLAST*, *Curated Gene List*, *Retrieval of Genes/Sequence by Coordinates*, *Region Viewer*, and *Organism Gene Lists*.

Of particular interest for the user is the *Region Viewer*, a tool that displays a region of a given assembly showing all genes in vicinity of a selected gene. The information on each of the genes in the viewer is shown upon placement of the cursor over a gene of interest. The region of interest can be expanded from the Viewer page, allowing the navigation along the selected assembly, and by the shrinking or enlargement of the selected area surrounding the gene of interest (Fig. 1.6).

The *Pathogenicity* section contains facts about amoebiasis, biosafety protocols, and a list of JCVI-curated virulence factors. From this page direct access to the latest '*Entamoeba* Pathogenicity'-related publications is also provided.

The *Entamoeba Protein Catalogue* provides individual lists of *Entamoeba* proteins including structural proteins, metabolism enzymes, and proteins involved in the cytoskeleton and vesicular trafficking. The protein catalogue lists are derived from a recent extensive review (Clark *et al.*, 2007) that based its content on the previously published *E. histolytica* genome annotation. The lists have been revised and mapped to the newly released *E. histolytica* annotation. Additionally, this page contains updated BLAST statistics for 96 lateral gene transfer candidates present in the *E. histolytica* genome.

As discussed in the 'Comparative genomics' section of this chapter, the Pathema-*Entamoeba* resource provides several whole-genome and multi-genome comparative tools to help identify regions and proteins of interest.

Pathema metabolic pathways

Integrated into Pathema are two suites of programs for visualization and comparative analysis of *Entamoeba* metabolic processes, the metabolic pathways from the Kyoto Encyclopaedia of Genes and Genomes (KEGG pathways) and the SRI Pathway Tools (Caspi *et al.*, 2008; Okuda *et al.*, 2008). KEGG pathways is a set of

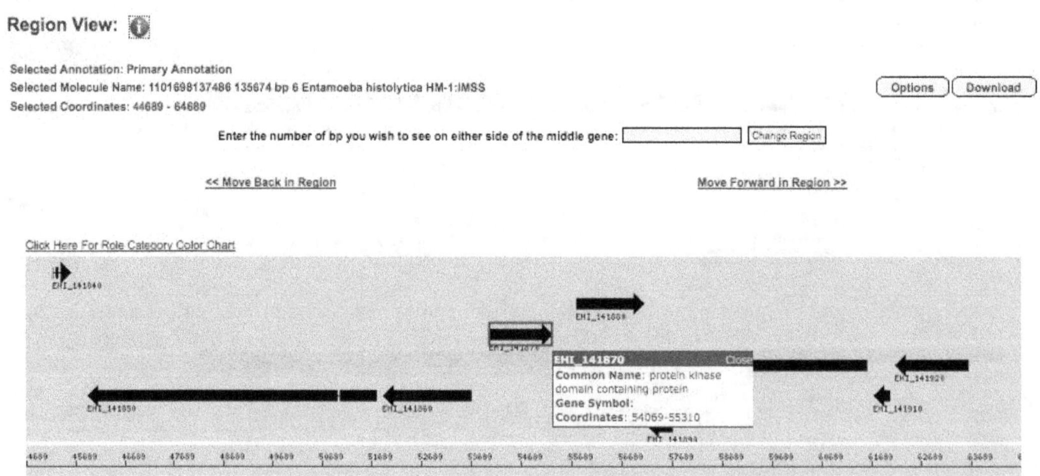

Figure 1.6 Region View allows the user to visualize the genomic context of a gene of interest. Black arrows represent protein coding genes. The page also provides information about the selected gene.

broad static pathway schemes representing the molecular interactions and enzymatic reactions for the collection of metabolic pathways, signal transduction and other cellular processes that have been described for different organisms. Proteins are assigned to reactions based on their EC number and then enzymatic reactions likely to occur in the organism are highlighted in colour on the selected metabolic map. In addition, the KEGG implementation in Pathema allows the simultaneous comparison of metabolic pathways or other cellular processes from two or more organisms. Pathema currently stores functional assignments for *E. histolytica, E. dispar* and *E. invadens* genes allowing comparative analysis among these three species to be performed.

Another Pathema resource available for the analysis of diverse cellular processes is SRI Pathway Tools (SRI-PT). In contrast to KEGG, metabolic pathways in SRI-PT are organized in smaller, more specific, diagrams. From the initial page it is possible to browse enzymes, reactions, pathways, proteins, genes and RNAs by either name, EC number or ontology. When looking at individual pathways, the program highlights pathway-specific reactions and those which require an enzyme that has not yet been identified in the organism of interest. This allows the user to discriminate metabolic pathways that are likely to be present in that organism from those that should be removed. The program also displays a cellular overview where all predicted metabolic reactions are depicted in a single cell-like diagram grouped by their ontology (e.g. carbohydrate degradation). Reactions are represented by lines while substrates and products are denoted by different symbols depending on their chemical nature. SRI-PT not only assigns proteins to reactions based on their EC number but also considers protein names, and in the next release of the program (12.5) it will incorporate GO terms. In addition, SRI-PT includes an *'omics viewer'* that allows the uploading and visualization of high-throughput data and other experimental results on top of the cellular overview using different colour-codes to represent changes in gene expression, protein/compound concentration, reaction flux, etc. The 'omics viewer' can be set up to visualize both relative or absolute data values and to display value changes over time or through different treatments. SRI-PT also lets the user run different comparative analyses among the three *Entamoebas*, from content of transporters, compounds and proteins to individual reactions and whole metabolic pathways.

TransportDB

Description

TransportDB (http://www.membranetransport.org/) (Ren *et al.*, 2007) is a relational database that describes all predicted cytoplasmic membrane transport proteins for organisms whose finished genome sequence is available. TransportDB contains the complete membrane transport complement identified for *E. histolytica, E. dispar* and *E. invadens* classified into protein families according to the transport classification system (TC, an IUBMB-approved classification system for membrane transport proteins), and also provides their functional predictions. The TC system is equivalent to the Enzyme Commission (EC) system for enzyme classification, but it also incorporates phylogenetic information. TCDB, which is used as part of the TransportDB pipeline, is a curated database of factual information from over 10,000 published references. It currently contains approximately 3000 protein sequences classified into over 550 transporter families based on the TC system (Saier *et al.*, 2006).

The complete lists of the transporters from *E. histolytica, E. dispar* and *E. invadens* are accessible from the pull down menus on the TransportDB home page (Fig. 1.7). Transport DB is also accessible through Pathema-*Entamoeba*.

When selecting *E. histolytica*, an overview of whole transporter content in the genome is summarized, including transporter families and types. For specific transporter types, a detailed list of possible transporters with their potential substrates is shown with links to individual *E. histolytica* predicted protein page, which contains peptide sequence and the most current annotation information.

TransportDB tools

TransportDB offers a suite of tools that facilitate analysis of the transporter complement of the

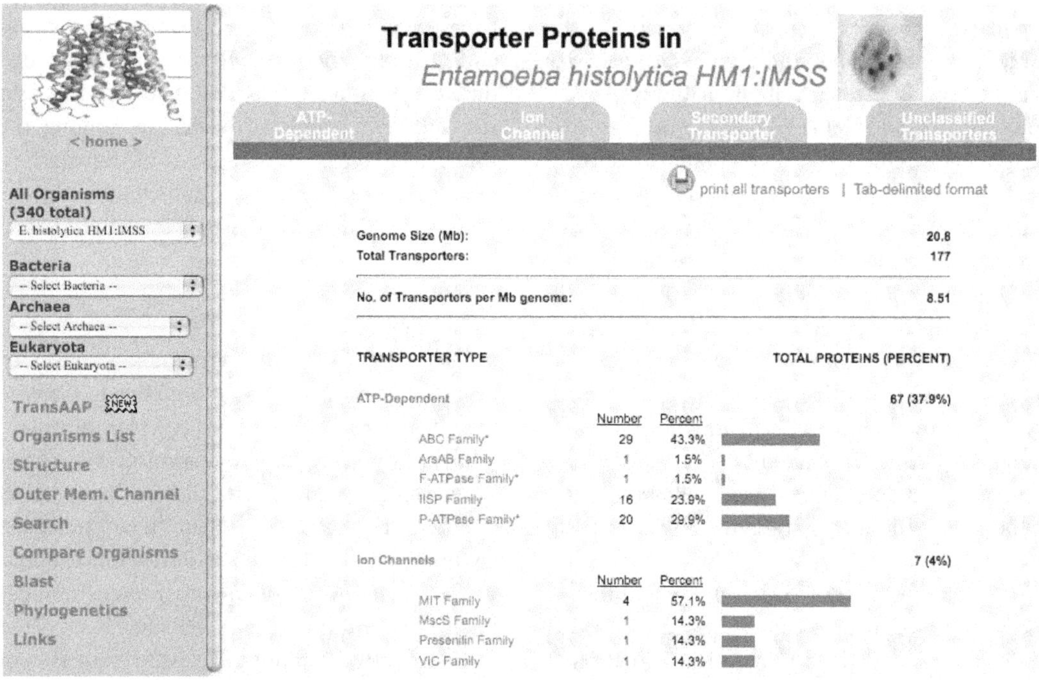

Figure 1.7 TransportDB display of transporter content in *E. histolytica*. The scroll down bar on the left side of the screen shows the selected organism, that can be chosen from the 'All organisms' scroll bar or from the specific kingdom groups. The right side of the screen shows a partial display of the transporter content, and the genome statistics.

organisms present in the database. The tools are displayed in the left side panel of the TransportDB webpage and include the following:

- ***Structure:*** Experimentally characterized 3D structures of membrane transporters from crystallization or NMR-based studies, derived by searching the entire collection of transport proteins against the protein data bank (PDB) (Berman, 2008), are incorporated into TransportDB. To date, 273 structures representing 98 transporters have been retrieved, and information on structure description, method and resolution are included.
- ***Search:*** TransportDB tools include a 'Search' engine set up to easily query the transporter database for any transporter type, family, individual proteins and their substrates.
- ***Compare Organisms***: Comparative analysis of the transporter system from the three species of amoebas, or to any other organisms in the TransportDB database, is possible by selecting the 'Compare Organisms' option on the left side of the webpage shown in Fig. 1.7. The page showing the comparative results is depicted in Fig. 1.8.
- ***Blast:*** This searches any protein sequence against the available TransportDB dataset, by selecting the 'Blast' option on the left panel of the webpage.
- ***Phylogenetics:*** TransportDB also provides phylogenetic analysis capabilities for the transporters. The 'Phylogenetics' option allows users to view the pre-computed neighbour-joining trees for each of the transporter families. This facilitates access to up-to-date phylogenetic trees of every transporter family and the ability to download them in FASTA or multiple sequence alignment formats.

With the rapid expansion of new and more efficient genomic sequencing technologies, a large number of genomes are now available for analysis of their transporter system. TransportDB developers at the J. Craig Venter Institute maintain the database up to date with the newly published genomes.

Comparision of Transporters from Different Organisms

Close Windows

	E. dispar SAW760	*E. histolytica HM1:IMSS*	*E. invadens IP1*
Genome Size(Mb):	22.9	20.8	40.8
Total Transporter Proteins:	186	177	209
No. of Transporters per Mb genome:	**6**	**8.51**	**6.53**
ATP-Dependent	**69 (37%)**	**67 (38%)**	**68 (33%)**
The ATP-binding Cassette (ABC) Superfamily	35	29	37
The Arsenite-Antimonite (ArsAB) Efflux Family	1	1	1
The Chloroplast Envelope Protein Translocase (CEPT or Tic-Toc) Family	0	0	1
The H+- or Na+-translocating F-type, V-type and A-type ATPase (F-ATPase) Superfamily	1	1	1
The H+-translocating Pyrophosphatase (H+-PPase) Family	0	0	0
The Type II (General) Secretory Pathway (IISP) Family	13	16	8
The Mitochondrial Protein Translocase (MPT) Family	1	0	1
The P-type ATPase (P-ATPase) Superfamily	18	20	19
The Septal DNA Translocator (S-DNA-T) Family	0	0	0

Figure 1.8 Comparative transporter display resulting page, for *E. histolytica*, *E. dispar* and *E. invadens*. The top part of the page shows a comparative overview of the organism genome size, transporter content and transporter density (transporters/Mb).

The *Entamoeba* Homepage

The *Entamoeba* Homepage (entamoeba.lshtm.ac.uk) provides links to information on *E. histolytica*. The site serves as a portal that contains basic *Entamoeba* facts and links to various information and sequencing centres, such as WHO, ATCC, CDC, FDA, JCVI, Sanger Institute and NCBI. In addition, the site provides direct links to *Entamoeba*-related publications and journals of interest to parasitologists.

The site also nucleates *Entamoeba*-related announcements of interest to the scientific community, such as reagent availability, conferences, workshops and job opportunities in the field.

The Sanger Institute

The Sanger Institute (www.sanger.ac.uk) is a genome research institute primarily funded by the Wellcome Trust (www.wellcome.ac.uk) that provides large-scale sequencing, informatics and analysis of genetic variation to further understand gene function in health and disease and to generate data and resources of lasting value to biomedical research. The protozoan sequencing efforts are concentrated on pathogens and model organisms, and all data from these projects are immediately and freely available.

Through their website, the Sanger Institute Pathogen Sequencing Unit (PSU) has limited read information available for two *Entamoeba* species in addition to *E. histolytica, E. dispar* and *E. invadens,* namely *Entamoeba terrapine* and *Entamoeba moshkovskii*. Although the current data are not annotated, the initial goal was to sequence both ends from 10,000 small insert clones from each species genome, with the objective of gathering preliminary data to study the genome evolution of these species, as well as providing a valuable resource for the scientific community who study these organisms. The raw sequence reads from this centre for all the organisms except *E. histolytica* are available for FTP download and BLAST searching. For *E. histolytica,* the initial annotation published by Loftus *et al.* is available through GeneDB (http://www.genedb.org/genedb/ehistolytica/). The GeneDB project is a central part of the Sanger Institute PSU's activities. Its primary goals are (i) to provide reliable access to the latest sequence data and annotation/curation for the whole range of organisms sequenced by the PSU; and (ii) to develop the website and other tools to aid the community in accessing and obtaining the maximum value from these data.

GeneDB currently provides access to 37 genomes at various stages of the sequencing curation pipeline, from early access to partial genomes with automatic annotation through to complete genomes with extensive manual curation.

Future directions

The sequencing and preliminary annotation of the *E. histolytica* genome provided an initial step in the understanding of parasite functionality and genome evolution. The improvement of the initial assembly and annotation was a necessary step to clear up a lot of the background noise that the initial semi-automated method of annotation generated. In addition, the completed sequencing and annotation of genomes in related species such as *E. dispar* and *E. invadens* adds significant value to our understanding of the evolutionary steps followed in the acquisition of pathogenesis traits in some of these species. Today's availability of websites such as Pathema that allow the scientific community to access the most up-to-date genomic data and analysis tools constitutes an invaluable resource for the future development of therapeutics, the identification and analysis of virulence factors, epitopes and other unsuspected characteristics of these organisms that are yet to be unraveled. In turn, the value of the web resources to the scientific community depends not only on the ease of use of the tools but more importantly on the feedback that the community supplies. This feedback can be provided, for example, in the form of third party annotation or suggestions for new tools, and is fundamental for the development, maintenance and improvement of the web resources. A good example of the combined effort from sequencing and annotation centres, developers and the scientific community is reflected in the Pathema-*Entamoeba* project, where a fluid exchange of information and critique exists between researchers working in specific fields and informatics experts.

Acknowledgements

Sybil was developed by the Bioinformatics department at The Institute for Genomic Research (TIGR) followed by the J. Craig Venter Institute and Institute for Genome Sciences (IGS) at the University of Maryland Baltimore's School of Medicine. The project has been funded by a variety of sources including: The National Institute of Allergy and Infectious Diseases (NIAID) as part of the NIAID Bioinformatics Resource Centres (BRC) for Biodefense and Emerging or Re-Emerging Infectious Diseases and PATH an organization supporting research in *Streptococcus* and *Neisseria* vaccine development.

References

Ali, I.K., Clark, C.G., and Petri, W.A., Jr. (2008). Molecular epidemiology of amebiasis. Infect. Genet. Evol. *8*, 698–707.

Ali, I.K., Hossain, M.B., Roy, S., Ayeh-Kumi, P.F., Petri, W.A., Jr., Haque, R., and Clark, C.G. (2003). *Entamoeba moshkovskii* infections in children, Bangladesh. Emerg. Infect. Dis. *9*, 580–584.

Ali, I.K., Mondal, U., Roy, S., Haque, R., Petri, W.A., Jr., and Clark, C.G. (2007). Evidence for a link between parasite genotype and outcome of infection with *Entamoeba histolytica*. J. Clin. Microbiol. *45*, 285–289.

Bakre, A.A., Rawal, K., Ramaswamy, R., Bhattacharya, A., and Bhattacharya, S. (2005a). The LINEs and SINEs of *Entamoeba histolytica*: comparative analysis and genomic distribution. Exp. Parasitol. *110*, 207–213.

Bakre, A.A., Rawal, K., Ramaswamy, R., Bhattacharya, A., and Bhattacharya, S. (2005b). The LINEs and SINEs of *Entamoeba histolytica*: comparative analysis and genomic distribution. Exp. Parasitol. *110*, 207–213.

Berman, H.M. (2008). The Protein Data Bank: a historical perspective. Acta Crystallogr. A *64*, 88–95.

Biessmann, H., Kasravi, B., Bui, T., Fujiwara, G., Champion, L.E., and Mason, J.M. (1994). Comparison of two active HeT-A retroposons of *Drosophila melanogaster*. Chromosoma *103*, 90–98.

Caspi, R., Foerster, H., Fulcher, C.A., Kaipa, P., Krummenacker, M., Latendresse, M., Paley, S., Rhee, S.Y., Shearer, A.G., Tissier, C., *et al.* (2008). The MetaCyc Database of metabolic pathways and enzymes and the BioCyc collection of Pathway/Genome Databases. Nucleic Acids Res. *36*, D623–631.

Clark, C.G., Ali, I.K., Zaki, M., Loftus, B.J., and Hall, N. (2006). Unique organisation of tRNA genes in *Entamoeba histolytica*. Mol. Biochem. Parasitol. *146*, 24–29.

Clark, C.G., Alsmark, U.C., Tazreiter, M., Saito-Nakano, Y., Ali, V., Marion, S., Weber, C., Mukherjee, C., Bruchhaus, I., Tannich, E., *et al.* (2007). Structure and content of the *Entamoeba histolytica* genome. Adv. Parasitol. *65*, 51–190.

Fotedar, R., Stark, D., Marriott, D., Ellis, J., and Harkness, J. (2008). *Entamoeba moshkovskii* infections in Sydney, Australia. Eur. J. Clin. Microbiol. Infect. Dis. *27*, 133–137.

Freedman, D.O., Weld, L.H., Kozarsky, P.E., Fisk, T., Robins, R., von Sonnenburg, F., Keystone, J.S., Pandey, P., and Cetron, M.S. (2006). Spectrum of disease and relation to place of exposure among ill returned travelers. N. Engl. J. Med. *354*, 119–130.

Ghosh, S., Frisardi, M., Ramirez-Avila, L., Descoteaux, S., Sturm-Ramirez, K., Newton-Sanchez, O.A., Santos-Preciado, J.I., Ganguly, C., Lohia, A., Reed, S., *et al.* (2000). Molecular epidemiology of *Entamoeba* spp.: evidence of a bottleneck (Demographic sweep) and transcontinental spread of diploid parasites. J. Clin. Microbiol. *38*, 3815–3821.

Haas, B.J., Salzberg, S.L., Zhu, W., Pertea, M., Allen, J.E., Orvis, J., White, O., Buell, C.R., and Wortman, J.R. (2008). Automated eukaryotic gene structure annotation using EVidenceModeler and the Program to Assemble Spliced Alignments. Genome Biol. *9*, R7.

Haas, B.J., Wortman, J.R., Ronning, C.M., Hannick, L.I., Smith, R.K., Jr., Maiti, R., Chan, A.P., Yu, C., Farzad, M., Wu, D., *et al.* (2005). Complete reannotation of the *Arabidopsis* genome: methods, tools, protocols and the final release. BMC Biol. *3*, 7.

Haque, R., Huston, C.D., Hughes, M., Houpt, E., and Petri, W.A., Jr. (2003). Amebiasis. N. Engl. J. Med. *348*, 1565–1573.

Huson, D.H., Reinert, K., Kravitz, S.A., Remington, K.A., Delcher, A.L., Dew, I.M., Flanigan, M., Halpern, A.L., Lai, Z., Mobarry, C.M., *et al.* (2001). Design of a compartmentalized shotgun assembler for the human genome. Bioinformatics *17* Suppl *1*, S132–139.

Jensen, S., Cavarec, L., Dhellin, O., and Heidmann, T. (1994). Retrotransposition of a marked *Drosophila* line-like I element in cells in culture. Nucleic Acids Res. *22*, 1484–1488.

Jurka, J., Kapitonov, V.V., Pavlicek, A., Klonowski, P., Kohany, O., and Walichiewicz, J. (2005). Repbase Update, a database of eukaryotic repetitive elements. Cytogenet. Genome Res. *110*, 462–467.

Loftus, B., Anderson, I., Davies, R., Alsmark, U.C., Samuelson, J., Amedeo, P., Roncaglia, P., Berriman, M., Hirt, R.P., Mann, B.J., *et al.* (2005). The genome of the protist parasite *Entamoeba histolytica*. Nature *433*, 865–868.

Lorenzi, H., Thiagarajan, M., Haas, B., Wortman, J., Hall, N., and Caler, E. (2008). Genome wide survey, discovery and evolution of repetitive elements in three *Entamoeba* species. BMC Genomics *9*, 595.

Lynch, M., and Conery, J.S. (2003). The origins of genome complexity. Science *302*, 1401–1404.

Mungall, C.J., and Emmert, D.B. (2007). A Chado case study: an ontology-based modular schema for representing genome-associated biological information. Bioinformatics *23*, i337–i346.

Nishihara, H., and Okada, N. (2008). Retroposons: genetic footprints on the evolutionary paths of life. Methods Mol. Biol. *422*, 201–225.

Nozaki, T., Kobayashi, S., Takeuchi, T., and Haghighi, A. (2006). Diversity of clinical isolates of *Entamoeba histolytica* in Japan. Arch. Med. Res. *37*, 277–279.

Okuda, S., Yamada, T., Hamajima, M., Itoh, M., Katayama, T., Bork, P., Goto, S., and Kanehisa, M. (2008). KEGG Atlas mapping for global analysis of metabolic pathways. Nucleic Acids Res. *36*, W423–426.

Parija, S.C., and Khairnar, K. (2005). *Entamoeba moshkovskii* and *Entamoeba dispar*-associated infections in Pondicherry, India. J. Health Popul Nutr *23*, 292–295.

Petri, W.A., Jr. (1996). Recent advances in amebiasis. Crit. Rev. Clin. Lab. Sci. *33*, 1–37.

Petri, W.A., Jr., Haque, R., Lyerly, D., and Vines, R.R. (2000). Estimating the impact of amebiasis on health. Parasitol. Today *16*, 320–321.

Pritham, E.J., Feschotte, C., and Wessler, S.R. (2005). Unexpected diversity and differential success of DNA transposons in four species of *Entamoeba* protozoans. Mol. Biol. Evol. *22*, 1751–1763.

Ren, Q., Chen, K., and Paulsen, I.T. (2007). TransportDB: a comprehensive database resource for cytoplasmic membrane transport systems and outer membrane channels. Nucleic Acids Res. *35*, D274–279.

Saier, M.H., Jr., Tran, C.V., and Barabote, R.D. (2006). TCDB: the Transporter Classification Database for membrane transport protein analyses and information. Nucleic Acids Res. *34*, D181–186.

Sharma, R., Bagchi, A., Bhattacharya, A., and Bhattacharya, S. (2001). Characterization of a retrotransposon-like element from *Entamoeba histolytica*. Mol. Biochem. Parasitol. *116*, 45–53.

Shire, A.M., and Ackers, J.P. (2007). SINE elements of *Entamoeba dispar*. Mol. Biochem. Parasitol. *152*, 47–52.

Sonnhammer, E.L., Eddy, S.R., Birney, E., Bateman, A., and Durbin, R. (1998). Pfam: multiple sequence alignments and HMM-profiles of protein domains. Nucleic Acids Res. *26*, 320–322.

Thompson, J.D., Higgins, D.G., and Gibson, T.J. (1994). CLUSTAL W: improving the sensitivity of progressive multiple sequence alignment through sequence weighting, position-specific gap penalties and weight matrix choice. Nucleic Acids Res. *22*, 4673–4680.

Van Dellen, K., Field, J., Wang, Z., Loftus, B., and Samuelson, J. (2002). LINEs and SINE-like elements of the protist *Entamoeba histolytica*. Gene *297*, 229–239.

WHO (1997). WHO/PAHO/UNESCO report. A consultation with experts on amoebiasis. Mexico City, Mexico 28–29 January, 1997. Epidemiol. Bull. *18*, 13–14.

Wicker, T., Sabot, F., Hua-Van, A., Bennetzen, J.L., Capy, P., Chalhoub, B., Flavell, A., Leroy, P., Morgante, M., Panaud, O., *et al.* (2007). A unified classification system for eukaryotic transposable elements. Nat. Rev. Genet. *8*, 973–982.

The Genome of *Giardia* and Other Diplomonads

2

Jan O. Andersson, Jon Jerlström-Hultqvist and Staffan G. Svärd

Abstract

Diplomonads are a group of mitochondrion-lacking, binucleated flagellates found in anaerobic or micro-aerophilic environments. Most research on diplomonads has focused on *Giardia*, which is a major cause of water-borne enteric disease in humans and other animals. The first diplomonad to have its genome sequenced was a *Giardia* isolate (WB) and the 11.7 million basepair genome is compact in structure and content with simplified basic cellular machineries and metabolism. Currently the genomes of several other *Giardia* isolates and diplomonads (the fish pathogens *Spironucleus vortens* and *S. salmonicida*) are being sequenced. Here we will review the current knowledge of the content, regulation and evolution of the genomes of diplomonads, with a main focus on the genus *Giardia*.

Introduction

The scientific story of *Giardia* and the other diplomonads begins in Delft, Holland, in the latter part of the 17th century. An energetic and curious tradesman, Antonie van Leeuwenhoek, started to construct microscopes using home-made lenses and investigated a world previously undetectable to the human eye. He opened the door to a world of unicellular organisms and made some spectacular findings by sampling his own personal environment. He observed 'animalcules' in water and in scrapings from his wife's teeth. He was the first to observe bacteria and red blood cells. Antonie van Leeuwenhoek documented all his findings in Dutch, from 1673 until his death in 1723. He sent the reports to the Royal Society in London. In one of these letters, he described *Giardia* trophozoites in his own diarrheic stool, making him the father of protozoology (Finlay and Esteban, 2001).

Since the initial discovery of *Giardia* by Leeuwenhoek several other organisms related to *Giardia* have been discovered and grouped into the diplomonads. Diplomonads are a group of mitochondrion-lacking, binucleated flagellates found in anaerobic or micro-aerophilic environments (Simpson, 2003). Most research on diplomonads has focused on *Giardia*, which is a major cause of water-borne enteric disease in humans and other animals in both industrialized and developing countries (Adam, 2001). However, there are important variations in lifestyle among diplomonads. For example, many species of *Spironucleus* are commensals or parasites associated with fish (Jørgensen and Sterud, 2006), *Trepomonas* and enteromonads are associated with various animals (Kolisko *et al.*, 2005) while *Hexamita* is considered free-living in aquatic environments rich in organic matter and deficient in oxygen (Brugerolle and Lee, 2002).

A genome project designed to obtain the complete genome sequence of a *Giardia* isolate (WB) officially started in February 1998, 300 years after the first observation of *Giardia*, and it was published in late 2007 (Morrison *et al.*, 2007). A high-throughput shotgun sequencing approach was used where the genome was fractionated and randomly sequenced until each position was covered on average by 11 individual sequence reads. The sequence reads have been assembled into 306 contiguous sequences (contigs) organized into 92 scaffolds (linked contigs). Together these cover 11.7 million basepairs (Mbp), which is estimated to represent 96% of the total genome size. More information about the genome project can be obtained from the *Giardia* genome database at http://www.giardiadb.org/giardiadb/.

Currently the genomes of several isolates from other *Giardia* genotypes (assemblages) are being sequenced and some of the data will be discussed here. Much less is known about the genomes of the other diplomonads. A genome survey project using expressed sequence tags (EST) and genomic survey sequences (GSS) has been completed in *Spironucleus salmonicida* (Andersson *et al.*, 2007), a full genome project of *S. vortens* was started a few years ago at the DOE-Joint Genome Institute, and complete genome sequencing of *S. salmonicida* will be performed by us. Here we will review our current knowledge of the content, regulation and evolution of the genomes of diplomonads, with a main focus on the genus *Giardia*.

Phylogeny and morphology of Diplomonads

Diplomonads were previously thought to form one of the earliest diverging lineages within the eukaryotic part of the universal tree of life (Sogin *et al.*, 1989). They were described as 'biological fossils', true eukaryotes but with many peculiarities, such as 'bacterial-like' metabolism, two nuclei, and a lack of aerobic mitochondria and introns (Cavalier-Smith, 1993; Upcroft and Upcroft, 1998). However, advances in genomics, molecular phylogenetics and cell biology during the last two decades strongly suggest that this view is incorrect.

Diplomonads in fact have most of the features previously thought to be lacking. Genes derived from the mitochondrial endosymbiont have been identified in the diplomonads' genomes (Horner and Embley, 2001; Roger *et al.*, 1998) and a double-membrane-bound organelle (mitosome) of likely mitochondrial origin has been identified in *G. lamblia* (Tovar *et al.*, 2003). The *S. salmonicida* genome survey identified genes coding for proteins associated with the mitosome in *Giardia*, suggesting that the mitosome might be a general feature of diplomonads (Andersson *et al.*, 2007). Mitochondria have indeed been found to be an ancestral feature of all present-day eukaryotes and all eukaryotes examined have been found to have retained the organelle in some form (Embley and Martin, 2006). Similarly, introns also have been found to be an ancestral feature of eukaryotes (Collins and Penny, 2005) and introns have now been reported in *Giardia* (Nixon *et al.*, 2002; Russell *et al.*, 2005). The 'bacterial-like' metabolism identified in *G. lamblia* is the result of lateral gene transfers (Andersson *et al.*, 2003; Andersson *et al.*, 2007; Field *et al.*, 2000; Morrison *et al.*, 2007), probably as a secondary adaptation to an anaerobic lifestyle (Andersson, 2006) rather than retention of an ancient eukaryotic metabolism.

The current interpretation of the phylogeny of eukaryotes lends no support for diplomonads being the earliest eukaryotic branch (Adl *et al.*, 2005; Simpson *et al.*, 2006); diplomonads are now classified within the taxon Excavata (Adl *et al.*, 2005). However, this eukaryotic super-group is based at present on cell ultrastructural data (Simpson, 2003). Multi-gene phylogenies have been inconclusive regarding its monophyly (Parfrey *et al.*, 2006; Simpson *et al.*, 2006), making it formally possible that the root of eukaryotes lie within Excavata. At any rate, monophyly of the diplomonads and parabasalids (*Trichomonas* etc.) is solid, which rules out the root being located on the branch leading to diplomonads (Andersson *et al.*, 2005; Simpson *et al.*, 2006).

Several unexpected results have been obtained using molecular phylogenetic methods on diplomonads and their close relatives (Fig. 2.1). The diplokaryotic diplomonads were recently classified within the taxon Fornicata together with the monokaryotic enteromonads, *Carpediemonas*, and retortamonads (Simpson, 2003). The first molecular data from *Retortamonas* surprisingly placed this group as a weakly supported sister group to *Giardia* within the diplomonads, although this was considered a likely artefact of the method at the time (Silberman *et al.*, 2002). Subsequent analyses with larger datasets confirmed *Retortamonas* as a sister to diplomonads (Kolisko *et al.*, 2008). The support for enteromonads, another monokaryotic group, being nested within diplomonads is solid, however (Jørgensen and Sterud, 2006; Kolisko *et al.*, 2005) (Fig. 2.1). This group had been considered a sister group to diplomonads

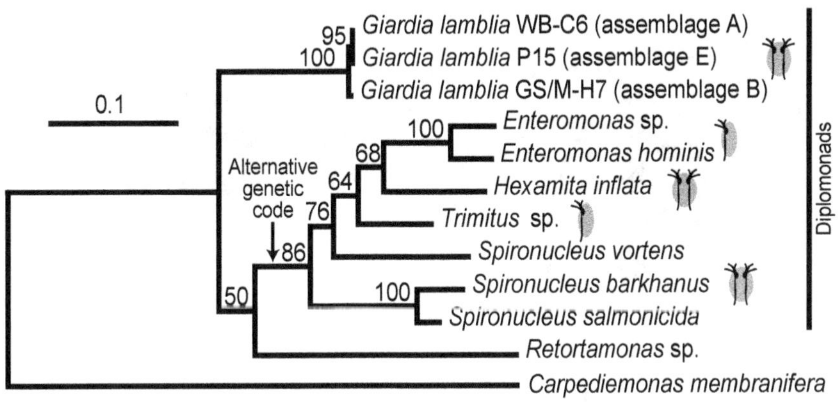

Figure 2.1 Phylogeny of Fornicata. Protein maximum likelihood tree based on 405 unambigously aligned amino acid positions of heat shock protein 90 from diplomonads, *Retortamonas* and *Carpediemonas*. The protein maximum likelihood analysis was performed with Phyml, version 2.4.5 (Guindon and Gascuel, 2003), using the WAG + Γ + I substitution model. Bootstrap support values are based on 100 replicates. Monokaryotic and diplokaryotic taxa within diplomonads are indicated.

based on ultrastructural data (Brugerolle and Lee, 2002). Unexpectedly, enteromonad sequences are clustered with diplomonads such as *Trepomonas, Hexamita* and *S. vortens* to the exclusion of *S. salmonicida* in molecular phylogenies (Jørgensen and Sterud, 2006; Kolisko *et al.*, 2005). The presence of an alternative genetic code in enteromonads (Kolisko *et al.*, 2008), in which TAA and TAG encode glutamine instead of being stop codons, that was previously reported from *Spironucleus, Trepomonas,* and *Hexamita* (Keeling and Doolittle, 1997) lends additional support for enteromonads being nested within diplomonads (Fig. 2.1). These results suggest either that the diplokaryotic state of diplomonads arose multiple times independently, or that the single nucleus of enteromonads is a derived feature (Kolisko *et al.*, 2005).

Currently, it is impossible to define the Fornicata on the basis of morphology, and monophyly of the group has only been demonstrated by molecular-phylogenetic studies, as described above. All Fornicata are flagellated and the ancestral number of flagella in the fornicate karyomastigont is probably four; however, the number of basal bodies and/or flagella has been reduced or duplicated independently in several lineages (Fig. 2.1). The diplomonads can be divided into diplozoic and unizoic diplomonads depending on the number of karyomastigonts per cell (Simpson, 2003). Most diplomonads are diplozoic (e.g. *Giardia*) and they have most cellular structures doubled. They appear like two cells fused back-to-back creating a cell with double symmetry with two nuclei, two flagellar apparatuses, and so on. Unizoic diplomonads (e.g. *Enteromonas* and *Trimitus*) possess only one karyomastigont and a cytostome (which has been lost in *Giardia*) per cell (Fig. 2.1). The cytostome forms a groove on the ventral surface of the cell and is used for up-take of nutrients (Simpson, 2003). Based on the presence or absence of cytostomes, diplozoic diplomonads have been divided into Hexamitinae and Giardiinae. Hexamitinae possess cytosomes that are either tube-like, passing through the whole cell and opening posteriorly (*Hexamita* and *Spironucleus*), or pocket-like grooves that also harbour three flagella (*Trepomonas*). Representatives of different Hexamitinae genera are easily recognizable. However, although *Hexamita* and *Spironucleus* are very similar they differ in the shape of the nuclei (oval in *Hexamita*, kidney-shaped in *Spironucleus*) and in the position of the basal bodies (Simpson, 2003). Giardiinae genera (*Octomitus, Brugerolleia* and *Giardia*) have lost the cytostomes with the associated cytoskeleton. *Octomitus* and *Brugerolleia* resemble diplozoic Hexamitinae diplomonads, but *Giardia* has a rather different morphology (Fig. 2.2).

It is likely that the life cycle of all diplomonads includes a trophozoite and a cyst form (Fig. 2.2). The cysts of both diplozoic and unizoic diplomonads are tetranucleated, oval and are surrounded by a thick cyst-wall (Fig. 2.2) (Simpson,

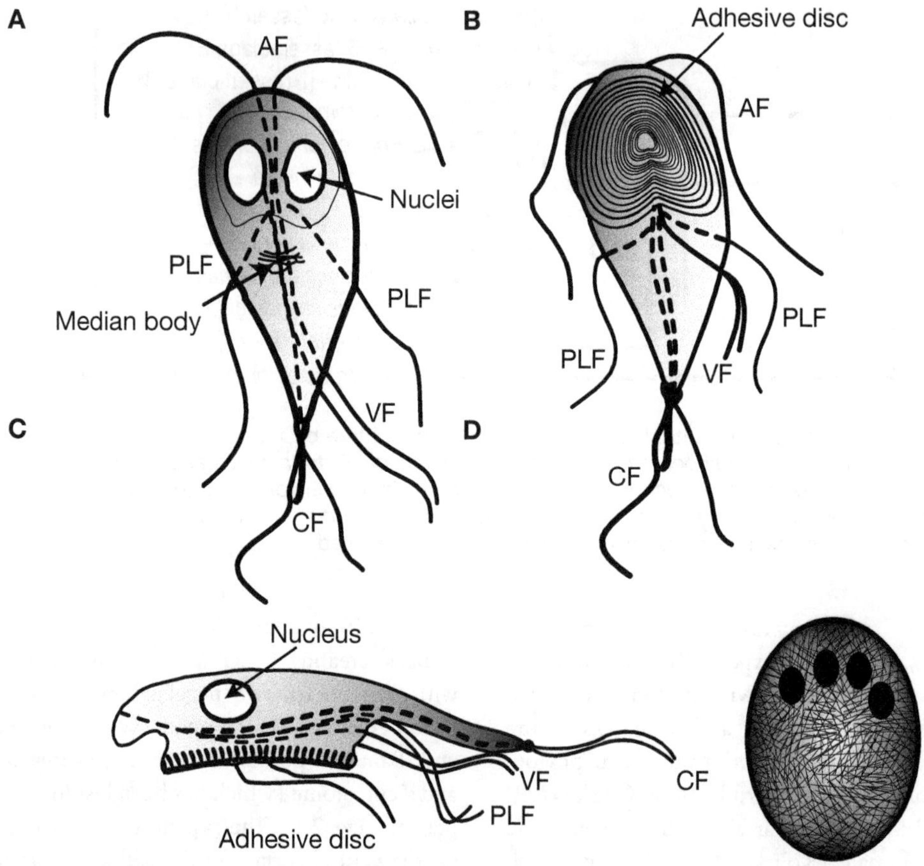

Figure 2.2 Ultrastructure of *Giardia.* (A) Trophozoite (dorsal view) with two nuclei and the median body just beneath. (B) Trophozoite (ventral view) showing the adhesive disc, an attachment structure specific for *Giardia*. (C) Trophozoite (lateral view) showing the flagella in the interior of the parasite. All the eight basal bodies localize between the two nuclei, (D) Cyst with four nuclei. AF, anterior flagella; CF, caudal flagella; PLF, posterior lateral flagella; VF, ventral flagella.

2003). The cysts are used in the transmission of diplomonads between hosts and for surviving unfavourable conditions in the case of free-living diplomonads. The best known and most widely studied diplomonad is the intestinal parasite *Giardia* and we will discuss the morphology of *G. lamblia* further here.

The giardial life cycle alternates between a vegetative trophozoite stage, which colonizes and attaches to the upper part of the small intestine, and the infective, dormant cyst stage (Fig. 2.2) via the differentiation processes of encystation and excystation. *G. lamblia* trophozoites measure 12–15 μm in length and 5–9 μm in width. The trophozoite display bilateral symmetry and it has the shape of a pear bisected lengthwise (Fig. 2.2), where the cut side of the pear represents the ventral surface of the cell and the convex side the dorsal surface. The anterior (front) ventral half of the parasite is covered by a unique attachment organelle, the adhesive disc (Fig. 2.2). This unique attachment structure is missing in all the other diplomonads and it most likely evolved in *Giardia* to make it possible for the trophozoites to attach efficiently to the intestinal epithelium. Another unique cytoskeletal structure is the median body (Fig. 2.2) but its function is still not known. Four pairs of flagella originate from the basal bodies between the two nuclei (Fig. 2.2) and the basal bodies are very close to the nuclear membrane. The nuclei have no distinct, visible nucleoli but recent data show that there is indeed a true nucleolus in each nucleus (Saraiya and Wang, 2008; Jimenez-Garcia *et al.*, 2008). As in other eukaryotes the nuclei are surrounded by an inner and an outer nuclear membrane. The outer nuclear membrane of a trophozoite has sparsely scattered ribosomes, whereas encysting cells have a nuclear envelope studded with ribosomes as well as a closely associated structure resembling endoplasmic reticulum (ER) (Benchimol, 2004). The nuclear membrane is different between the two nuclei; one nucleus has more nuclear pores than the other (Benchimol, 2004, 2005). The nuclei appear to be partitioned equationally at cytokinesis of trophozoites (Yu *et al.*, 2002) and each nucleus appears to be enclosed by a strong net of microtubules that is important during the partitioning of the nuclei during cell division (Sagolla *et al.*, 2006).

Trophozoites divide by binary fission and colonize the intestinal epithelium from duodenum to mid-jejunum. They attach to the epithelial lining for 12–19 days post-infection, producing symptoms consisting of diarrhoea and abdominal pain. As the parasite travels further down the gastrointestinal (GI) tract, a number of environmental stimuli in the jejunum trigger differentiation into the cyst form (Fig. 2.2). The cysts are morphologically simple with an oval cyst wall, fragmented flagella and adhesive disc, and four nuclei (Fig. 2.2). Here the localization of the nuclei seems to be more flexible and they can move and localize to different parts of the cysts (Poxleitner *et al.*, 2008). This could be due to the disassembly of the cytoskeleton during encystation. Parasites are flushed out from the host's GI system in the form of water-resistant cysts and are immediately ready to infect a new host. Cell numbers favours the probability of infecting a new host, since a patient can shed 10^8 viable cysts per gram of faecal material and the infectious dose has been reported to be as low as ten to a hundred cysts (Rendtorff, 1954).

Genotyping and classification of *Giardia*

Giardia infecting humans and other vertebrates have been rediscovered and renamed many times since Leeuwenhoek's initial observations. Today it is known that the genus *Giardia* infects a wide range of different animals (Caccio and Ryan, 2008). The nomenclature of the parasite was first based upon morphology of the trophozoite and host-specificity (Adam, 2001). A landmark study published by Filice (1952) suggested a classification into only three species based on differences in morphology: *G. lamblia, G. agilis* and *G. muris*. *G. lamblia* is pear-shaped and has one or two claw-shaped median bodies; *G. agilis* is long and slender and has a teardrop-shaped median body and the *G. muris* trophozoite is shorter and rounder with a small, rounded median body. Three additional *Giardia* species have been described on the basis of electron microscopic and DNA sequence differences: *G. psittaci, G. ardeae* and *G. microti* (Table 2.1). The use of different names for *Giardia* species infecting humans persist (Caccio and Ryan, 2008). Thus, *Giardia* affecting humans have three synonymous names; *G. lamblia, G. duodenalis* and *G. intestinalis*. For simplicity, we will consistently use the name *G. lamblia* in this review. However, it would be very good if the *Giardia* community could agree on the names of different *Giardia* variants in order to avoid unnecessary confusion; the use of alternative names is a serious complication for communication with medical personnel, researchers outside the field and the general public.

Studies of *G. lamblia* isolates recovered from faecal cysts extracted from *Giardia*-infected humans, using a range of phenotypic and genetic criteria, show that they fall into two major genetic groups or genotypes (also called assemblages), A and B (Caccio and Ryan, 2008). Genotype A isolates are easier to grow *in vitro* and most early studies of *Giardia* were performed on Genotype A trophozoites. However, epidemiological studies indicate that Genotype B isolates currently are more common in humans world-wide (Caccio and Ryan, 2008). These two genotypes are found in other animals as well, for example livestock, cats, dogs and rats (Table 2.1). There are also isolates found in a variety of animals and belonging to the morphological group of *G. lamblia* that are genetically distinct from the human-associated *Giardia* Genotypes A and B (Genotypes C-G) (Caccio and Ryan, 2008). Genotypes C and D *Giardia* parasites have been identified in dogs, wolves, coyotes and cats, Genotype E in cattle, sheep, pigs, goats and water buffalo, Genotype F in cats and Genotype G in rats (Table 2.1). It is worth noting that even though these seven genotypes are all classified as *G. lamblia*, they vary substantially and may ultimately be designated as separate species.

Table 2.1 *Giardia* classification and host range

Species	Genotype	Host range
G. lamblia	Assemblage A	Humans, livestock, cats, dogs, beavers and guinea pigs
	Assemblage B	Humans, dogs, beavers, rats
	Assemblage C	Dogs, wolves, coyotes and cats
	Assemblage D	Dogs, wolves, coyotes and cats
	Assemblage E	Cattle, sheep, pigs, goats and water buffaloes
	Assemblage F	Cats
	Assemblage G	Rats
G. agilis		Amphibians
G. muris		Rodents
G. psittaci		Psittacine birds
G. ardeae		Heron birds
G. microti		Voles and muskrats

One major aim in *Giardia* research is to connect genotype with symptoms (Homan and Mank, 2001). Clinical manifestations of giardiasis are very variable and range from asymptomatic infections to acute and chronic diarrhoea, abdominal pain, nausea, vomiting, weight loss and even arthritis (Buret, 2007). The severity of infection is determined by the interaction between the parasite and the host, but studies on the possible connection between genotype of the parasite and symptoms have all failed to give a conclusive picture. Although isolates from the same genotype infect both humans and other animals, the issue of whether zoonotic transmission occurs is another unsolved question in *Giardia* research (Caccio and Ryan, 2008; Thompson *et al.*, 2008). Molecular epidemiological studies involving genotyping of isolates using PCR amplification of *Giardia* genes in DNA extracted from *Giardia*-containing faeces have addressed this issue. Only a few different molecular genotyping approaches have been used in the published studies (Caccio and Ryan, 2008). These differ in the PCR target gene(s) (small subunit ribosomal RNA-*ssu*, beta-giardin-*bg*, glutamate dehydrogenase-*gdh*, triose phosphate isomerase-*tpi* and elongation factor 1 alpha-*ef1a*), the region of the gene examined (*Giardia*-specific, *G. lamblia*-specific or genotype-specific regions) and method of analysis (PCR product size polymorphisms, restriction fragment length polymorphisms (RFLP) or DNA sequencing of PCR products). The genes used show different levels of polymorphisms with *tpi* and *gdh* genes being the most variable, followed by *bg*, *ef1a* and the *ssu* gene, which is the most highly conserved. Different research groups have been promoting their respective methods for different genotyping applications but the picture that is emerging is that at least three, and probably five, different PCR targets are needed in a multi-locus genotyping (MLG) approach in order to get enough resolution between and within different genotypes (Caccio *et al.*, 2008). Many of the Genotype A and B subtype isolates that have been identified in animals are not identical to isolates from humans, which argues against zoonotic transmission as a major route of infection, but more variable loci are needed in order to resolve this issue. More variable loci are also needed in order to resolve the relationship between different human isolates, especially within Genotype A where the DNA sequences of the various isolates tend to be very similar. We believe the field of genotyping will benefit from the data from the ongoing *Giardia* genome projects. With the availability of several complete genomes, the selection of genes for genotyping can be evaluated on objective criteria. Other genotype-specific genomic traits, such as lineage-specific genes, may also be useful for genotyping purposes.

Chromosomes and genome ploidy

Pulsed-field gel electrophoresis (PFGE) analysis of *Giardia* identified five major chromosomes between 1.6 and 3.8 Mbp, resulting in a haploid genome size estimate of about 12 Mbp (Adam, 1992; Adam *et al.*, 1988; Le Blancq and Adam, 1998). The size of the five *Giardia* chromosomes varies between different isolates, mainly due to changes in the subtelomeric regions (Adam, 1992; Hou *et al.*, 1995). Tandem repeats of ribosomal RNA genes (rDNA) are adjacent to the telomeric repeats (Adam, 1992; Hou *et al.*, 1995) and are frequently involved in subtelomeric rearrangements. A recent report suggested that there are differences in chromosomal numbers between the nuclei of isolates from different genotypes (Tumova *et al.*, 2007) but more data are needed in order to resolve this issue. This is also true for other diplomonads since virtually nothing is known about chromosome numbers or ploidy in other diplomonads. A surprisingly low level of allelic sequence divergence, less than 0.01%, was reported from the analysis of the WB genome (Morrison *et al.*, 2007), and no allelic variations were found in the *S. salmonicida* genome survey (Andersson *et al.*, 2007). However, the allelic sequence divergence seems to be higher in other *Giardia* genotypes and *Spironucleus* species (see Chapter 8). The cellular genome ploidy varies between 4N and 8N in trophozoites, and cysts have a ploidy of 16 N (Bernander *et al.*, 2001). The knowledge of ploidy differences during differentiation is limited from other diplomonads, mainly due to problems with encystation *in vitro*. This is most likely due to lack of knowledge of suitable experimental conditions rather than an absence of cystic forms in their life cycle since cysts can be found *in vivo* (Baker *et al.*, 1998; von Allmen *et al.*, 2004). Indeed, homologues of three proteins associated with the cyst stage in *G. lamblia* were identified in *S. salmonicida* (Andersson *et al.*, 2007), and cysts have been described from *Spironucleus muris* (Januschka *et al.*, 1988).

Variable G+C content and codon usage in diplomonads

An analysis of the G+C content in the contigs that have been assembled as part of the WB genome project has yielded an overall G+C content of 49%, excluding the rRNA genes (Morrison *et al.*, 2007). The intergenic regions are much more A+T-rich, especially the promoter regions (Morrison *et al.*, 2007). Recently, we have obtained a draft genome sequence of the Genotype B isolate GS using 454 sequencing. 3 separate runs generated 808,172 high-quality reads with an average length of 227 basepairs (bp), totaling 182 Mbp (16× genome coverage). This was distributed on 3118 contigs with an average length of 3535 bp (Table 2.2). The average G+C-content in the GS contigs is 46.5% in coding regions and 37% in intergenic regions, compared to 49.0% in the WB genome (Table 2.2).

Early molecular phylogenetic studies of diverse diplomonads indicated a large variation of G+C contents within the group (Keeling and Doolittle, 1997; Rozario *et al.*, 1996). Genes from *Trepomonas* showed very low G+C content, whereas *G. lamblia* genes had intermediate values. In these studies genes from isolates later described as *S. salmonicida* (Jørgensen and Sterud, 2006) had high G+C content (Keeling and Doolittle, 1997; Rozario *et al.*, 1996). In contrast, the average G+C content of the GSS sequences in the *S. salmonicida* genome survey was 35.9% (Andersson *et al.*, 2007). However, the G+C content was found to be very variable along the genome with, for example, stretches of around 30% G+C followed by regions with around 70%. The size of these G+C-rich regions seems to be at least a couple of thousand basepairs, although the shortage of longer contigs from the genome prevent any detailed study of their distribution (Andersson *et al.*, 2007). Most of the *S. salmonicida* genome consists of low G+C regions, but no drastic differences in gene density and content were observed between the high and low G+C regions. Obviously, the first few genes sequenced from the *S. salmonicida* genome (Keeling and Doolittle, 1997; Rozario *et al.*, 1996) were from the high G+C regions.

Similarly puzzling patterns have been found from analysis of codon usage in diplomonads (Andersson *et al.*, 2007; Lafay and Sharp, 1999). The codon usage in *G. lamblia* is rather predictable; a subset of codons is preferred in highly expressed genes, whereas weakly expressed genes show a much more random use of codons, most likely affected by mutational biases (Andersson

Table 2.2 Comparison of eukaryotic microbial genome content and organization

	Saccharomyces cerevisiae	*Encephalitozoon cuniculi*	*Entamoeba histolytica*[c]	*Cryptosporidium parvum*	*Giardia* WB	*Giardia* GS
Genome size (Mbp)	12.5	2.3	20.8	9.1	11.7	11.1[a]
% G+C content	38.3	47.6	24.2	30.3	49.0	46.5
Proteins encoded	5770	1997	8160	3952	4787[b]	4794
Mean CDS (bp)	1424	1077	1261	1720	1283	1260
Mean intergenic distance (bp)	515	129	709	585	372	158
Gene density, per kbp	0.48	0.97	0.39	0.44	0.58	0.42
Introns	272	2	2536	5% of genes	4	4
tRNAs	275	44	Subtelomeric arrays	45	63	69

a Number of base-pairs in the 3011 contigs of assembled data.

b Transcriptionally proven ORFs.

c Data from chapter by Caler and Lorenzi, this book.

et al., 2007; Lafay and Sharp, 1999). Probably there is a selection for translation efficiency in this parasite (Lafay and Sharp, 1999), as has been suggested for many other organisms (Knight *et al.*, 2001). Most genes in *S. salmonicida* follow this general pattern. Highly expressed genes use a reduced set of codons and weakly expressed genes tend to show codon usage in agreement with a mutational bias towards A+T (Andersson *et al.*, 2007). However, there is a third set of genes that strongly disagree with this pattern, including genes from both the highly and the weakly expressed class. These genes have higher G+C content and each gene tends to use a subset of the codons, although the favoured codons differ between genes (Andersson *et al.*, 2007). These genes with enigmatic codon usage are localized to the genomic regions with high G+C content in the *S. salmonicida* genome.

Codon usage by highly expressed genes is probably affected by mutational biases in addition to translational selection. The favoured codons in highly expressed genes differ between the two diplomonads analysed (Andersson *et al.*, 2007). In *G. lamblia*, codons ending in 'C' or 'G' are favoured in highly expressed genes, whereas in *S. salmonicida* preferred codons end in 'U' and 'A' in the case of nine and one amino acids, respectively (Andersson *et al.*, 2007). This is probably a reflection of the lower overall G+C content in the *S. salmonicida* genome. An even more drastic change of the translation machinery is the reassignment of codons in the *Spironucleus* lineage, where the canonical termination codons UAA and UAG instead encode glutamine (Keeling and Doolittle, 1997). These drastic differences clearly show that the selection acting on the translational process differs significantly between the two diplomonads.

The *Giardia* proteome

Diplomonad genomes are very compact. Both *G. lamblia* and *S. salmonicida* have gene-dense genomes with examples of putatively over-lapping genes (Andersson *et al.*, 2007; Iwabe and Miyata, 2001; Morrison *et al.*, 2007). Automatic open reading frame (ORF) identification programs identified 6470 ORFs with a mean intergenic distance of 372 base pairs in the *G. lamblia* WB genome (Morrison *et al.*, 2007). Serial analysis of gene expression (SAGE) and cDNA sequences provided transcriptional evidence for 4787 of these ORFs (Table 2.2). Automatic ORF identification in the *G. lamblia* GS genome identified 6768 ORFs and manual curation of the data resulted in a set of 4576 intact and 218 broken ORFs with a mean intergenic distance of 158 bp (Table 2.2). Approximately 74% of the genome is coding, which is similar to what was observed in the WB genome. Although the haploid genome size and the total number of ORFs in *G. lamblia* are similar to that of *Saccharomyces cerevisiae*, many eukaryotic processes appear simpler in the diplomonad than in the yeast. Analyses of the *G. lamblia* WB genome indicated that this organism encodes rudimentary forms of many cellular processes with fewer and more basic subunits in cellular machineries, and a limited metabolic repertoire with many introduced bacterial-like enzymes. DNA synthesis, transcription, RNA processing and cell division machineries are examples of simplified cellular machineries in *G. lamblia* WB (Morrison *et al.*, 2007). The 'missing' proteins in these processes were not detected in the *G. lamblia* GS genome; however, they may be highly divergent and difficult to recognize. Others may be non-essential or functionally redundant with other proteins in the same or another pathway. Complete genome sequences of other diplomonads, as well as a diversity of other protists, will be useful to determine whether these proteins have never been present in the lineages leading to diplomonads, whether they were secondarily lost within this lineage, or whether they simply are too divergent to be recognized.

Cell cycle regulation is essential for infection and induction of symptoms during giardiasis. The two *Giardia* genomes (WB and GS) encode at least three CDK1-like kinases and two mitotic cyclins, which suggests they use a eukaryotic type of cell cycle regulation. The machinery that initiates DNA replication in *Giardia* is made up of typically eukaryotic proteins and is very similar to that of trypanosomes (El-Sayed *et al.*, 2005). However, certain aspects of the composition of the replication machinery are more similar to archaea than to higher eukaryotes (Morrison *et al.*, 2007). There are only two origin recognition complex (ORC) proteins in *Giardia*, Orc1/Cdc6 and Orc4. The role of Orc1/Cdc6 in archaea appears to be both in origin recognition and loading of the MCM helicase (Bernander, 2007). However, the MCM DNA helicase in *Giardia* is composed of 6 different subunits, as in other eukaryotes. *Giardia* is also similar to archaea in the apparent absence of regulatory initiation proteins, such as Cdt1, Dpb11, Cdc45, MCM10, and Gemini. Furthermore, no single strand binding protein complex like RPA or the bacterial SSB was identified in any of the *Giardia* genomes.

The Mitotic Exit Network (MEN) coordinates nuclear division with cytokinesis and regulates actomyosin ring constriction in *S. cerevisiae.* The basic function of MEN is degradation of mitotic cyclins and/or inhibition of the mitotic CDK–cyclin complex (Piatti *et al.*, 2006). During mitotic exit in anaphase, the FEAR (CdcFourteen Early Anaphase Release) proteins release the phosphatase Cdc14 from the nucleolus (Fig. 2.3). Cdc5 allows the RAS-like GTPase Tem1 (Fig. 2.3) to accumulate in the GTP-bound form. One of the substrates of Cdc14 is another kinase, Cdc15, activating MEN signalling. This keeps Cdc14 out of the nucleolus and activates Dbf2-Mob1, which brings about cytokinesis. Cdc14 increases the level of a CDK inhibitor (Sic1) and promotes mitotic cyclin destruction. This will inhibit the mitotic CDK-cyclin activity and allow actin ring formation and cytokinesis. Regulation of cytokinesis in *Giardia* may be similar to that of yeast (Fig. 2.3). Homologues of actin, CDKs and the mitotic cyclins A and B were detected in both *Giardia* isolates. Furthermore, we identified giardial homologues of several MEN proteins, Tem1, Cdc5, Cdc14, Cdc15, Bub2 and Mob1, as well as two members of the FEAR complex, the kinase Cdc5 and the protease Esp1 (Fig. 2.3). The functions of these proteins in regulation of cytokinesis in *Giardia* remain to be determined but these results suggest an actomyosin-like cell-division mechanism in *Giardia.*

Influx of genes and metabolic adaptation in diplomonads

The metabolism of diplomonads has been described as bacterial-like (Biagini *et al.*, 2003;

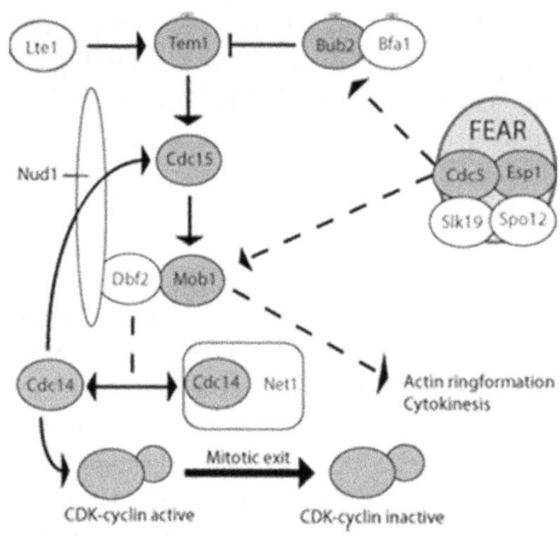

Figure 2.3 Regulatory cytokinesis proteins in *Giardia* and *S. cerevisiae*. White indicates non-identified proteins in *Giardia* and grey identified orthologues. A signalling cascade results in the assembly of an actinomyosin ring.

Upcroft and Upcroft, 1998). Phylogenetic analyses have indicated that many genes indeed have been acquired from anaerobic bacteria (Andersson *et al.*, 2003; Andersson *et al.*, 2007; Field *et al.*, 2000; Morrison *et al.*, 2007). Many of these genes encode functions associated with an anaerobic lifestyle, such as A-type flavoprotein, acetyl-CoA synthetase and alcohol dehydrogenases. These anaerobic proteins are typically also found in other anaerobic protists, such as *Entamoeba histolytica* (Andersson *et al.*, 2006; Andersson *et al.*, 2003; Field *et al.*, 2000). These protists usually have acquired these genes from different anaerobic bacteria in independent events. This suggests that two lineages independently acquired a similar set of genes from the environment during the evolutionary transition from an aerobic to an anaerobic lifestyle. This is indeed a general pattern for eukaryotic adaptation to environments deficient in oxygen (Andersson, 2006).

Phylogenomic studies of *S. salmonicida* identified 84 lateral gene transfer (LGT) candidates, corresponding to 12% of the analysed genes (Andersson *et al.*, 2007). A similar number of candidates were found in the *G. lamblia* WB genome (Morrison *et al.*, 2007). The vast majority of these genes encode proteins with either metabolic or uncharacterized functions, suggesting metabolic adaptation probably was the selective force for their maintenance in the diplomonad genomes. Most of the gene transfer events probably occurred before the diversification of diplomonads as only 19 were unique to *Spironucleus*. Twelve LGT candidates were shared with *Trichomonas*, suggesting an early introduction into the Excavata lineage before the split of diplomonads and parabasalids (Andersson *et al.*, 2007). Furthermore, 17 genes appear to have been transferred between eukaryotic lineages after an initial acquisition from prokaryotes. Most of these are exchanges with members of Amoebozoa, such as the lineage leading to *E. histolytica*. These findings indicate that there is a continuous exchange of genes between diplomonads and all domains of life (Andersson *et al.*, 2007). This is in agreement with the increased understanding of the importance of LGT in eukaryotic genome evolution (Andersson, 2009; Keeling and Palmer, 2008).

The genome sequence from a second isolate of *G. lamblia* enabled us to address whether gene acquisition is also important for more short-scale evolution within diplomonads. Surprisingly, only three *G. lamblia* WB genes, not belonging to large gene families, were found to be completely missing from the GS genome, all coding for hypothetical proteins. On the other hand, 28 genes are uniquely present in GS. Four of these are of bacterial origin, and four show similarities to a gene family associated with rolling circle replication mostly found in viruses. The remaining lineage-specific genes code for hypothetical proteins without any sequence similarities in the databases. Interestingly, one of the lineage-specific genes is flanked by two pseudogenes of recent bacterial origin in the GS genome. This suggests that a bacterial fragment coding for at least three genes was introduced recently into this lineage and that one of these genes presently is under purifying selection. Thus, comparative genomics has indicated that there is an ongoing influx of genes into the *Giardia* genome, although the biological significance of this process needs to be tested by functional and further comparative studies.

Giardia promoters

The two nuclei in *Giardia* are nearly identical in appearance and both are transcriptionally active as determined by uridine incorporation into nuclear RNA (Kabnick and Peattie, 1990). However, recent studies on the distribution and clustering of pore complexes suggest that they are morphologically different and that this may reflect differences in metabolic activity and protein synthesis (Benchimol, 2004). Cytogenetic studies also showed differences in the number of chromosomes in the two nuclei (Tumova *et al.*, 2007) and a precursor to a microRNA was only found in one of the nuclei (Saraiya and Wang, 2008). Further studies are needed to improve our understanding of nuclear transcription and whether there is a tight communication between the nuclei in order to coordinate transcription. The *Giardia* genome is very compact with the genes tightly packed and short, A+T-rich intergenic regions (Morrison *et al.*, 2007). Promoter analyses indicated that less than 50 bp upstream of an open reading frame are needed for transcriptional regulation (Elmendorf *et al.*, 2001; Knodler *et al.*, 1999; Sun *et al.*, 2002). The 5′ untranslated regions (UTRs) of mRNAs are generally less than 20 nucleotides (nt) and often only 1 to 5 nt. Thus, compared to other studied eukaryotes, *Giardia* has very short promoter sequences and extremely short 5′UTRs. Short intergenic regions appear to be a general feature of diplomonads (Andersson *et al.*, 2007). Sterile transcripts that are not translated into proteins are abundant in *Giardia*. It has been estimated that 20% of the transcripts in *Giardia* are sterile (Elmendorf *et al.*, 2001). Whether this is due to 'sloppy' transcription or antisense regulation of the transcriptome is not clear. However, this seems to be a *Giardia*-specific trait; only ~1% of the *S. salmonicida* transcripts were found to be sterile (Andersson *et al.*, 2007). The overall A+T richness in the most proximal ~50 bp region of the promoter is crucial for recognition by the parasite's transcriptional machinery and determination of transcriptional efficiency. *Giardia* protein-encoding genes have no TATA-boxes but have an initiator-like A+T-rich sequence at the transcriptional start site. Similar observations were made from the few 5′ regions of *S. salmonicida* genes (Andersson *et al.*, 2007). An artificial promoter consisting of an A+T-rich stretch of nucleotides can induce transcription of reporter genes (Elmendorf *et al.*, 2001) and many promoters seem to display bidirectional transcriptional activity (Teodorovic *et al.*, 2007).

Genomic comparative analyses of RNA polymerases and transcription factors showed that *Giardia* possesses a similar but smaller set of transcription factors compared to higher eukaryotes (Best *et al.*, 2004); only eight of the twelve general transcription factors were identified in the *Giardia* genome. TATA-binding protein is highly divergent and no TFIIB was found. This can explain the relaxed requirements for initiation of transcription in *Giardia* and the large amount of sterile transcripts. Few upstream or downstream distal regulatory elements, activators or repressors have been reported in *G. lamblia* and most specific regulatory elements have been reported for developmentally regulated genes (Davis-Hayman *et al.*, 2003; Knodler *et al.*, 1999).

The *G. lamblia* WB genome project database has been searched for eukaryotic transcription factors associated with differentiation. Two proteins belonging to the Myb transcription factor family were first identified by Sun *et al.* (2002). Myb proteins are tightly connected to activation of genes related to cell cycle progression and differentiation in other eukaryotes. Normally these proteins contain two or more Myb repeats that can bind to specific sequences in the promoter region of a target gene, thereby activating transcription. One of the Myb proteins (Myb2) found in the *G. lamblia* genome was shown to be up-regulated during encystation. Interestingly, Myb2 exhibited affinity to a double stranded DNA probe representing the −90/−46 part of the promoter region of the cyst wall protein 1 gene (*cwp1*). Further band shift experiments demonstrated that the Myb2 binding was specific to a sequence within the −90/−69 region of the *cwp1* promoter. Scanning mutagenesis of the target sequence (CTACAG) was performed to elucidate nucleotides important for binding and Myb2 was shown to bind the sequence C(T/A) AC(A/T)(G/C). Computer analyses of the WB genome showed that the sequence C(T/A) ACAG was found in the promoter region of five known *Giardia* genes: *cwp1*, *cwp2*, *cwp3*, *G6PI-B* (the protein product responsible for synthesis of a carbohydrate component of the cyst wall) and *myb2* itself. All these genes are induced during encystation and the target sequence was not found in other *Giardia* genes. Band shift analysis confirmed that Myb2 could bind to the promoter region of these genes. Only 65 bp of upstream sequence is needed for encystation-specific expression of the G6PI-B gene and the promoter region contains the Myb target sequence at position −46/−41. To determine if this region is involved in transcriptional regulation during *in vivo* encystation, a series of deletion constructs in the G6PI-B promoter was generated using luciferase as the reporter gene. Deletions down to −45 had no drastic effect on expression during encystation. However, deletion to −42 caused a fivefold reduction of the expression. Finally, the target sequence was fused to multiple copies of the G6PI-B promoter, which further increased the induction of expression during encystation. A constitutive promoter (*ran*) conferred encystation inducibility when multiple copies of the target sequence were added (Sun *et al.*, 2002).

Recent experiments have shown that over-expression of Myb2 results in an increased expression of encystation-specific genes (Huang *et al.*, 2008) and Myb2 over-expressing trophozoites have an increased capacity to differentiate into cysts. Chromatin immunoprecipitation assays confirmed that Myb2 binds to promoters with its binding sites *in vivo* and *myb2* antisense constructs reduced the levels of cyst-wall transcripts and cyst formation. In conclusion, these data strongly suggests that Myb2 is involved in gene regulation during encystation of *Giardia*. These studies have been complemented by other similar studies, and a GARP-like protein and an A+T–rich interaction domain (ARID) protein have been shown to bind to promoters of cyst wall protein and other genes (Sun *et al.*, 2006; Wang *et al.*, 2007). Thus, within the relatively short encystation-specific promoters (65bp) of *Giardia* several different transcription factors bind and regulate gene expression. This compaction most likely reflects the pressure to reduce the size of intergenic regions in *Giardia*.

mRNA capping, introns and polyadenylation in diplomonads

The 5′m^{7}Gppp5′N cap is a distinctive feature of RNA pol II transcripts in higher eukaryotes. It is important for translation, stability and processing of mRNAs and is produced early during the transcription of mRNAs, when the nascent mRNA 5′end is exiting the RNA polymerase. The presence of caps on giardial mRNAs has been controversial. However, recent data have identified RNA triphosphatase, guanyltransferase and RNA (guanine-N7) methyltransferase genes in *Giardia* (Hausmann *et al.*, 2005; Hausmann and Shuman, 2005), indicating that *Giardia* actually do have capped RNAs. The triphosphatase belongs to the tunnel family of metal-dependent phosphohydrolases and it is encoded separately from the guanyltransferase gene (Hausmann *et al.*, 2005). The guanyltransferase and 7-methyltransferase proteins interact with the C-terminal domain (CTD) tail of RNA polymerase II of higher eukaryotes and catalyse the two last steps of capping, coupling capping and transcription. The giardial catalytic subunit of RNA polymerase II has a short, divergent CTD tail, which makes it uncertain if these transferase interactions occur in *Giardia* (Seshadri *et al.*, 2003). Most native giardial mRNAs have blocked 5′ ends and 7-methylguanosine caps promote translation in *Giardia* (Hausmann *et al.*, 2005; Li and Wang, 2005). Extending the capped 5′UTR to more than 9 nt reduced the efficiency of translation (Li and Wang, 2004, 2005). Immunoprecipitation assays suggested that there are several small nuclear RNAs (snRNAs) with 5′m^{2,2,7}Gppp5′N caps (Niu *et al.*, 1994) in *Giardia*. Two enzymes (Tgs1 and 2) that potentially could be involved in production of these caps from the ordinary 5′m^{7}Gppp5′N caps that have been identified in *Giardia* (Hausmann *et al.*, 2007). Furthermore, two eukaryotic translation initiation factor 4E homologues were identified: one (eIF4E-1) that binds the 5′m^{2,2,7}Gppp5′N caps of snRNAs and one (eIF4E-2) that binds the 5′m^{7}Gppp5′N of mRNAs (Li and Wang, 2005).

Four introns have been identified so far in *Giardia* (Morrison *et al.*, 2007; Nixon *et al.*, 2002; Russell *et al.*, 2005). First, a single 35 nt intron was found in the 2Fe-2S ferredoxin gene with a canonical 3′ splice site (AG|GG) and branch point (CUAAC) but with a non-canonical 5′splice site (AA|CU) (Nixon *et al.*, 2002). During the analysis of the WB genome, three more introns were identified in genes encoding the ribosomal protein L7A, a dynein light-chain protein, and an unknown protein (Morrison *et al.*, 2007; Russell *et al.*, 2005). The last three show canonical GT/AG splice sites and contain an AC-repeat motif, [AC]CT[GA]AC[AC]CACAG. The AC-repeat motif is very similar to the corresponding motif in *Trichomonas* introns, suggesting similarities in the splicing mechanisms. A genomic survey showed that many of the around 100 proteins involved in splicing in higher eukaryotes are missing in *Giardia* (Svärd, unpublished). There are very large differences in the complexes that bind the U RNAs that participate in mRNA splicing; in the WB and GS genomes we only detected Prp39 among the U1 snRNP proteins, SF3b49 and Sf3b155 of U2 snRNP, PRP8, U5–200kD and U5–15kD of U5 snRNP and 15.5 tri-snRNP of U4/U6 snRNP. This shows that splicing in *Giardia* is different compared to the process in higher eukaryotes. No gene coding for proteins associated with splicing was found in the *S. salmonicida* genome, nor were there any indications of splicing from EST sequences when compared to genome sequences (Andersson *et al.*, 2007). Thus, introns seem to be rare in diplomonads, and might indeed have been completely lost in some lineages. Two introns were reported in one of the first genes sequenced from the *Carpediemonas* (Simpson *et al.*, 2002), a sister lineage of diplomonads within Excavata (Fig. 2.1). This suggests that diplomonads probably are secondarily intron-poor; intron-loss is a common evolutionary phenomenon (Jeffares *et al.*, 2006).

The 3′ UTRs of giardial mRNAs are short, typically 10–30 nt in length. Transfection of mRNAs with and without polyA tails into *Giardia* showed a need for the polyA tail in the expression of the mRNAs (Hausmann *et al.*, 2005). A putative consensus polyadenylation signal (AGTRAAY) has been identified but it is highly degenerate in certain genes. A peculiar pattern of polyadenylation was found in *S. salmonicida* (Andersson *et al.*, 2007). The 3′ UTRs are very short: in 122 of the 134 cases the distance between the termination codon and the first A in the polyA tail was between 9 and 14 nt. No strong consensus sequence for polyadenylation was found outside the unique termination codon TGA, but A seems to be preferred two bases upstream (Andersson *et al.*, 2007). The central three 'TRA' bases are the most conserved positions in the consensus polyadenylation signal identified in *Giardia* (Que *et al.*, 1996), suggesting that the termination codons in *S. salmonicida* might also function as a short but effective polyadenylation signal in the fish pathogen (Andersson *et al.*, 2007). It remains to be determined whether this is a lineage-specific trait, or if all diplomonads that use the alternative genetic code also use the termination codon as a polyadenylation signal. At least one differentially polyadenylated gene, ENC6, has been identified in *G. lamblia* (Que *et al.*, 1996). It is weakly expressed as a long transcript in trophozoites and as a more abundant, shorter transcript in encysting cells. A genomic survey of polyadenylation factors showed that *Giardia* has a very rudimentary set (Morrison *et al.*, 2007). We only found homologues of 5 (CPSF-30, -73, PAP, PABP-1 and Glc7) of the 40 proteins involved in polyadenylation in yeast. This can explain the relaxed specificity of polyadenylation in *Giardia*.

Small RNAs in *Giardia*

Small nuclear RNA molecules are involved in the splicing of nuclear pre-mRNA (snRNAs), processing of pre-rRNA (snoRNAs) and tRNA (RNase P RNA) and in the transport of proteins to the ER (SRP RNA). Most of these small RNAs from other eukaryotes have a trimethylguanosine cap at the 5′ end. In the first study of small RNAs in *Giardia* a number of candidate small RNAs from *G. lamblia* were immunoprecipated by anti-trimethylguanosine antiserum; caps were confirmed by the susceptibility of the antibody reactivity to the decapping effect of tobacco acid pyrophosphatase (Niu *et al.*, 1994). This study has been followed by several others using sequencing of clones from libraries of small RNA (Saraiya and Wang, 2008; Ullu *et al.*, 2005; Yang *et al.*, 2005), bioinformatic searches (Chen *et al.*, 2008) or a combination of both (Chen *et al.*, 2007). Currently more than 20 verified and 60 putative snoRNAs are known in *Giardia*. Most snoRNAs probably bind to rRNA and participate in RNA modification but a few of the identified snoRNAs do not bind rRNA and two have been shown to be micro RNA (miRNA) precursors (Saraiya and Wang, 2008). The miRNA precursors are processed by the enzyme Dicer, a cytoplasmic endoribonuclease III, generating 26 nt miRNAs that can inhibit translation of specific reporter mRNAs in *Giardia* (Saraiya and Wang, 2008). Interestingly, the miRNA precursor GlsR17 was found in the nucleolus of only one of the nuclei and the mature miRNA miR2 was found in the cytoplasm (Saraiya and Wang, 2008). Many mRNAs (296) potentially bind the miR2 miRNA and among these genes are a family of 22 VSP genes, suggesting a role in control of antigenic variation.

Spliceosomal RNAs have not been identified in the cloning experiments mentioned above but putative spliceosomal RNAs were identified recently by computer searches of the WB genome (Chen *et al.*, 2008). Comparisons with GS showed that only the U4 to U6 RNAs are conserved (82 to 93% sequence identity). In contrast, U1 and U2 RNAs showed less than 50% identity indicating that these are not under strong purifying selection. This is surprising considering that all other verified small RNAs showed a high level of sequence identity. It has been suggested that *Giardia* lack a 5S RNA gene (Chen *et al.*, 2008). We detected several 5S RNA genes in the WB genome and the genes showed 100% identity to 5S RNA genes in the GS genome. Their expression could be verified using Northern blot analysis (Svärd, unpublished). The 5S RNA genes are often at the ends of contigs in WB and GS, suggesting that they are repeated in the genome, and the genes are also often found close to VSP genes in G+C-rich islands in the genome. Sixty nine tRNAs could be identified in the GS isolate, compared to 63 in WB (Table 2.2). However, 45 different isotypes are found in both genomes. Most tRNAs are encoded by one gene in the GS genome. Surprisingly, the $tRNA^{Gln(TGG)}$ gene, was found in 8 copies, 7 of which contained an intron. Many tRNAs localize close to the start of downstream ORFs (3–10 bp). One such example is a $tRNA^{Met(CAT)}$ that ends only 9 bp upstream of the ATG startcodon of the DNA replication initiation protein Orc1/Cdc6 in the two genomes. It will be interesting to see how this tRNA is co-regulated with Orc1 during the *Giardia* cell cycle since it was recently shown that over-expression of $tRNA^{Met}$ in mouse fibroblasts induces cell proliferation (Marshall *et al.*, 2008).

Ribonuclease P is a ubiquitous cellular enzyme responsible for the 5′-processing of tRNA precursors in all kingdoms. The chemical composition and the enzymatic properties differ in various organisms. Bacterial RNase P is composed of a catalytic RNA (M1 RNA) and a single protein (C5). The RNA is catalytic in the absence of the protein subunit *in vitro*. Both human and *S. cerevisiae* RNase P consist of one RNA molecule and at least 10 proteins (Jarrous and Reiner, 2007). However, it was recently shown that the RNA part and two proteins (Rpp21 and Rpp29) are sufficient for 5′-processing of human tRNA precursors *in vitro* (Mann *et al.*, 2003). The giardial RNase P RNA has been cloned (Marquez *et al.*, 2005) and it is the most divergent eukaryotic RNase P RNA identified so far. Recently the RNA component of human and *Giardia* RNAse P was shown to be catalytically active *in vitro* in the absence of the proteins, as in bacteria (Kikovska *et al.*, 2007). The RNase P RNA from GS showed 91% identity to the WB RNase P RNA. Interestingly, in the two *Giardia* genomes we detected only two RNAse P protein subunits, Rpp21 and 29. This suggests that giardial RNase P has fewer protein components and the RNA is probably more important for catalysis *in vivo* than in other eukaryotes.

The signal recognition particle (SRP) is universally conserved in protein transport (Pool, 2005). The human and yeast SRP complex consists of 6 different proteins and one RNA (Pool, 2005). Only three of the six SRP proteins (SRP 19, 54 and 68) were detected in the two *Giardia* genomes. We have identified the giardial SRP RNA (Svärd, unpublished) and it is very divergent compared to other eukaryotic SRP RNAs. Thus, the SRP system in *Giardia* is very different from the system in other studied eukaryotes, but this is not so surprising since the ER-Golgi system is also different in *Giardia*.

Lineage-specific gene families

VSP and HCMp genes

The best characterized virulence factors in *Giardia* are the variant-specific surface proteins (VSPs) (Nash, 2002). The outer plasma membrane of the parasite is completely covered by a thick coat of VSPs, and genome sequencing of WB has recently shown that there are approximately 225 genes encoding different VSPs scattered across the 5 chromosomes (Morrison *et al.*, 2007). The protein products are cysteine-rich (11–12%), acidic (pI around 4.5), range in size between 30 and 200 kilodaltons, and display characteristic structures such as zinc-finger motifs and a semi-conserved hydrophobic C-terminus (Nash, 2002). Only one VSP protein is expressed per trophozoite and the cell has the ability to switch surface coats. *In vitro*, trophozoites switch their coats every 6–12 generations depending on their initial VSP (Nash, 2002). The VSPs are very immunoreactive and antibodies against VSPs can be found in serum, intestinal secretions and breast milk from *Giardia* infected humans and mice (Palm *et al.*, 2003; Stager *et al.*, 1998; Tellez *et al.*, 2005). Antibodies from the host targeting VSP have been shown to be lethal to the parasite *in vitro* and *in vivo*, suggesting that the parasite switches antigens to avoid immune responses (Nash, 2002).

The VSP 3′ UTRs have the typical *G. lamblia* polyadenylation signal, AGTRAAY, immediately preceded by the sequence ACTYAGRT, generating an extended polyadenylation signal (Svärd *et al.*, 1998). In addition to their marked 3′ similarity, the VSP genes demonstrate different degrees of identity in other regions. In some cases, there is a high degree of similarity throughout the entire VSP gene, whereas in other cases certain regions are identical and other regions completely different. This suggests that the VSP gene repertoire has been expanded by gene duplication and/or recombination followed by sequence divergence. The sequences of one VSP from two geographically separated isolates (WB from Afghanistan and G3M from Peru) were identical over the complete region that was compared (Adam, 2001), suggesting that VSP divergence is not an extremely rapid process, at least not in certain VSPs and in Genotype A isolates. Several VSP genes related to the classical VSP TSA417 have been identified in Genotype A-1 and A-2 isolates (Ey and Darby, 2002; Ey *et al.*, 1998). Comparisons have shown that the differences between sub-genotypes are smaller than the differences between members of this VSP gene family within an isolate, suggesting that the divergence among the various members of the TSA417 gene family occurred before the divergence between genotypes A-1 and A-2.

Southern blot analyses originally showed that different *Giardia* isolates have markedly divergent VSP repertoires (Nash and Mowatt, 1992). We identified 181 unique, complete VSP genes in the WB genome (Svärd, unpublished). A search for similar VSP proteins in GS using the tblastn algorithm on all DNA sequences generated in the GS genome project showed that the average amino-acid identity was only 55.0% and no identical VSP proteins were identified (spread of identity 70–33%). The fraction of the reads containing VSPs suggested that there are around 200 VSP genes in GS. The only conserved regions in the proteins were the CXXC motifs, the membrane-spanning region and the CRGKA cytoplasmic tail. This shows that the VSP repertoires are completely different in the two strains and this may reflect observed differences in infectivity rates and clinical outcome.

Allele-specific expression of the VSP genes and the absence of sequence alteration or DNA rearrangements (Yang and Adam, 1994) associated with antigenic variation suggest the possibility of an epigenetic form of control for VSP gene expression. Nash *et al.* showed that VSP genes that are activated change the pattern of histone acetylation in their promoter region (Kulakova *et al.*, 2006), suggesting an epigenetic mechanism of antigenic variation in *Giardia*. Lujan *et al.* have recently shown that Dicer, RNA-dependent RNA polymerase and Argonaute are needed to repress the non-expressed VSPs in the genome (Prucca *et al.*, 2008). VSP switching seems to occur randomly and the authors suggested that local variations in the concentration of different VSP transcripts circumvent the silencing system. However, it remains to be seen how allele-specific expression is achieved in *Giardia*.

During the analysis of VSP genes in the WB genome, a new class of cysteine-rich membrane proteins was identified: the high-cysteine membrane proteins (HCMps) (Morrison *et al.*, 2007). There are 61 HCMps (with 10% or more cysteine residues and 20 or more CXXC or CXC motifs) in the WB genome. HCMps lack the CRGKA tail, and their single membrane-spanning domain diverges from the VSPs. One of the HCMps named HCNCp, is up-regulated during encystation and localizes to the cyst wall (Davids *et al.*, 2006).

One hundred and forty nine of the genes identified in the *S. salmonicida* genome had a cysteine content of >10% (Andersson *et al.*, 2007). Most of the cysteines in these were found as CXXC motifs, similar to *G. lamblia* VSPs. However, no conserved membrane-spanning region or CRGKA cytoplasmic tail was identified (Andersson *et al.*, 2007), making them superficially more similar to the *G. lamblia* HCMps. Interestingly, the majority *S. salmonicida* cysteine-rich proteins showed higher sequence similarity to ciliate homologues than those in *Giardia*. This argues against a close relationship between the protein families in the two diplomonads; rather, they seem to be lineage-specific expansions (Andersson *et al.*, 2007). Several of the cysteine-rich proteins in *S. salmonicida* had a structural organization with a transmembrane domain and cysteine-rich domains facing the extracellular milieu. This suggests that at least some of the cysteine-rich proteins in *S. salmonicida* are surface proteins, perhaps with a role similar as the VSPs in *G. lamblia* (Andersson *et al.*, 2007).

NEK kinases

G. lamblia WB has 276 putative protein kinases including members of 43 of the 61 primordial kinase subfamilies present in widely diverged eukaryotes (Morrison *et al.*, 2007). Most notable is that 180 (70%) of the putative giardial protein kinases belong to the NIMA (Never in Mitosis Gene A)-Related Kinase (NEK) family, and that 137 of them are predicted to be catalytically inactive. By contrast, most organisms have fewer than 10 NEK kinases. The rest of the giardial kinome is very rudimentary and the non-NEK kinome is the most compact known from any eukaryote. *Giardia* has no tyrosine-specific or histidine kinases. Broad-spectrum signal transduction proteins gain specificity by localization to specific cellular target structures. *Entamoeba histolytica* has more than 80 transmembrane kinases (Loftus *et al.*, 2005) but only four predicted giardial kinases have transmembrane domains. Most giardial kinases have either ankyrin repeats or coiled-coil domains. These domains can be important for specific localization within the cell and this might reduce the need for membrane-spanning regions. Protein dephosphorylation is also critical in signal transduction networks. *Giardia* has 32 predicted protein phosphatases (Morrison *et al.*, 2007), but only one is predicted to be membrane associated.

The annexin (alpha giardin) gene family

The *Giardia* trophozoite must either attach to intestinal epithelium or swim upstream, or it will face elimination by peristalsis. Since the *Giardia* cytoskeleton is involved in both attachment and motility it has been classified as a virulence factor. The cytoskeleton is a mixture of classical eukaryotic protein homologues and *Giardia*-specific proteins and structures. Its major parts are the ventral disc, four pairs of flagella and the median body. The giardins originally described by Crossley and Holberton in the early 1980s are cytoskeletal proteins unique to *Giardia* (Crossley and Holberton, 1983). There are four types of giardins: alpha, beta, gamma and delta giardins. The alpha giardins are annexin homologues (Gerke *et al.*, 2005), beta giardin (Palm *et al.*, 2005) and delta giardin (Jenkins *et al.*, 2009) are striated fibre (SF)-assemblin homologues and the gamma giardin (Nohria *et al.*, 1992) is a protein without notable homologues. Although the giardins belong to four separate gene families they are all associated with the cytoskeleton and unique to *Giardia*.

The name annexin means 'bring/hold together' and describes the principal property of annexins, namely binding and possibly holding membranes together. Annexins are Ca^{2+}-dependent, phospholipid-binding proteins with a wide range of biological functions (Gerke *et al.*, 2005). About 160 unique annexin proteins exist and are present in more than 65 different eukaryotic species (Gerke *et al.*, 2005). Despite detailed biochemical and structural knowledge of annexins, a clear physiological function for most annexins still needs to be assigned. The presence of 21 annexin homologues in *Giardia* (Weiland *et al.*, 2005) is surprising since other eukaryotes, like *S. cerevisiae, Dictyostelium discoideum, Neurospora crassa* and *Caenorhabditis elegans,* have either none or only two to four homologues. This points towards a prolific expansion by gene duplication accompanied by selection based on functions related to the specific needs of the parasite.

To understand how the alpha giardin gene family has evolved, we looked at the distribution of the alpha giardin genes in the WB genome by using PCR with different combinations of gene-specific primers and by mapping their placement within the current *Giardia* genome sequence assembly. In previous studies, alpha-1 and -2 giardin, as well as gamma giardin, were localized to chromosome III by pulse field gradient separation and hybridization techniques. By looking at all alpha giardin genes we found that several of the genes localized to the same chromosomal region (Weiland *et al.*, 2005). Alpha-2, -4, -5, and -6 giardin genes are located in a cluster within a ~4 kilobasepair (kbp) region, and on the same scaffold alpha-1, -8, and -9 giardin genes are found in a second cluster. Alpha-1 and -8 giardin genes were found within a 6.5 kbp region. Since alpha-1 and -2 giardin localize to chromosome III (Morrison *et al.*, 2007), most alpha giardins are probably located on this chromosome. In a separate region of the assembly is a cluster containing alpha-10, -11, -15, -16, -17, and -18 giardin genes located within a ~23 kbp region. The third cluster of genes encodes alpha-7.3 and -12 giardins located within a ~15.5 kbp stretch of the genome. The remaining alpha giardins are distributed throughout the assembly (Weiland *et al.*, 2005).

To investigate expression of the alpha giardins, reverse transcriptase PCR analysis was performed to determine the alpha giardin gene-specific mRNA levels in trophozoites, encysting cells, and cysts (Weiland *et al.*, 2005). All alpha giardin genes had detectable mRNA levels and two genes, alpha-1 and -11 giardin, showed an expression level even higher than the highly expressed gene beta giardin (Weiland *et al.*, 2005). Alpha-2, -3, -6, -7.2, -7.3 and -9 giardin all had moderate expression levels, while the remaining alpha giardins had low expression levels. Most alpha giardins were expressed constantly throughout encystation, with only alpha-4, -6, -9, and -18 giardin showing a tendency towards higher expression late in encystation.

The significant homology between alpha giardins and annexins suggests that these proteins have a cell membrane adhesive property. Alpha-14 giardin (Annexin XXI) has been suggested to function as a Ca^{2+}-regulated structural element linking the phospholipid bi-layer to the underlying axoneme (Szkodowska *et al.*, 2002; Vahrmann *et al.*, 2008). Alpha-14 giardin localizes to cytoskeletal components like the median body and the eight flagella (Szkodowska *et al.*, 2002) and it binds tubulin and phospholipids (Vahrmann *et al.*, 2008). In order to determine the cellular localization of all the different alpha giardins transfections of *Giardia* trophozoites were performed with epitope tagged (AU-1) alpha giardins (Weiland *et al.*, 2005). All *Giardia* proteins were expressed with their own putative promoter to reduce the risk of mislocalization due to unnatural over-expression. Most alpha giardins showed a plasma membrane localization and/or localization to different flagella, indicating their close association to the cytoskeleton. A number of alpha giardins (alpha-2, -5, -9, -10, -14, and -17 giardin) localized to the flagella, indicating their role in flagella-dependent processes. The flagella of *Giardia*, in addition to enabling the parasite to swim, have been implicated in a number of different processes including attachment, detachment and food uptake (Elmendorf *et al.*, 2003). A few other alpha giardins (alpha-3, -5, and -17 giardin) localized to the adhesive disc. Interestingly, transfection experiments with constructs encoding alpha-4, -7.1, -8 or -11 giardin were lethal to the parasite and transfection with alpha-7.2 giardin disturbed cell division and caused parasites to encyst. A polyclonal antibody against alpha-11 giardin, produced in rabbits using recombinant protein, clearly showed specific localization to the basal bodies (Svärd, unpublished). Localization of alpha-15 and -16 giardin showed a patchy distribution along the plasma membrane, resembling vesicles. This and alpha-7.2 giardin's localization close to the cyst wall, may point towards their participation in the encystation process. The fact that epitope-tagging and over-expression of some alpha giardin proteins was lethal to the parasite, with impaired cell division and/or differentiation, indicates that the alpha giardin gene family is likely to play an important role in cell division of *Giardia*. The close association of the proteins with cytoskeletal elements probably indicates their importance for overall cell stability.

The 21 members of the annexin-like alpha giardin gene family are likely to serve different functions within the cell. Further studies will show their precise function and link to cytoskeletal processes. Similar to other annexins, one possible function for the alpha giardins could be to serve as lectins. The alpha giardins may interact with glycosaminoglycans on the epithelial lining, mediating the intriguing initial contact with the microvilli. Certain alpha giardins probably interact directly with the microfilament or microtubule system of the parasite. The two alpha giardins (alpha-11 and -14) with determined structures (Pathuri *et al.*, 2009; Pathuri *et al.*, 2007) have slightly different structures, different lipid-binding and calcium binding activities, and they localize to different regions of the flagella. This shows that the two genes have diverged after putative gene duplication and now encode proteins with different functions. The importance of alpha giardins in *Giardia* biology and their high divergence compared to homologues from other eukaryotes could be useful. A drug directed against the unique alpha giardins may impair the parasite cytoskeleton and other cellular functions and thereby be lethal for the parasite, but at the same time not disturb the cytoskeleton of the host.

Prospects for the future

Much information about the genomes of diplomonads has accumulated during recent years. However, this is only the start of an intensive research period where diplomonads will be used as model system to understand genome function and evolution, gene expression, cell biology and intestinal immune responses. Several *Giardia* genomes from isolates of different genotypes are currently being sequenced, as are genomes of other diplomonads. These genome data will make it possible to develop methods for transfection, gene expression studies and functional genomics. We think this will rapidly turn diplomonads into a model system representing a branch of eukaryotic diversity that currently is very poorly studied. Such systems from diverse eukaryotes are needed to complement the more established systems like yeast and *C. elegans* for an understanding of the basic principles for a eukaryotic cell, and to gain insights into the fascinating diversity of microbial eukaryotes.

References

Adam, R.D. (1992). Chromosome-size variation in *Giardia lamblia*: the role of rDNA repeats. Nucleic Acids Res. *20*, 3057–3061.

Adam, R.D. (2001). Biology of *Giardia* lamblia. Clin. Microbiol. Rev. *14*, 447–475.

Adam, R.D., Nash, T.E., and Wellems, T.E. (1988). The *Giardia lamblia* trophozoite contains sets of closely related chromosomes. Nucleic Acids Res. *16*, 4555–4567.

Adl, S.M., Simpson, A.G., Farmer, M.A., Andersen, R.A., Anderson, O.R., Barta, J.R., Bowser, S.S., Brugerolle, G., Fensome, R.A., Fredericq, S., *et al.* (2005). The new higher level classification of eukaryotes with emphasis on the taxonomy of protists. J. Eukaryot. Microbiol. *52*, 399–451.

von Allmen, N., Bienz, M., Hemphill, A., and Müller, N. (2004). Experimental infections of neonatal mice with cysts of *Giardia lamblia* clone GS/M-83-H7 are associated with an antigenic reset of the parasite. Infect. Immun. *72*, 4763–4771.

Andersson, J.O. (2006). Genome evolution of anaerobic protists: metabolic adaptation via gene acquisition. In Genomics and Evolution of Microbial Eukaryotes, L.A. Katz, and D. Bhattacharya, eds. (Oxford, Oxford University Press), pp. 109–122.

Andersson, J.O. (2009). Gene transfer and diversification of microbial eukaryotes. Annu. Rev. Microbiol. *63*, 177–193.

Andersson, J.O., Hirt, R.P., Foster, P.G., and Roger, A.J. (2006). Evolution of four gene families with patchy phylogenetic distributions: influx of genes into protist genomes. BMC Evol. Biol. *6*, 27.

Andersson, J.O., Sarchfield, S.W., and Roger, A.J. (2005). Gene transfers from Nanoarchaeota to an ancestor of diplomonads and parabasalids. Mol. Biol. Evol. *22*, 85–90.

Andersson, J.O., Sjögren, Å.M., Davis, L.A.M., Embley, T.M., and Roger, A.J. (2003). Phylogenetic analyses of diplomonad genes reveal frequent lateral gene transfers affecting eukaryotes. Curr. Biol. *13*, 94–104.

Andersson, J.O., Sjögren, Å.M., Horner, D.S., Murphy, C.A., Dyal, P.L., Svärd, S.G., Logsdon Jr, J.M., Ragan, M.A., Hirt, R.P., and Roger, A.J. (2007). A genomic survey of the fish parasite *Spironucleus salmonicida* indicates genomic plasticity among diplomonads and significant lateral gene transfer in eukaryote genome evolution. BMC Genomics *8*, 51.

Baker, D.G., Malineni, S., and Taylor, H.W. (1998). Experimental infection of inbred mouse strains with *Spironucleus muris*. Vet. Parasitol. *77*, 305–310.

Benchimol, M. (2004). *Giardia lamblia*: behavior of the nuclear envelope. Parasitol. Res. *94*, 254–264.

Benchimol, M. (2005). The nuclei of *Giardia lamblia*: new ultrastructural observations. Arch. Microbiol. *183*, 160–168.

Bernander, R. (2007). The cell cycle of *Sulfolobus*. Mol. Microbiol. *66*, 557–562.

Bernander, R., Palm, J.E., and Svärd, S.G. (2001). Genome ploidy in different stages of the *Giardia lamblia* life cycle. Cell. Microbiol. *3*, 55–62.

Best, A.A., Morrison, H.G., McArthur, A.G., Sogin, M.L., and Olsen, G.J. (2004). Evolution of eukaryotic transcription: insights from the genome of *Giardia lamblia*. Genome Res. *14*, 1537–1547.

Biagini, G.A., Yarlett, N., Ball, G.E., Billetz, A.C., Lindmark, D.G., Martinez, M.P., Lloyd, D., and Edwards, M.R. (2003). Bacterial-like energy metabolism in the amitochondriate protozoon *Hexamita inflata*. Mol. Biochem. Parasitol. *128*, 11–19.

Brugerolle, G., and Lee, J.J. (2002). Order Diplomonadida. In An Illustrated Guide to the Protozoa, 2nd edn, J.J. Lee, G.F. Leedale, and P. Bradbury, eds. (Lawrence, Kansas, Society of Protozoologists), pp. 1125–1135.

Buret, A.G. (2007). Mechanisms of epithelial dysfunction in giardiasis. Gut *56*, 316–317.

Caccio, S.M., Beck, R., Lalle, M., Marinculic, A., and Pozio, E. (2008). Multilocus genotyping of *Giardia duodenalis* reveals striking differences between assemblages A and B. Int. J. Parasitol. *38*, 1523–1531.

Caccio, S.M., and Ryan, U. (2008). Molecular epidemiology of giardiasis. Mol. Biochem. Parasitol. *160*, 75–80.

Cavalier-Smith, T. (1993). Kingdom Protozoa and its 18 phyla. Microbiol. Mol. Biol. Rev. *57*, 953–994.

Chen, X.S., Rozhdestvensky, T.S., Collins, L.J., Schmitz, J., and Penny, D. (2007). Combined experimental and computational approach to identify non-protein-coding RNAs in the deep-branching eukaryote *Giardia intestinalis*. Nucleic Acids Res. *35*, 4619–4628.

Chen, X.S., White, W.T., Collins, L.J., and Penny, D. (2008). Computational identification of four spliceosomal snRNAs from the deep-branching eukaryote *Giardia intestinalis*. PLoS ONE 3, e3106.

Collins, L., and Penny, D. (2005). Complex spliceosomal organization ancestral to extant eukaryotes. Mol. Biol. Evol. *22*, 1053–1066.

Crossley, R., and Holberton, D.V. (1983). Characterization of proteins from the cytoskeleton of *Giardia lamblia*. J. Cell Sci. *59*, 81–103.

Davids, B.J., Reiner, D.S., Birkeland, S.R., Preheim, S.P., Cipriano, M.J., McArthur, A.G., and Gillin, F.D. (2006). A new family of giardial cysteine-rich non-VSP protein genes and a novel cyst protein. PLoS ONE *1*, e44.

Davis-Hayman, S.R., Hayman, J.R., and Nash, T.E. (2003). Encystation-specific regulation of the cyst wall protein 2 gene in *Giardia lamblia* by multiple cis-acting elements. Int. J. Parasitol. *33*, 1005–1012.

El-Sayed, N.M., Myler, P.J., Bartholomeu, D.C., Nilsson, D., Aggarwal, G., Tran, A.N., Ghedin, E., Worthey, E.A., Delcher, A.L., Blandin, G., *et al.* (2005). The genome sequence of *Trypanosoma cruzi*, etiologic agent of Chagas disease. Science *309*, 409–415.

Elmendorf, H.G., Dawson, S.C., and McCaffery, J.M. (2003). The cytoskeleton of *Giardia lamblia*. Int. J. Parasitol. *33*, 3–28.

Elmendorf, H.G., Singer, S.M., and Nash, T.E. (2001). The abundance of sterile transcripts in *Giardia lamblia*. Nucleic Acids Res. *29*, 4674–4683.

Embley, T.M., and Martin, W. (2006). Eukaryotic evolution, changes and challenges. Nature *440*, 623–630.

Ey, P.L., and Darby, J.M. (2002). VSP417-6, a variant-specific surface protein encoded at a sixth locus within the vsp417 gene subfamily of *Giardia intestinalis*. Int. J. Parasitol. *32*, 425–436.

Ey, P.L., Darby, J.M., and Mayrhofer, G. (1998). Comparison of tsa417-like variant-specific surface protein (VSP) genes in *Giardia intestinalis* and identification of a novel locus in genetic group II isolates. Parasitology *117*, 445–455.

Field, J., Rosenthal, B., and Samuelson, J. (2000). Early lateral transfer of genes encoding malic enzyme, acetyl-CoA synthetase and alcohol dehydrogenases from anaerobic prokaryotes to *Entamoeba histolytica*. Mol. Microbiol. *38*, 446–455.

Filice, F.P. (1952). Studies on the cytology and life history of a *Giardia* from the laboratory rat. Univ. Calif. Publ. Zool. *57*, 53–146.

Finlay, B.J., and Esteban, G.F. (2001). Exploring Leeuwenhoek's legacy: the abundance and diversity of protozoa. Int. Microbiol. *4*, 125–133.

Gerke, V., Creutz, C.E., and Moss, S.E. (2005). Annexins: linking Ca2+ signalling to membrane dynamics. Nat. Rev. Mol. Cell Biol. *6*, 449–461.

Guindon, S., and Gascuel, O. (2003). A simple, fast, and accurate algorithm to estimate large phylogenies by maximum likelihood. Syst Biol. *52*, 696–704.

Hausmann, S., Altura, M.A., Witmer, M., Singer, S.M., Elmendorf, H.G., and Shuman, S. (2005). Yeast-like mRNA capping apparatus in *Giardia lamblia*. J. Biol. Chem. *280*, 12077–12086.

Hausmann, S., Ramirez, A., Schneider, S., Schwer, B., and Shuman, S. (2007). Biochemical and genetic analysis of RNA cap guanine-N2 methyltransferases from *Giardia lamblia* and *Schizosaccharomyces pombe*. Nucleic Acids Res. *35*, 1411–1420.

Hausmann, S., and Shuman, S. (2005). *Giardia lamblia* RNA cap guanine-N2 methyltransferase (Tgs2). J. Biol. Chem. *280*, 32101–32106.

Homan, W.L., and Mank, T.G. (2001). Human giardiasis: genotype linked differences in clinical symptomatology. Int. J. Parasitol. *31*, 822–826.

Horner, D.S., and Embley, T.M. (2001). Chaperonin 60 phylogeny provides further evidence for secondary loss of mitochondria among putative early-branching eukaryotes. Mol. Biol. Evol. *18*, 1970–1975.

Hou, G., Le Blancq, S.M., E., Y., Zhu, H., and Lee, M.G. (1995). Structure of a frequently rearranged rRNA-encoding chromosome in *Giardia lamblia*. Nucleic Acids Res. *23*, 3310–3317.

Huang, Y.C., Su, L.H., Lee, G.A., Chiu, P.W., Cho, C.C., Wu, J.Y., and Sun, C.H. (2008). Regulation of cyst wall protein promoters by Myb2 in *Giardia lamblia*. J. Biol. Chem. *283*, 31021–31029.

Iwabe, N., and Miyata, T. (2001). Overlapping genes in parasitic protist *Giardia lamblia*. Gene *280*, 163–167.

Januschka, M.M., Erlandsen, S.L., Bemrick, W.J., Schupp, D.G., and Feely, D.E. (1988). A comparison of *Giardia microti* and *Spironucleus muris* cysts in the vole: an immunocytochemical, light, and electron microscopic study. J. Parasitol. *74*, 452–458.

Jarrous, N., and Reiner, R. (2007). Human RNase P: a tRNA-processing enzyme and transcription factor. Nucleic Acids Res. *35*, 3519–3524.

Jeffares, D.C., Mourier, T., and Penny, D. (2006). The biology of intron gain and loss. Trends Genet. *22*, 16–22.

Jenkins, M., O'Brien, C.N., Murphy, C., Schwarz, R., Miska, K.B., Rosenthal, B.M., and Trout, J. (2009). Antibodies to the ventral disc protein delta-giardin prevent *in vitro* binding of *Giardia lamblia* trophozoites. J. Parasitol. *95*, 895–899.

Jimenez-Garcia, L.F., Zavala, G., Chavez-Munguia, B., Ramos-Godinez Mdel, P., Lopez-Velazquez, G., Segura-Valdez Mde, L., Montanez, C., Hehl, A.B., Arguello-Garcia, R., and Ortega-Pierres, G. (2008). Identification of nucleoli in the early branching protist *Giardia duodenalis*. Int. J. Parasitol. *38*, 1297–1304.

Jørgensen, A., and Sterud, E. (2006). The marine pathogenic genotype of *Spironucleus barkhanus* from farmed salmonids redescribed as *Spironucleus salmonicida* n. sp. J. Eukaryot. Microbiol. *53*, 531–541.

Kabnick, K.S., and Peattie, D.A. (1990). In situ analyses reveal that the two nuclei of *Giardia lamblia* are equivalent. J. Cell Sci. *95*, 353–360.

Keeling, P.J., and Doolittle, W.F. (1997). Widespread and ancient distribution of a noncanonical genetic code in diplomonads. Mol. Biol. Evol. *14*, 895–901.

Keeling, P.J., and Palmer, J.D. (2008). Horizontal gene transfer in eukaryotic evolution. Nat. Rev. Genet. *9*, 605–618.

Kikovska, E., Svärd, S.G., and Kirsebom, L.A. (2007). Eukaryotic RNase P RNA mediates cleavage in the absence of protein. Proc. Natl. Acad. Sci. U.S.A. *104*, 2062–2067.

Knight, R.D., Freeland, S.J., and Landweber, L.F. (2001). A simple model based on mutation and selection explains trends in codon and amino-acid usage and GC composition within and across genomes. Genome Biol. *2*, 0010.

Knodler, L.A., Svärd, S.G., Silberman, J.D., Davids, B.J., and Gillin, F.D. (1999). Developmental gene regulation in *Giardia lamblia*: first evidence for an encystation-specific promoter and differential 5′ mRNA processing. Mol. Microbiol. *34*, 327–340.

Kolisko, M., Cepicka, I., Hampl, V., Kulda, J., and Flegr, J. (2005). The phylogenetic position of enteromonads: a challenge for the present models of diplomonad evolution. Int. J. Syst. Evol. Microbiol. *55*, 1729–1733.

Kolisko, M., Cepicka, I., Hampl, V., Leigh, J., Roger, A.J., Kulda, J., Simpson, A.G., and Flegr, J. (2008). Molecular phylogeny of diplomonads and enteromonads based on SSU rRNA, alpha-tubulin and HSP90 genes: implications for the evolutionary history of the double karyomastigont of diplomonads. BMC Evol. Biol. *8*, 205.

Kulakova, L., Singer, S.M., Conrad, J., and Nash, T.E. (2006). Epigenetic mechanisms are involved in the control of *Giardia lamblia* antigenic variation. Mol. Microbiol. *61*, 1533–1542.

Lafay, B., and Sharp, P.M. (1999). Synonymous codon usage variation among *Giardia lamblia* genes and isolates. Mol. Biol. Evol. *16*, 1484–1495.

Le Blancq, S.M., and Adam, R.D. (1998). Structural basis of karyotype heterogeneity in *Giardia lamblia*. Mol. Biochem. Parasitol. *97*, 199–208.

Li, L., and Wang, C.C. (2004). Capped mRNA with a single nucleotide leader is optimally translated in a primitive eukaryote, *Giardia lamblia*. J. Biol. Chem. *279*, 14656–14664.

Li, L., and Wang, C.C. (2005). Identification in the ancient protist *Giardia lamblia* of two eukaryotic translation initiation factor 4E homologues with distinctive functions. Eukaryot. Cell *4*, 948–959.

Loftus, B., Anderson, I., Davies, R., Alsmark, U.C., Samuelson, J., Amedeo, P., Roncaglia, P., Berriman, M., Hirt, R.P., Mann, B.J., *et al.* (2005). The genome of the protist parasite *Entamoeba histolytica*. Nature *433*, 865–868.

Mann, H., Ben-Asouli, Y., Schein, A., Moussa, S., and Jarrous, N. (2003). Eukaryotic RNase P: role of RNA and protein subunits of a primordial catalytic ribonucleoprotein in RNA-based catalysis. Mol. Cell *12*, 925–935.

Marquez, S.M., Harris, J.K., Kelley, S.T., Brown, J.W., Dawson, S.C., Roberts, E.C., and Pace, N.R. (2005). Structural implications of novel diversity in eucaryal RNase P RNA. RNA *11*, 739–751.

Marshall, L., Kenneth, N.S., and White, R.J. (2008). Elevated tRNA(iMet) synthesis can drive cell proliferation and oncogenic transformation. Cell *133*, 78–89.

Morrison, H.G., McArthur, A.G., Gillin, F.D., Aley, S.B., Adam, R.D., Olsen, G.J., Best, A.A., Cande, W.Z., Chen, F., Cipriano, M.J., *et al.* (2007). Genomic minimalism in the early diverging intestinal parasite *Giardia lamblia*. Science *317*, 1921–1926.

Nash, T.E. (2002). Surface antigenic variation in *Giardia lamblia*. Mol. Microbiol. *45*, 585–590.

Nash, T.E., and Mowatt, M.R. (1992). Characterization of a *Giardia lamblia* variant-specific surface protein (VSP) gene from isolate GS/M and estimation of the VSP gene repertoire size. Mol. Biochem. Parasitol. *51*, 219–227.

Niu, X.H., Hartshorne, T., He, X.Y., and Agabian, N. (1994). Characterization of putative small nuclear RNAs from *Giardia lamblia*. Mol. Biochem. Parasitol. *66*, 49–57.

Nixon, J.E., Wang, A., Morrison, H.G., McArthur, A.G., Sogin, M.L., Loftus, B.J., and Samuelson, J. (2002). A spliceosomal intron in *Giardia lamblia*. Proc. Natl. Acad. Sci. U.S.A. *99*, 3701–3705.

Nohria, A., Alonso, R.A., and Peattie, D.A. (1992). Identification and characterization of gamma-giardin and the gamma-giardin gene from *Giardia lamblia*. Mol. Biochem. Parasitol. *56*, 27–37.

Palm, D., Weiland, M., McArthur, A.G., Winiecka-Krusnell, J., Cipriano, M.J., Birkeland, S.R., Pacocha, S.E., Davids, B., Gillin, F., Linder, E., *et al.* (2005). Developmental changes in the adhesive disk during *Giardia* differentiation. Mol. Biochem. Parasitol. *141*, 199–207.

Palm, J.E., Weiland, M.E., Griffiths, W.J., Ljungstrom, I., and Svärd, S.G. (2003). Identification of immunoreactive proteins during acute human giardiasis. J. Infect. Dis. *187*, 1849–1859.

Parfrey, L.W., Barbero, E., Lasser, E., Dunthorn, M., Bhattacharya, D., Patterson, D.J., and Katz, L.A. (2006). Evaluating support for the current classification of eukaryotic diversity. PLoS Genet. *2*, e220.

Pathuri, P., Nguyen, E.T., Ozorowski, G., Svärd, S.G., and Luecke, H. (2009). Apo and calcium-bound crystal structures of cytoskeletal potein alpha-14 giardin (Annexin E1) from the intestinal protozoan parasite *Giardia lamblia*. J. Mol. Biol. *385*, 1098–1112.

Pathuri, P., Nguyen, E.T., Svärd, S.G., and Luecke, H. (2007). Apo and calcium-bound crystal structures of Alpha-11 giardin, an unusual annexin from *Giardia lamblia*. J. Mol. Biol. *368*, 493–508.

Piatti, S., Venturetti, M., Chiroli, E., and Fraschini, R. (2006). The spindle position checkpoint in budding yeast: the motherly care of MEN. Cell Div. *1*, 2.

Pool, M.R. (2005). Signal recognition particles in chloroplasts, bacteria, yeast and mammals (review). Mol. Membr. Biol. *22*, 3–15.

Poxleitner, M.K., Carpenter, M.L., Mancuso, J.J., Wang, C.J., Dawson, S.C., and Cande, W.Z. (2008). Evidence for karyogamy and exchange of genetic material in the binucleate intestinal parasite *Giardia intestinalis*. Science *319*, 1530–1533.

Prucca, C.G., Slavin, I., Quiroga, R., Elias, E.V., Rivero, F.D., Saura, A., Carranza, P.G., and Lujan, H.D. (2008). Antigenic variation in *Giardia lamblia* is regulated by RNA interference. Nature *456*, 750–754.

Que, X., Svärd, S.G., Meng, T.C., Hetsko, M.L., Aley, S.B., and Gillin, F.D. (1996). Developmentally regulated transcripts and evidence of differential mRNA processing in *Giardia lamblia*. Mol. Biochem. Parasitol. *81*, 101–110.

Rendtorff, R.C. (1954). The experimental transmission of human intestinal protozoan parasites. II. *Giardia lamblia* cysts given in capsules. Am. J. Hyg. *59*, 229–220.

Roger, A.J., Svärd, S.G., Tovar, J., Clark, C.G., Smith, M.W., Gillin, F.D., and Sogin, M.L. (1998). A mitochondrial-like chaperonin 60 gene in *Giardia lamblia*: evidence that diplomonads once harbored an endosymbiont related to the progenitor of mitochondria. Proc. Natl. Acad. Sci. U.S.A. *95*, 229–234.

Rozario, C., Morin, L., Roger, A.J., Smith, M.W., and Müller, M. (1996). Primary structure and phylogenetic relationships of glyceraldehyde-3-phosphate dehydrogenase genes of free-living and parasitic diplomonad flagellates. J. Eukaryot. Microbiol. *43*, 330–340.

Russell, A.G., Shutt, T.E., Watkins, R.F., and Gray, M.W. (2005). An ancient spliceosomal intron in the ribosomal protein L7a gene (Rpl7a) of *Giardia lamblia*. BMC Evol. Biol. *5*, 45.

Sagolla, M.S., Dawson, S.C., Mancuso, J.J., and Cande, W.Z. (2006). Three-dimensional analysis of mitosis and cytokinesis in the binucleate parasite *Giardia intestinalis*. J. Cell Sci. *119*, 4889–4900.

Saraiya, A.A., and Wang, C.C. (2008). snoRNA, a novel precursor of microRNA in *Giardia lamblia*. PLoS Pathog. *4*, e1000224.

Seshadri, V., McArthur, A.G., Sogin, M.L., and Adam, R.D. (2003). *Giardia lamblia* RNA polymerase II: amanitin-resistant transcription. J. Biol. Chem. *278*, 27804–27810.

Silberman, J.D., Simpson, A.G., Kulda, J., Cepicka, I., Hampl, V., Johnson, P.J., and Roger, A.J. (2002). Retortamonad flagellates are closely related to diplomonads: implications for the history of mitochondrial function in eukaryote evolution. Mol. Biol. Evol. *19*, 777–786.

Simpson, A.G. (2003). Cytoskeletal organization, phylogenetic affinities and systematics in the contentious taxon Excavata (Eukaryota). Int. J. Syst. Evol. Microbiol. *53*, 1759–1777.

Simpson, A.G.B., Inagaki, Y., and Roger, A.J. (2006). Comprehensive multi-gene phylogenies of excavate protists reveal the evolutionary positions of 'primitive' eukaryotes. Mol. Biol. Evol. *23*, 615–625.

Simpson, A.G.B., MacQuarrie, E.K., and Roger, A.J. (2002). Eukaryotic evolution: early origin of canonical introns. Nature *419*, 270.

Sogin, M.L., Gunderson, J.H., Elwood, H.J., Alonso, R.A., and Peattie, D.A. (1989). Phylogenetic meaning of the kingdom concept: an unusual ribosomal RNA from *Giardia lamblia*. Science *243*, 75–77.

Stager, S., Gottstein, B., Sager, H., Jungi, T.W., and Müller, N. (1998). Influence of antibodies in mother's milk on antigenic variation of *Giardia lamblia* in the murine mother-offspring model of infection. Infect. Immun. *66*, 1287–1292.

Sun, C.H., Palm, D., McArthur, A.G., Svärd, S.G., and Gillin, F.D. (2002). A novel Myb-related protein involved in transcriptional activation of encystation genes in *Giardia lamblia*. Mol. Microbiol. *46*, 971–984.

Sun, C.H., Su, L.H., and Gillin, F.D. (2006). Novel plant-GARP-like transcription factors in *Giardia lamblia*. Mol. Biochem. Parasitol. *146*, 45–57.

Svärd, S.G., Meng, T.C., Hetsko, M.L., McCaffery, J.M., and Gillin, F.D. (1998). Differentiation-associated surface antigen variation in the ancient eukaryote *Giardia lamblia*. Mol. Microbiol. *30*, 979–989.

Szkodowska, A., Müller, M.C., Linke, C., and Scholze, H. (2002). Annexin XXI (ANX21) of *Giardia lamblia* has sequence motifs uniquely sdhared by giardial annexins and is specifically localized in the flagella. J. Biol. Chem. *277*, 25703–25706.

Tellez, A., Palm, D., Weiland, M., Aleman, J., Winiecka-Krusnell, J., Linder, E., and Svärd, S. (2005). Secretory antibodies against *Giardia intestinalis* in lactating Nicaraguan women. Parasite Immunol. *27*, 163–169.

Teodorovic, S., Walls, C.D., and Elmendorf, H.G. (2007). Bidirectional transcription is an inherent feature of *Giardia lamblia* promoters and contributes to an abundance of sterile antisense transcripts throughout the genome. Nucleic Acids Res. *35*, 2544–2553.

Thompson, R.C., Palmer, C.S., and O'Handley, R. (2008). The public health and clinical significance of *Giardia* and *Cryptosporidium* in domestic animals. Vet. J. *177*, 18–25.

Tovar, J., Leon-Avila, G., Sanchez, L.B., Sutak, R., Tachezy, J., van der Giezen, M., Hernandez, M., Müller, M., and Lucocq, J.M. (2003). Mitochondrial remnant organelles of *Giardia* function in iron-sulphur protein maturation. Nature *426*, 172–176.

Tumova, P., Hofstetrova, K., Nohynkova, E., Hovorka, O., and Kral, J. (2007). Cytogenetic evidence for diversity of two nuclei within a single diplomonad cell of *Giardia*. Chromosoma *116*, 65–78.

Ullu, E., Lujan, H.D., and Tschudi, C. (2005). Small sense and antisense RNAs derived from a telomeric retroposon family in *Giardia intestinalis*. Eukaryot. Cell *4*, 1155–1157.

Upcroft, J., and Upcroft, P. (1998). My favorite cell: *Giardia*. Bioessays *20*, 256–263.

Vahrmann, A., Saric, M., Koebsch, I., and Scholze, H. (2008). alpha14-Giardin (annexin E1) is associated with tubulin in trophozoites of *Giardia lamblia* and forms local slubs in the flagella. Parasitol. Res. *102*, 321–326.

Wang, C.H., Su, L.H., and Sun, C.H. (2007). A novel ARID/Bright-like protein involved in transcriptional activation of cyst wall protein 1 gene in *Giardia lamblia*. J. Biol. Chem. *282*, 8905–8914.

Weiland, M.E., McArthur, A.G., Morrison, H.G., Sogin, M.L., and Svärd, S.G. (2005). Annexin-like alpha giardins: a new cytoskeletal gene family in *Giardia lamblia*. Int. J. Parasitol. *35*, 617–626.

Yang, C.Y., Zhou, H., Luo, J., and Qu, L.H. (2005). Identification of 20 snoRNA-like RNAs from the primitive eukaryote, *Giardia lamblia*. Biochem. Biophys. Res. Commun. *328*, 1224–1231.

Yang, Y., and Adam, R.D. (1994). Allele-specific expression of a variant-specific surface protein (VSP) of *Giardia lamblia*. Nucleic Acids Res. *22*, 2102–2108.

Yu, L.Z., Birky, C.W., Jr., and Adam, R.D. (2002). The two nuclei of *Giardia* each have complete copies of the genome and are partitioned equationally at cytokinesis. Eukaryot. Cell *1*, 191–199.

The Genome of *Trichomonas vaginalis* 3

Jane M. Carlton, Shehre-Banoo Malik, Steven A. Sullivan, Thomas Sicheritz-Ponten, Petrus Tang and Robert P. Hirt

Abstract

The genome of *Trichomonas vaginalis* – the first from a parabasalid – was published in 2007, the culmination of a project marked by several surprises and not a little distress. The genome proved to be approximately 10 times larger and far more repetitive than expected, presenting new challenges to the standard genome sequencing pipeline, as well as to continued funding. Here we present an overview of the project, including both a historical synopsis of how the project was conceived and carried out, as well as a summary of the fascinating biology revealed by analysis of the genome. To illustrate the utility of the complete genome sequence, specific areas of the parasite's biology – including membrane trafficking and mode of reproduction – are expanded and brought up-to-date.

Introduction

Trichomonas vaginalis belongs to the Parabasalia, a highly diverged lineage within the eukaryotic supergroup Excavata (for recent reviews concerning eukaryotic parasite systematics, see Adl *et al.*, 2005; Burki and Pawlowski, 2006; Dacks *et al.*, 2008; Edgcomb *et al.*, 1998; Hampl *et al.*, 2005; Moreira *et al.*, 2007; Simpson, 2003). Excavates are free-living, commensal, symbiotic or parasitic protists characterized by a pair of flagella attached anteriorly to an 'excavated' ventral feeding groove (Simpson, 2003; Simpson and Patterson, 1999). Parabasalids together with Fornicata (*Carpediemonas*, retortamonads, and diplomonads such as *Giardia lamblia* syn. *intestinalis*) form a sub-group within excavates named the Metamonada (Arisue *et al.*, 2005; Burki and Pawlowski, 2006; Edgcomb *et al.*, 1998; Moreira *et al.*, 2007; Simpson, 2003). Morphological and molecular phylogenetic data, while controversial in details, divide parabasalids into two groups, the hypermastigids (gut symbionts of wood-feeding cockroaches and termites, e.g. Trichonymphida) and the trichomonads (symbiotic, parasitic or free-living flagellates, e.g. Trichomonadidae) (Carpenter and Keeling, 2007; Gerbod *et al.*, 2004; Hampl *et al.*, 2004; Hampl *et al.*, 2006; Kleina *et al.*, 2004). Recent phylogenetic analyses of parabasalid 18S ribosomal (r)RNA genes summarized in Fig. 3.1 illustrate the current view of relationships among parabasalids.

Trichomonads are microaerophilic and flagellated commensals, symbionts or parasites of vertebrates. They have an apparently simple life cycle dominated by an extracellular trophozoite stage (typically a pear-shaped cell with visible flagella), which is transmitted from host to host and can differentiate into amoeboid forms upon contact with host tissue. The trophozoites are characterized morphologically by multiple anterior flagella, a parabasal body (the Golgi complex), and an 'axostyle–pelta complex' composed of a sheet of microtubules extending along the longitudinal axis of the cell. Some trichomonads appear to form 'pseudocysts' under unfavourable environmental conditions (Pereira-Neves *et al.*, 2003). Trichomonads lack 'classic' mitochondria (of alpha-proteobacterial origin that are sites of oxidative phosphorylation) and peroxisomes but have specialized organelles (hydrogenosomes) that conduct anaerobic metabolism (Müller,

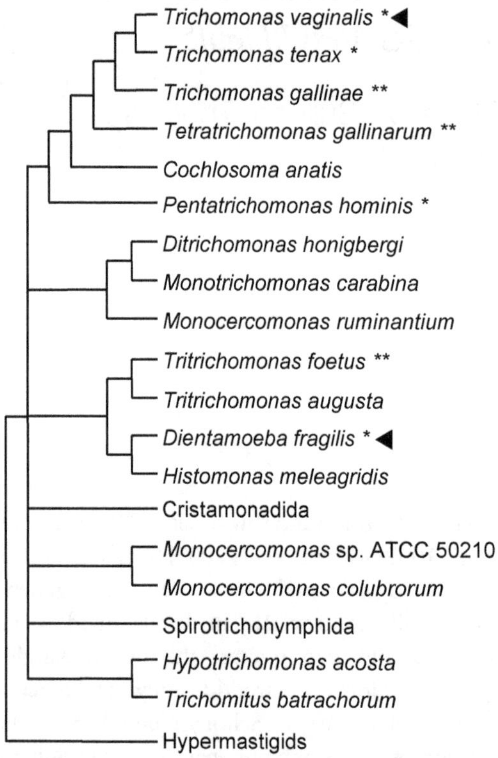

Figure 3.1 Cartoon phylogeny of parabasalids with emphasis on trichomonads. The phylogeny was redrawn from recent phylogenetic analyses of 18S rRNA genes (Carpenter and Keeling, 2007; Hampl *et al.*, 2004; Noël *et al.*, 2007). Trichomonad species found in humans are indicated with one asterisk and two asterisks indicate species considered to infect other animals but that were recently identified in human patients too (Duboucher *et al.*, 2008). Arrowheads indicate the two species currently recognized as human pathogens.

1993) and are now considered to be modified versions of mitochondria (Bradley *et al.*, 1997; Hrdy *et al.*, 2004). Trichomonads also notably carry out a peculiar type of 'closed' mitosis called cryptopleuromitosis, characterized by the persistence of the nuclear envelope and the presence of an extranuclear spindle (Ribeiro *et al.*, 2002).

Most studies on trichomonads have focused on the biology of two sexually transmitted species, the human parasite *Trichomonas vaginalis* and the bovine parasite *Tritrichomonas foetus*, because of their medical and veterinary importance, respectively (Schwebke and Burgess, 2004). However, several other trichomonads are of interest to human and animal health. *Trichomonas tenax*, a close sister taxon to *T. vaginalis* (Fig. 3.1), is found in patients with poor oral hygiene; while usually considered a harmless commensal of the human mouth, it can also be involved in pulmonary trichomoniasis (see below). *Dientamoeba fragilis* was only recently recognized as a human pathogen and a common cause of chronic diarrhoea (Johnson *et al.*, 2004; Stark *et al.*, 2006). *Pentatrichomonas hominis*, considered to be a commensal of the vertebrate digestive tract, infects cats, dogs, guinea pigs and primates including humans, and is thought to cause feline diarrhoea (Gookin *et al.*, 2005). *Trichomonas gallinae* is the causative agent of avian trichomoniasis or 'canker', a lethal and epizootic disease found primarily in pigeons and several other avian orders worldwide (Gerhold *et al.*, 2008). Another common avian trichomonad is *Tetratrichomonas gallinarium*, found in chickens and other birds (Cepicka *et al.*, 2005). Futhermore *T. foetus* is now considered to be synomymous to *Tritrichomonas suis*, isolated from pigs, and was also shown to infect other mammals including cats and dogs, where it can cause disease (Lun *et al.*, 2005; Stockdale *et al.*, 2009). More recently, several of these species have been identified by molecular and immunological methods in the human upper or lower respiratory tract of patients with chronic respiratory conditions, demonstrating that these trichomonads, including the veterinarily important *T. foetus* and *T. gallinarum*, are neither site- nor host-specific, and indicating the likely zoonotic origins of such infections (Duboucher *et al.*, 2008). These respiratory tract infections are actually surprisingly common in patients with *Pneumocystis* sp. pneumonia or respiratory distress syndrome (Duboucher *et al.*, 2008).

Trichomonas vaginalis causes trichomoniasis, arguably the most common non-viral sexually transmitted infection (STI) with an estimated 174 million new cases per year worldwide, including ~5 million cases in the United States (World Health Organization, 2001). Prevalence varies widely (typically 3.2–34% in women, 6.3–11% in men, and 7.6–12.6% for both sexes) but is greatest in resource-limited settings and among 'high-risk' groups in industrialized countries (e.g. African and Native Americans, prison inmates, drug users, and sex workers) (Johnston and Mabey, 2008). Up to one third of cases in females are asymptomatic, although the majority of infected women develop vaginal discharge, abnormal vaginal odor, and vulval itching. In men, the infection is thought to be mainly asymptomatic and self-limiting, although complications such as urethritis and prostatitis are recognized and a positive association with prostate cancer was also recently suggested (Sutcliffe *et al.*, 2006). This striking variation in pathogenesis between the sexes implies different host–parasite interactions and has led to the disease being labelled a female 'nuisance' infection, resulting in a lack of scientific and medical attention, and scant interest from public health officials in trichomoniasis control programs (Bowden and Garnett, 1999; McClelland, 2008; Van der Pol, 2007).

Acute infections in women are associated with pelvic inflammatory disease and adverse pregnancy outcomes, such as pre-term delivery and low birth weight. More significantly, *T. vaginalis* is emerging as one of the most important cofactors in amplifying HIV transmission in women (Sorvillo and Kerndt, 1998); individuals infected with *T. vaginalis* have a significantly increased incidence of HIV transmission (Miller *et al.*, 2008; Van Der Pol *et al.*, 2008). *Trichomonas vaginalis* infection elicits an aggressive local cellular immune response with a heavy influx of target cells for HIV. This response to *T. vaginalis* infection may increase the chance of portal entry of the virus in HIV naïve individuals. In an HIV-seropositive individual, punctate haemorrhages that are frequently associated with *T. vaginalis* infection increase shedding and subsequent transmission of the virus (Cohen *et al.*, 1997; Galvin and Cohen, 2004). Conversely, treating *T. vaginalis* infections reduces HIV shedding (Kissinger *et al.*, 2009). The demonstrated uptake by endocytosis, and potential subsequent release, of HIV particles by *T. vaginalis* could also facilitate viral transmission from dually infected patients to non-infected individuals (see membrane trafficking section below). The ability to identify methods to decrease the incidence of *T. vaginalis* could thus ultimately contribute to reduced HIV infection rates worldwide (Galvin and Cohen, 2004; McClelland, 2008; Van der Pol, 2007). Two approved 5-nitroimidazole drugs, metronidazole and tinidazole, are used to treat *T. vaginalis* infections, but parasites resistant to both drugs have been documented (Schwebke and Barrientes, 2006) and resistance is likely to become problematic if trends observed in other pathogens are considered (Clatworthy *et al.*, 2007).

Little information was available regarding the characteristics of the *T. vaginalis* genome prior to the large-scale sequencing project. A publication describing the electrophoretic karyotype of fourteen *T. vaginalis* strains derived from different geographic regions proposed a genome size of approximately 16 million basepairs (Mb) (Lehker and Alderete, 1999), similar to that of *G. lamblia* (~12 Mb, 5 chromosomes) and *Entamoeba histolytica* (~20 Mb, 14 chromosomes). Analysis of cultured, condensed, metaphase chromosomes showed that there were six per cell, each with one pair of sister chromatids, indicating that the genome was haploid (Yuh *et al.*, 1997). Genomic DNA libraries of *T. vaginalis* strains G3 and C-1:NIH had been used by investigators in the community to clone >100 genes, none of which were found to contain introns (Liston and Johnson, 1999), indicating that *cis*-splicing might be rare or non-existent; there was also no evidence of *trans*-splicing. Studies on the structure and expression of a number of *T. vaginalis* genes (Quon *et al.*, 1994) showed genes arranged in single transcription units. Furthermore, all examined genes had been shown to possess a conserved promoter element (the initiator element Inr) surrounding the start site of transcription that must be bound by a novel initatior binding protein to promote gene expression (Liston *et al.*, 2001; Schumacher *et al.*, 2003). The genome had an estimated A+T content of ~36% (Wang and Wang, 1985).

The *T. vaginalis* genome sequencing project

The project to sequence the genome of *T. vaginalis* began as a collaborative venture between the *T. vaginalis* research community, led by Patricia Johnson (University of California at Los Angeles), and the genome sequencing centre TIGR (The Institute for Genomic Research). In the early part of 2000, *T. vaginalis* was identified by NIAID/NIH (National Institute of Allergy and Infectious Diseases/National Institutes of Health) as a target for genome sequencing funding via a Blue Ribbon Panel on Genomics report that proposed the ten most important eukaryotic human pathogens. An initial application to the funding agency outlining a plan to sequence the parasite was declined, but a subsequent successful proposal resulted in a two year project commencing under the leadership of Jane Carlton at TIGR in September 2002.

The nuts and bolts of the genome project

The *T. vaginalis* isolate G3, from a patient in Kent, England (Coombs, 1982), was chosen as the genome reference strain due to its widespread use by investigators in studies of virulence, biochemistry, and cell and molecular biology, its ability to grow axenically in culture to high density, its lack of infection with the dsRNA virus that has been discovered in some *T. vaginalis* strains, and the absence of contamination with *Mycoplasma hominis*. Genomic DNA and frozen stocks of parasites were sent to TIGR, and small (2–3 kilobasepair (kb)) and medium (10–12 kb) insert-size libraries were constructed for whole genome shotgun sequencing by standard Sanger capillary methods.

The initial project plan was for the assumed 16 Mb genome to be shotgun sequenced to 8× 'gold standard' coverage, assembled into contigs and scaffolds, and degapped using PCR-based methodologies and transposon tagging. A total of 30,000 ESTs would provide information regarding 5′ and 3′ non-coding regions and could then be used for training of specialized *T. vaginalis* gene finders. Annotation of the genome would include identifying genes through similarity searches of current databases (particularly sequence data from the *E. histolytica* and *G. lamblia* genome projects), characterization of gene families, identification of tRNA and rRNA units, analysis of sequences for signal peptide, transmembrane and other motifs, and identification of putative regulatory elements, GC-rich isochores and GC-skew across chromosomes. Under the terms of the project funding, all sequence data were to be made available through several publicly-available websites, including a series of TIGR *T. vaginalis* project and GenBank pages.

Several months into the project in January 2003, the results of an initial assembly of ~77,000 reads caused the project to grind to a halt. Under the Lander-Waterman algorithm used to estimate genome assembly statistics (Lander and Waterman, 1988), a draft assembly of ~4000 contigs produced with ~77,000 reads of a 16Mb genome would be expected to provide 3× coverage. However, the actual number of contigs (~8800) and their much smaller length (~880 basepairs (bp) as opposed to an expected 3.7 kb) indicated a significantly larger genome. Considerable validation was undertaken to exclude the possibility that a contaminant genome was being sequenced, or that the assembly software was at fault, but no such errors were apparent. Instead, as more genome sequence reads were produced, it became evident that not only was the genome almost ten times the original estimated size (~160 Mb compared to ~16 Mb), but that it contained a large number of repeat families, members of which were very similar to each other (see below). A modification in sequencing strategy was attempted through end-sequencing of large 35–40 kb insert *Mbo*I BAC and fosmid libraries constructed at Children's Hospital Oakland Research Institute (http://bacpac.chori.org/libraries.php). The *Mbo*I restriction recognition site is the 17th rarest of 136 different four-base motifs in the *T. vaginalis* genome, and was found to occur more frequently in unique reads (50.4%) than in repetitive reads (30.6%), thereby anchoring more of the reads in unique regions of the genome. Both the genome size and repetitive nature of the genome made the *T. vaginalis* project the most challenging that had ever been attempted at TIGR.

The first assembly at 2.7× coverage was released in October 2003, followed by a 3.3× assembly in March 2004 and a 3.8× assembly with a draft annotation in September 2004. At this point, funds for the two-year project had been depleted, and an additional 18 months of funding via NIAID's Microbial Sequencing Centres contract (http://www3.niaid.nih.gov/research/resources/mscs/default.htm) to TIGR was obtained thanks to vigorous lobbying by the *T. vaginalis* research community. The total funded period was thus three-and-a-half years and the total cost $3.7 million. A final genome assembly at 7.2× coverage was released in February 2005, and a final annotation shortly thereafter in April 2005. An intense period of genome analysis by the community culminated in the submission of a genome paper to *Science* in July 2006 and publication as the front cover article on January 12, 2007 (Carlton *et al.*, 2007).

Interactions with the *T. vaginalis* research community were maintained throughout the project via e-mail updates and data release to the TIGR *T. vaginalis* web site. In addition, two specialized workshops were held in 2003 and 2004 respectively. 'The Genome and Biology of the Trichomonads' took place in December 2003 at the annual meeting of the American Society of Tropical Medicine and Hygiene in Philadelphia, and included a day-long session for discussion of the *T. vaginalis* genome sequencing project followed by a two-hour symposium 'Mind the Gap: Bridging the Divide Between Clinical and Molecular Genomic Studies of the Trichomonads' within the conference itself (Lyons and Carlton, 2004). The two-day workshop 'Genomics and Biology of the Amitochondriates' took place in September 2004 prior to the Molecular Parasitology Meeting at Woods Hole, and brought together ~70 members of the *Trichomonas, Giardia*, and *Entamoeba* research communities to discuss the current status and future directions of the genome projects for these organisms, as well as to identify common goals and opportunities for amitochondriate research.

The *T. vaginalis* genome is highly repetitive

A list of the characteristics of the *T. vaginalis* genome is shown in Table 3.1. The current assembly contains 1.4 million shotgun reads assembled into >17,000 scaffolds at ~7.2x coverage. The abundance of frequently-occurring 22mers in the sequencing reads indicates the extremely repetitive nature of the genome; at least 65% of the *T. vaginalis* genome is repeated. Despite this, estimates of the genome size could be made, and using two different types of analysis (of the start position offsets for overlapping reads and analysis of the genome assembly), a genome size of between 158.1Mb and 165.6Mb was determined. Thus *T. vaginalis* has one of the largest genomes identified for a parasitic protist, with a degree of repetitiveness that is unprecedented among such organisms.

Analysis of the most common ~60 repeat families characterized them as follows: (1) length from ~300 bp to ~11 kb (average length 1.5 kb); (2) present in hundreds of copies in the genome (average copy number ~660, cumulative length ~38 Mb); (3) located on small 1–5 kb contigs or at contig ends; and (4) each repeat family is extraordinarily homogenous, despite the large number of copies, with an average polymorphism of ~2.5%. The repeats could be classified into three types: (1) virus-like repeats having significant similarity to poxviruses, mimiviruses and phage proteins, the presence of which may be explained by the ability of *T. vaginalis* to ingest and retain viruses from the environment (Pindak *et al.*, 1989); (2) transposon-like repeats, including ~1000 copies of the first *mariner* element identified outside animals (Silva *et al.*, 2005); (3) retrotransposon-like repeats consisting of just two families, indicating that retrotransposons have difficulties invading the genome of *T. vaginalis* in contrast to other human parasites, in which retrotransposons predominate over transposons (El-Sayed *et al.*, 2005); and (4) unclassified repeats, which have no similarity to known sequences in current public databases and account for ~25 Mb of the genome.

Interestingly, the lack of a strong correlation between copy number and average pairwise difference between copies (Carlton *et al.*, 2007) suggests that the repeat families underwent a sudden expansion. By estimating the polymorphism and divergence in synonymous sites among several geographically diverse *T. vaginalis* strains and *T. tenax* (a close sister taxon to *T. vaginalis;* Fig. 3.1) for several protein-coding loci, the repeat family expansion was inferred to have occurred after the split between the two species, although several families appear to have undergone multiple expansions. Sequences with high sequence similarity to the repeat families in *T. vaginalis* could not be found in *T. tenax*, but were present in *T. vaginalis* isolates of distinct origin, indicating that the repeat family spread happened before the diversification of *T. vaginalis*. Moreover, the *T. vaginalis* genome appears to have undergone a single or several large-scale genome duplication event(s). Thus, the large genome size, high repeat copy number, low repeat polymorphism and indication of repeat expansion occurring after the split between *T. vaginalis* and *T. tenax*, suggests that

Table 3.1 Summary of the *T. vaginalis* genome sequence data

Feature	Value
The genome	
Estimated genome size (Mb)	~158–166
No. genomic DNA libraries	17
Total no. good reads	1,403,509
Genome coverage (fold)	7.4
Total G+C content (%)	32.7
No. scaffolds	17,290
N_{50} scaffold size (bp)[a]	68,338
Occurrence of 22mers > 100 times in all reads (%)	53
Protein-coding genes	
No. predicted genes	59,681
No. evidence-supported genes[b]	25,949
Mean gene length (bp)	928.6
Gene G+C content (%)	35.5
Gene density per kb[c]	0.60
No. genes with introns	65
No. genes with predicted UTR	7708
Avg. UTR length (bp)	80
No. genes with predicted signal peptide	3051
No. genes with predicted transmembrane domain	4947
No. genes with Pfam/TIGRfam match	37,486
No. paralogous gene families	4141
Mean length intergenic regions (bp)	1165
Intergenic G+C content (%)	28.8
Non protein-coding genes	
No. predicted tRNA genes	479
No. predicted 5.8S, 18S & 28S rRNA units	~250

a N_{50} scaffold size is the minimum scaffold length such that more than half the genome is contained in scaffolds of at least that length.

b The number of evidence-supported genes with either significant similarity to a known protein ($E < 1 \times 10^{-10}$, >25% length of protein) or similarity to an EST (>95% identity over >90% length of the gene).

c Total number of all predicted genes including transposable elements (97,873) divided by estimated 162 Mb genome size.

T. vaginalis has undergone a very recent and substantial increase in genome size. What are the potential explanations for this? One hypothesis put forward is that massive repeat expansion occurred following a reduction in the parasite's effective population size when it moved from an enteric environment (the habitat of most trichomonads) to the urogenital tract (Carlton *et al.*, 2007). Since there is a positive correlation between genome size and cell volume, an increase in cell volume favoured by selection will lead to, or be facilitated by, an increase in genome size. The increased genome size of *T. vaginalis* could have been achieved through rapid, selection-driven fixation of recently arisen repeat copies, driven by selection for a larger cell size. Since *T. vaginalis* is a highly predatory parasite, which phagocytoses bacteria, erythocytes and other host cells (Pereira-Neves and Benchimol, 2007; Rendon-Maldonado *et al.*, 1998) and is itself ingested by macrophages, an increase in cell size could tip the phagocytosis war in favour of the parasite, as well as providing an increased cell surface area for colonization of host vaginal epithelial cells. Support for this hypothesis comes from the cell volume of *T. vaginalis*, which is larger than that of *T. tenax* and the intestinal species *P. hominis* (Honigberg and Brugerolle, 1989) and *T. gallinae* (Shuter, 1983).

Since publication of the *T. vaginalis* genome paper, several new papers have thrown additional light on the size of trichomonad genomes and their repeat content. An analysis of nine trichomonad species (*T. vaginalis, T. tenax, T. gallinarum, P. hominis, T. foetus, Tritrichomonas augusta, Monocercomonas colubrorum, Trichomitus batrachorum* and *Hypotrichomonas acosta*) by chromosome spreading and flow cytometry found chromosome numbers to range from 4–6 per genome and genome sizes to range from 85–177 Mb (Zubacova *et al.*, 2008). The largest genomes were typically observed in the genera *Trichomonas* and *Tritrichomonas* (133–177 Mb), and the *T. vaginalis* strain T1 was assigned a genome size of 154 ± 9 Mb, very close to the size determined by analysis of the *T. vaginalis* G3 genome sequence. The genome sizes correlated with their respective cell volumes, and it was concluded that no relationship existed between genome size and the site of infection or trichomonad phagocytic ability. Determining the repeat content of these genomes will go some way towards inferring how the genomes from these different lineages have evolved. In further studies on the *T. vaginalis* repeats, a new class of transposable element designated *Mavericks*, previously identified in *Caenorhabditis elegans*, *C. briggsae* and the zebrafish *Danio rerio*, was found among the repeats (Pritham *et al.*, 2007). *Maverick* elements are 15–22 kb in length and encode a retroviral-like c-integrase along with a conserved set of 5 to 9 genes that are predicted to encode proteins with homology to replication and packaging proteins of some bacteriophages and eukaryotic double-stranded DNA viruses. There are an estimated 3000 *Maverick* copies in the *T. vaginalis* strain G3 genome, constituting perhaps as much as ~37% (~60 Mb) of the 160 Mb genome.

Genomic insights into *T. vaginalis* biology

A staggering number of genes (97,883) were predicted from the genome sequence using gene finding algorithms trained on a set of ~300 manually-curated *T. vaginalis* genes. Of these, 38,213 were reclassified as repeats by screening with the ~60 repeat families described above and excluded from the gene set. The remaining 59,681 gene models were considered the core set of putative protein coding genes, with 25,949 (43%) having good supporting evidence (Table 3.1). A total of 65 genes were identified as containing introns, increasing the number of intron-containing genes over those documented using preliminary sequence data (Vanacova *et al.*, 2005). tRNAs for all 20 amino acids were identified and ~250 rRNA units were found on small contigs. Interestingly, all rRNA units appear to be located on only one of the six *T. vaginalis* chromosomes (Carlton *et al.*, 2007). A comparison of the Pfam protein domains in the *T. vaginalis* genome revealed 80% of the domains to be shared with *E. histolytica* and 45% with other parasitic protists (*Plasmodium, Trypanosoma, Leishmania* and *Cryptosporidium*), while 33%, 32%, and 16% are shared with Archaea, plants, and Bacteria respectively.

What biological insights have been provided by analysis of the genome sequence? A considerable number of novel findings were made that are too numerous to report here, and the reader is referred to the genome paper for further reading (Carlton *et al.*, 2007). Some of the more notable findings are summarized below.

Metabolism

Trichomonas vaginalis uses carbohydrate as a main energy source in the classical Embden-Meyerhof pathway of glycolysis in the cytosol, and metabolism is fermentative under both aerobic and anaerobic conditions. Prior to the genome project, little was known concerning *T. vaginalis* amino acid metabolism, but analysis of the sequence identified a central role for aminotransferases and glutamate dehydrogenase in amino acid catabolism. The large number of aminotransferases, which allow interconversion of amino acids, suggests that *T. vaginalis* utilizes a variety of amino acids as energy substrates but, with the exception of cysteine, the parasite appears unable to synthesize amino acids *de novo*. Analysis of the genome sequence identified a pathway for *de novo* synthesis of cysteine, which may occur via cysteine synthase, also a potential lateral gene transfer candidate. Analysis of the genome also revealed enzymes involved in synthesis of lipophosphoglycan (LPG), a major component of the *T. vaginalis* cell surface that forms a dense glycocalyx and is involved in attachment to human cervical cells *in vitro* (Bastida-Corcuera *et al.*, 2005; Gilbert *et al.*, 2000) and for which the human receptor was recently identified as galectin-1 (Okumura *et al.*, 2008). *Trichomonas vaginalis* LPG molecules have a sphingolipid anchor and a polysaccharide chain that contains various monosaccharides including rhamnose and galactose. Homologues of the main enzymes for the synthesis of the LPG anchor and chain and the monosaccharide rhamnose were identified. This monosaccharide is of particular interest since it is absent in the human host and so the biosynthetic pathway could be the target of drug therapy.

Massively expanded gene families

Analysis of the *T. vaginalis* G3 genome sequence identified massive expansion of many gene families on a scale not previously seen for a parasitic protist. The largest 31 families (out of a total of 4141 predicted protein families; Table 3.1) make up 42% (24,981 proteins) of the predicted proteome, and contain *T. vaginalis*-specific domains, predicted signal peptides, and transmembrane domains. These families have no known homologues and no functional data exist, except that members of 21 of the families are transcriptionally active. Known gene families that appear massively amplified include the parasite's protein kinase repertoire, which in *T. vaginalis* consists of ~880 genes encoding distinct eukaryotic protein kinases (ePKs) and ~40 atypical protein kinases (aPKs), one of the largest kinomes known to exist among eukaryotes. In contrast to *E. histolytica*, which has 90 receptor ePKs (Loftus *et al.*, 2005), *T. vaginalis* has no such proteins, indicating that the two species differ significantly in their mechanisms of cell signalling despite similar anaerobic environments. No receptor protein kinases were identified that could act to transduce extracellular signals. However, the parasite has heterotrimeric G proteins (Hirt *et al.*, 2003) and several components of the Mitogen-Activated Protein Kinase (MAPK) pathway, suggesting signal transduction mechanisms that may be more similar to yeast than other protists. Another unusual feature of the *T. vaginalis* kinome is the presence of ~120 cytosolic tyrosine kinase-like (TKL) genes, which are important for activating signalling molecules. They are found in *E. histolytica* but are absent from other protozoa such as the trypanosomatids (El-Sayed *et al.*, 2005). The *T. vaginalis* 'degradome' was found to contain 439 peptidases and to be more complex than those of several other parasitic protists. Several peptidases were identified as of particular interest. First, *T. vaginalis* has a single aspartic peptidase in the same family as retropepsin of human immunodeficiency virus 1 (HIV-1). Several HIV-1 drugs target inhibition of retropepsin, so this protein presents itself as a putative candidate for anti-HIV peptidase inhibitors. Second, ~50 papain-like cysteine peptidases were identified, highlighting the diversity and importance of this family of enzymes, which are implicated as virulence factors and in the ability to induce apoptosis in human vaginal epithelial cells (Klemba and Goldberg, 2002). Finally, *T. vaginalis* has nine NlpC/P60-like members, which play a role in the degradation of bacterial cell wall components and show evidence of lateral gene transfer from a bacterial donor into the trichomonad lineage.

Molecular mechanisms of pathogenesis

Surface proteins of *T. vaginalis* are implicated in the initial adherence of the parasite to mucosal tissue and may play an important role in the long-term survival of the pathogen on mucosal surfaces. Eight protein families containing ~3300 proteins that represent candidate surface molecules likely to be involved in *T. vaginalis* pathogenicity were identified from the genome sequence, all but one of which (P270 proteins (Musatovova and Alderete, 1999)) were novel. One family, the BspA-like proteins, are characterized by the leucine-rich repeat TpLRR, and are found on the surface of pathogenic and other bacteria which commonly live on mucosal surfaces, where they mediate cell aggregation and binding to ECM proteins (Ikegami *et al.*, 2004). The ~600 *T. vaginalis* copies exhibited a marked diversity in sequence features, such as variation in the number of TpLRR repeats (from 3 to >70), most likely generated by gene duplication and recombination events; such mechanisms could generate *T. vaginalis* antigenic variation during infections and phenotypic plasticity between different strains. The mucosal pathogen *E. histolytica* and commensal *E. dispar* are the only other eukaryotes currently known to encode BspA-like proteins (Loftus *et al.*, 2005). A second family, the GP63-like proteins, are homologues of the most abundant surface peptidase of *Leishmania major*, 'leishmanolysin', which contributes to virulence and pathogenicity through diverse functions in both insect and mammalian hosts (Yao *et al.*, 2003). GP63 proteins are metalloproteinases and are broadly distributed among eukaryotes. The several other *T. vaginalis* surface protein families share domains with *Chlamydia* polymorphic membrane proteins, *G. lamblia* variant surface proteins and *E. histolytica* immuno-dominant variable surface antigens. A review of the various surface protein families of *T. vaginalis* encoded by its genome has been published recently (Hirt *et al.*, 2007).

In addition to surface molecules that may be involved in pathogenesis, genes encoding cytolytic effectors, which may be released upon contact between the parasite and host membranes, were identified. In particular, twelve SAPLIP genes were found, which are probably the 'trichopores' that mediate lysis of host red blood cells by *T. vaginalis* (Fiori *et al.*, 1993). *E. histolytica* also secretes pore-forming proteins called amoebapores (Leippe and Herbst, 2004). The presence of a Lys75 basic residue in all 12 trichopores and the triggering of haemolysis by acidic pH suggest a pH-dependent protonation mechanism underlying oligomerization in *T. vaginalis*, similar to that of amoebapores (Fiori *et al.*, 1996).

The hydrogenosome

T. vaginalis lacks typical mitochondria but possesses double-membrane-bound organelles called hydrogenosomes, which produce ATP and molecular hydrogen through fermentation. Analysis of the *T. vaginalis* genome illuminated several areas of hydrogenosomal biology. First, few genes encoding homologues of mitochondrial transporters, translocons and soluble proteins were identified in the genome, indicating that *T. vaginalis* resembles other parasitic eukaryotes whose mitochondrial proteomes also show reductive evolution (e.g. *Cryptosporidium* and *Plasmodium*). Second, a 5–14 residue presequence containing ML(S/T/A) $X_{(1..15)}$R (N/F/E/XF) or MSLX$X_{(1..15)}$R(N/F/E/XF) motifs was identified and used to detect ~145 genes that are probably targeted to the matrix of hydrogenosomes. Recent investigations of the hydrogenosomal 'mitochondrial processing peptidase' (MPP)-like enzyme, which processes the hydrogenosomal N-terminal targeting signal of matrix proteins upon their import into the organelle (Brown *et al.*, 2007), have demonstrated that it functions as a classic α/β heterodimer metalloprotease, as does the mitochondrial homologue, and that it processes various hydrogenosomal protein N-terminal targeting signals (Smid *et al.*, 2008). Careful phylogenetic analyses demonstrated that the *T. vaginalis* α and β subunits are orthologues of their respective mitochondrial subunits (Smid *et al.*, 2008). Third, analysis of the genome sequence enabled prediction of a new function for hydrogenosomes to be determined and tested experimentally, that of amino acid metabolism (Mukherjee *et al.*, 2006a,b). Finally, several hints as to possible mechanisms of metronidazole resistance were found in the genome, in particular the presence of three *NimA*-like genes implicated in metronidazole resistance in bacteria (Leiros *et al.*, 2004). A recent study that shows recombinant NIM proteins conferring high resistance to transformed bacteria appears to vindicate these initial findings (Pal *et al.*, 2009).

The *T. vaginalis* transcriptome and transcriptional regulation

Several remarkable observations concerning the genome size and the number of genes in *T. vaginalis* G3 resulted from the completion of the genome sequencing project of this extremely divergent eukaryote lineage. Primary among them is that the ~160 Mb genome harbours more than 25,000 protein coding genes supported by good evidence for their existence, and another ~35,000 genes predicted by sophisticated algorithms. While many of the latter genes ultimately may be reclassified as constituents of transposable elements during the ongoing re-annotation of the genome, and thus will not be 'true' *T. vaginalis* protein-coding genes, these findings nevertheless raise an interesting question: how does the parasite regulate and coordinate the expression of such a large number of genes under different growth and environmental conditions?

The *T. vaginalis* transcriptome

Large-scale gene expression studies carried out by different laboratories have generated close to 100,000 *T. vaginalis* expressed sequence tags (ESTs), which have been used for gene finder output verification (Carlton *et al.*, 2007) and for comparative transcriptomics (P. Tang, unpublished). The transcriptome data for *T. vaginalis* are available from the genome sequence database TrichDB (http://www.trichdb.org/) and from the Chang Gung University TvXpress Database (http://TvXpress.cgu.edu.tw). Table 3.2 is a summary of the *T. vaginalis* cDNA libraries from which the majority of ESTs have been generated.

The TvXpress project profiled gene expression patterns from cDNA libraries prepared from parasites cultured under seven defined conditions related to cell cycle, growth, starvation and pathogenesis. A total of 71,428 EST sequences were assembled from these libraries; 61,363 (86%) were collapsed into 6309 'tentative consensus' (TC) sequences while the remaining 10,165 (14%) EST sequences were 'singletons'. Together these 16,474 'unigenes' (TCs plus singleton sequences) account for almost 37% of the predicted proteome, assuming a gene count of ~100,000 (i.e. including transposable elements). Analysis of the EST data has revealed several significant features of the *T. vaginalis* transcriptome. First, although the overall coverage per library is modest (average: ~3500 unigenes/library), most of the unigenes (94%) derive from just one or a very few ESTs (i.e. either singletons or TCs comprised of <10 ESTs). Moreover, only 239 of the unigenes are shared by all seven libraries. This expression profile indicates that *T. vaginalis* may possess certain unique features of gene expression, such as stringent, differential transcription rate controls even for housekeeping genes. This hypothesis is supported by the fact that the synthesis of mRNAs encoding ribosomal proteins was suppressed when cells were grown in nutrient-poor media but the ratio of the transcripts coding for proteins in ribosome subunits 60S:40S in different libraries varied independently from 1:3 to 2:1, based on 8158 ESTs (9% of the total ESTs). In addition, data from low glucose culture (TvL) and fibronectin-mediated cytoadherence culture (TvF) libraries showed the greatest redundancy and lowest number of singletons (9% and 12%, respectively), compared with the other libraries (range: 23% to 37%). However, this is not simply correlated with the expression of housekeeping genes (such as ribosomal proteins), since the TvF library was also found to contain a high level of such transcripts.

Another example of support for differential gene expression of *T. vaginalis* under different physiological conditions is the modulation of protein kinase gene expression at the mRNA level. Protein kinases are one of the major classes of enzymes involved in signal transduction and signalling networks and therefore mirror the response of cells to their external environments. As mentioned above, the *T. vaginalis* 'kinome' is composed of more than 880 ePKs and 40 aPKs, making it one of the largest kinomes identified so far (Carlton *et al.*, 2007). A total of 451 kinase genes (51% of the kinome) were identified in the *T. vaginalis* EST dataset. Expression profiling of the kinase genes showed that the calmodulin-dependent kinases (CaMK) were the major kinase family expressed, followed by the cyclic-nucleotide- and calcium/phospholipid-dependent kinases (AGC) and the CMGC group, which consists of the cyclin-dependent-, mitogen-activated-, glycogen-synthase- and CDK-like kinases (Fig. 3.2). ESTs from the unsynchronized normal culture (TvE) library expressed 160

Table 3.2 Summary of significant EST sequencing libraries.

cDNA Libraries	TvE[1]	TvG[2]	TvC[3]	TvV[4]	TvF[5]	TvI[6]	TvL[7]	TIGR ESTs[8]
No. of ESTs	10,749	10,470	9865	9904	9955	10,506	9979	17,969
No. of TCs	1341	1399	1326	1059	810	1477	1421	1585
No. of ESTs in TCs	7655	7649	6956	7908	8881	7427	9144	ND
% ESTs in TCs	71.22	73.06	70.51	79.85	89.21	70.69	91.63	ND
No. of singletons	3094	2821	2909	1996	1074	3079	835	280
% ESTs in singletons	28.78	26.94	29.49	20.15	10.79	29.31	8.37	1.56
No. of unigenes	4435	4220	4235	3055	1884	4556	2256	1865
% TCs	30.24	33.15	31.31	34.66	42.99	32.42	62.99	84.99
% Singletons	69.76	66.85	68.69	65.34	57.01	67.58	37.01	15.00

The seven cDNA libraries, constructed from several different *T. vaginalis* isolates, were not normalized. Libraries 1–7 are from the Chang Gung University TvXpress project, where each clone was sequenced from a single end, and Library 8 is from the TIGR EST project where clones were sequenced from both ends. Tentative Consensus sequences (TCs) refer to ESTs collapsed into virtual transcripts, which may contain full or partial cDNA sequences; unigenes are unique genes.

1 Unsynchronized trophozoites of isolate ATCC30236 (JH 31A#4) grown in nutrient-rich medium;

2 G2-phase synchronized trophozoites of isolate ATCC30001 (C-1:NIH);

3 Pseudocyst stage of isolate ATCC30001 (C-1:NIH) induced by cold-stress;

4 Vaginal epithelial cell (VEC)-induced amoeboid form of isolate TO16;

5 Fibronectin-induced amoeboid form of isolate TO16;

6 Trophozoites of isolate T1 grown in low iron medium;

7 Trophozoites of isolate ATCC30236 (JH 31A#4) grown in low glucose culture;

8 Trophozoites of isolate T1 grown under standard conditions. ND, not determined

kinases, with an average of 1.31 transcripts per gene. The remaining libraries expressed 66–115 kinases with an average of 1.14–1.28 transcripts per gene. The exception to the pattern was the low-glucose culture (TvL), which contained ESTs for only 79 kinase genes but with an average of 2.57 transcripts per gene. The large number of kinase genes expressed differentially in the EST libraries together with the low number of transcripts per gene suggest that the functions of most *T. vaginalis* kinases are specific to physiological changes in response to changes in the external environment.

The association between GC content of *T. vaginalis* genes and their expression level is intriguing. The average GC content of the total *T. vaginalis* genome is 32.7% (Fig. 3.3A), but the GC content of coding regions is partitioned into two peaks of 36% and 51% (Fig. 3.3B, C). When the GC contents of the complete set of ESTs (which is enriched for highly-expressed genes with multiple copies of ESTs) and the set of unigenes (which is enriched for singletons) are plotted separately for all the libraries, the highly-expressed, redundant ESTs tends to cluster in the 51% GC peak while the singleton ESTs tend to cluster in the 36% GC peak (this is even more striking when the ratio of ESTs and unigenes are plotted as a function of GC content; Fig. 3.3D). The highly-expressed genes constitute a small fraction (~10%) of the total number of genes. Some of them are differentially expressed, and exist as several hundreds of ESTs in some cDNA libraries but are completely absent (possibly shut down) in others. This transcription-associated GC content variation is has been proposed to be related to the level of gene expression and transcription-coupled DNA repair mechanisms that are universal to all cellular organisms (DeRose-Wilson and Gaut, 2007; Vinogradov, 2005).

The *T. vaginalis* microRNAome

Double-stranded RNA has regulatory roles in a range of eukaryotes(Du and Zamore, 2005; Fire *et al.*, 1998; Mello and Conte, 2004; Ruvkun, 2001; Ullu *et al.*, 2004). Such regulatory RNAs fall into two major classes: (1) microRNAs (miRNAs), which are derived from endogenous mRNA transcripts and are processed initially within the nucleus; and (2) small interfering RNAs (siRNAs), which are derived from endogenous or exogenous transcripts. In animals, stem-loop structures of primary miRNA (pri-mRNA) transcribed by RNA polymerase II are processed within the nucleus into 60–70 nucleotide precursor miRNAs (pre-miRNAs) that are exported from the nucleus by the Exportin-5 and Ran-GTPase proteins. Cytosolic Dicer and Argonaute proteins further process pre-miRNAs into miRNAs (or siRNA precursors into siRNAs). In plants, Dicer-1 (DCL1) and Argonaute-1 (AGO1) mediate nuclear pri-miRNA processing, and pre-miRNAs are exported to the cytosol by HASTY, an Exportin-5 homologue. Plant siRNA formation relies on several Argonaute homologues and RNA-dependent RNA polymerases (Rdp1/Rdr2) as well as Dicer homologues DCL2 and DCL3, while *trans*-acting siRNA formation relies upon DCL4.

Both miRNAs and siRNAs may cause degradation of their target mRNAs or suppress translation (Du and Zamore, 2005). Interestingly, the RNA interference (RNAi) machinery is also involved in transcriptional gene silencing in animals, fungi, plants (AFP) and ciliates, involving histone H3 lysine-9 methyltransferase Clr4 or DNA cytosine methylation (reviewed in Matzke and Birchler, 2005). RNAi has become a powerful research tool for investigating the effects of genes in development by introducing exogenous siRNAs into cells to interfere with gene expression. While this phenomenon is widely used in the study of animals and plants, and more recently ciliates (Aronica *et al.*, 2008; Howard-Till and Yao, 2006), *Trypanosoma brucei* (Ullu *et al.*, 2004) and *G. lamblia* (Prucca *et al.*, 2008), RNAi in protists is still not fully understood (Shabalina and Koonin, 2008; Ullu *et al.*, 2004). However, it promises to be a powerful tool to investigate protist biology, particularly if sequence similarity-based methods can be used to (i) predict miRNAs and siRNAs, (ii) identify components of the miRNA and siRNA processing pathways and RNAi machinery, and (iii) predict mRNA targets.

As part of the *T. vaginalis* genome project, sequence similarity-based methods identified Dicer-like and Argonaute genes (Carlton *et al.*, 2007). Re-analysis of the sequence shows that the parasite genome encodes Exportin-5/HASTY, Ran-GTPase, Dicer, Argonaute, and

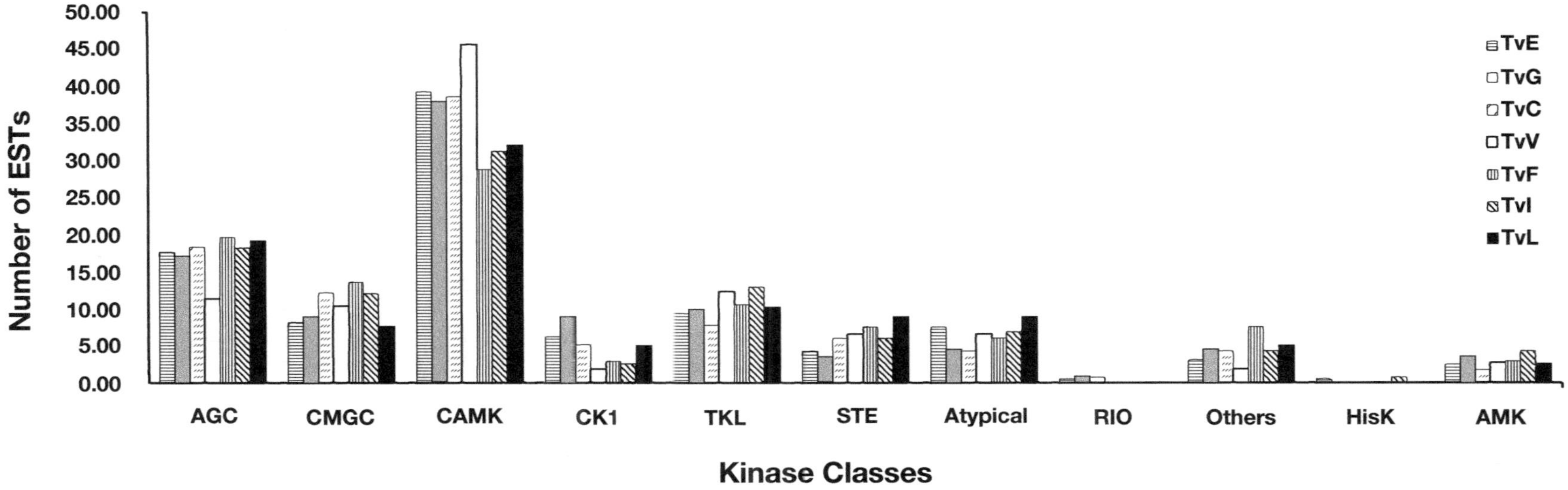

Figure 3.2 Differential expression of kinase genes in *T. vaginalis* under different growth conditions. A total of 880 eukaryotic protein kinases (ePKs) protein kinases identified in the genome were classified into 11 classes (x-axis). The total number of ESTs for each class of kinases expressed in different growth conditions was used to estimate the number of transcripts per kinase gene (y-axis). Growth conditions of the different EST libraries (TvE, TvG, TvC, TvV, TvF, TvV, TvI and TvL; abbreviated in the inset box) are described in Table 3.2.

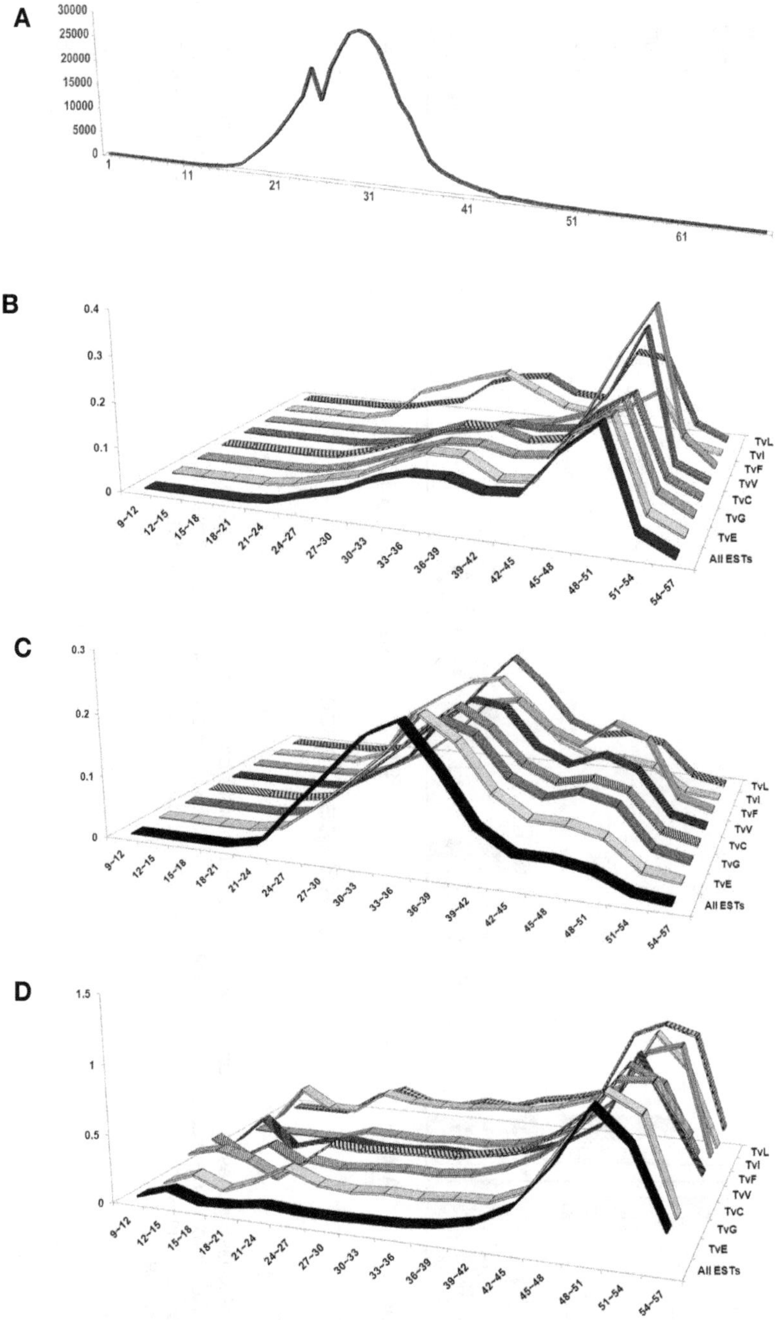

Figure 3.3 Analysis of GC content in the TvXpress EST dataset. A) GC content (x-axis) in 500 bp windows of the *T. vaginalis* genome (y-axis). (B) GC content (x-axis) of ESTs plotted as a proportion of each library (y-axis) that makes up the total number of ESTs. (C) GC content (x-axis) of unigenes plotted as a proportion of each library (y-axis) that makes up the total number of ESTs. (D) Ratio of number of ESTs to number of unigenes (y-axis) plotted as a function of GC content (x-axis).

Clr4 orthologues (Table 3.3), indicating that it is equipped to process small RNAs in the nucleus, transport them to the cytosol, and potentially use miRNAs and siRNAs to target transcripts for degradation and prevent transcription (through chromatin modification via siRNA). For example, small RNAs potentially could play a role in regulating some of the massively expanded gene

Table 3.3 Inventory of miRNA and siRNA effector proteins in diverse eukaryotes.

	Exportin5 (Hasty)	Ran-GTPase	Dicer	Argonaute	Clr4	Rdp1 (Rdr2)
Excavata						
Trichomonas vaginalis	EAY02963 EAY18864	EAY23202 EAY16220	EAY14451 EAX89071 EAY03005	EAY17553	EAY23675	
Giardia lamblia		1	1	1	1	1
Trypanosoma brucei	1	2	2	2		
Leishmania major	1	1	–	–		
Unikonta						
Entamoeba histolytica		1		2		1
Dictyostelium discoideum	1	1	1	5	1	2
Homo sapiens	1	1	1	4	2	
Schizosaccharomyces pombe	1	1	1	1	1	1
Alveolata						
Plasmodium falciparum/ chabaudi	1	1	–	–	1	
Cryptosporidium parvum/ hominis	1	1	–	–	2	
Tetrahymena thermophila		1	3			
Stramenopila						
Phaeodactylum tricornutum		1	1		1	
Phytophthora ramorum/sojae		1	3	3	1	
Archaeplastida						
Arabidopsis thaliana	1	3	4	10		6
Cyanidioschyzon merolae	1	1				

Data are summarized from BLASTP/PSI-BLASTP searches of the NCBI nr, stramenopile (http://genome.jgi-psf.org/; Bowler *et al.*, 2008), and *Cyanidioschyzon merolae* (http://merolae.biol.s.u-tokyo.ac.jp/blast/blast.html; Matsuzaki *et al.*, 2004) databases, phylogenetic analyses (not shown) correcting for among-site amino acid substitution rate heterogeneity with the WAG+I+8γ model (Whelan and Goldman, 2001) implemented in MrBayes3.1.2 (Ronquist and Huelsenbeck, 2003), and the following papers: (Carlton *et al.*, 2007; Cerutti and Casas-Mollano, 2006; Flynt and Lai, 2008; Morrison *et al.*, 2007; Murphy *et al.*, 2008; Nakayashiki *et al.*, 2006; Shabalina and Koonin, 2008). Minus indicates confident absence; numbers indicate copy number. *T. vaginalis* homologues are indicated by GenBank accession numbers.

families in the parasite's highly repetitive genome. An expanded phylogenomic survey of the homologues across diverse eukaryotes including *G. lamblia* and *E. histolytica* indicates that some miRNA transport and transcriptional gene silencing proteins are broadly conserved (Table 3.3; S.-B. Malik, unpublished). Some key components of the siRNA and miRNA processing machinery such as Dicer and Argonaute are ancient, since they are broadly distributed among diverse eukaryotes (Murphy *et al.*, 2008; Shabalina and Koonin, 2008). However, Table 3.3 shows that Exportin-5/Hasty and Ran-GTPase orthologues are as widely dispersed among diverse eukaryotes as Dicer and Argonaute, indicating that eukaryotes also share conserved machinery for transporting small dsRNAs from the nucleus to the cytosol. Notably, Dicer orthologues are absent from *E. histolytica,* kinetoplastids, apicomplexans and the red alga *Cyanidioschyzon merolae,* although a recent genetic study reported the presence of two RNAse III homologues in *T. brucei* that exhibit Dicer-like biochemical activity (Shi *et al.*, 2004). Several, if not most, major lineages have more than one Dicer-like homologue that appear to have evolved by lineage-specific gene duplications. For example, while most animals have one Dicer, arthropods and fungi have two independently duplicated copies, and *T. vaginalis* has three recently duplicated Dicer homologues. *G. lamblia* has one Dicer homologue (Morrison *et al.*, 2007) that performs like a canonical Dicer even though it is highly diverged from other eukaryotic orthologues (Macrae *et al.*, 2006). Since lineage-specific roles for Dicer evolved independently several times in model AFP organisms, these data fall short of establishing universal models for possible roles of independently-duplicated Dicer homologues during protist life cycles. Argonaute homologues are also highly variable and similarly widespread among eukaryotes, while subject to lineage-specific rather than ancient gene duplications. Clr4 is conserved in *T. vaginalis, G. lamblia* and other eukaryotes, indicating that RNAi-mediated transcriptional silencing by heterochromatin formation via Clr4 might be an ancient characteristic of eukaryotes. Substantial evolutionary plasticity of miRNA and siRNA processing proteins is indicated by independent gene duplications (and possible losses); their persistence rather than loss in the face of plasticity throughout eukaryotic evolution may be a relic of their adaptive value.

As part of ongoing functional studies of miRNA in *T. vaginalis,* a genome-wide screen of the ESTs, coding regions and intergenic DNA identified candidate miRNAs (Table 3.4; Lin *et al.*, 2009). Screening for putative hairpin structures, a characteristic of all known miRNAs, identified 558 candidates, the majority of which (555) were identified in intergenic regions, with the remainder in coding sequences. In contrast, in mammalian systems approximately 50% of miRNAs are expressed from introns of protein-coding genes – although of course *T. vaginalis* is effectively intron-less. A few of the candidate *T. vaginalis* miRNAs showed sequence similarity to known miRNAs though searching miRBase, which contains over 8000 mature miRNAs from 87 species. Experimental data from these preliminary studies suggest that the miRNA families in *T. vaginalis* may be significantly divergent from those of other systems (Lin *et al.*, 2009).

Table 3.4 Summary of putative miRNA candidates identified by a miRNA bioinformatic analysis pipeline

	Number of miRNA candidates
From scaffolds	22,019
From ESTs	4537
From ORFs	783
Putative miRNA candidates[a]	3754
Unique miRNA candidates	558
Exonic miRNA	3
Intergenic miRNA	555
Sequence similarity with known miRNAs	27
Novel miRNAs	531

a Putative miRNA candidates have at least one matching EST record and no matching ORF record.

Unraveling *T. vaginalis* reproduction

Sexual reproduction is unique to eukaryotes and is comprised of two fundamental parts: the reduction of ploidy by meiosis, followed by the restoration of ploidy by syngamy. Meiosis is a specialized cell division process that differs from mitosis in that replication is followed by a reductional division that halves ploidy and then an equational (mitosis-like) division. The reductional division is a key eukaryotic innovation and utilizes meiosis-specific proteins that have evolved from mitotic paralogues (Villeneuve and Hillers, 2001; Malik *et al.*, 2008). During meiotic prophase I, DNA double-strand breaks (DSBs) are created by a meiosis-specific recombination initiation complex and then repaired by DNA recombination and repair machinery that shares some components with mitotic DNA repair machinery (reviewed by Egel, 2008; Ehmsen and Heyer, 2008). The general model for DSB repair in meiotic recombination is based primarily upon empirical studies in AFP (Fig. 3.4) (reviewed by Haber, 2008; Jones and Franklin, 2008). Timing is everything: meiotic recombination follows a critical series of coordinated events in AFP.

Recent studies that compare the phylogenies of meiotic genes with one another and with a current consensus phylogeny of eukaryotes inferred that meiotic genes are derived from early gene duplications followed by gene loss (Lin *et al.*, 2006; Malik *et al.*, 2007, 2008; Ramesh *et al.*, 2005). Homologues of genes encoding proteins essential for the completion of meiosis in AFP exist in diverse protists and some have prokaryotic orthologues (Malik *et al.*, 2008), suggesting that the most recent common ancestor of eukaryotes may have had meiosis. In contrast, the conservation of proteins involved in gamete fusion or the restoration of ploidy across diverse eukaryotes is far less well understood; where studied, the proteins tend to evolve quickly (Vacquier, 1998). The twenty-eight 'core meiotic genes' conserved in diverse eukaryotic genome sequences include ten coding for proteins only known to function in meiosis and 18 repair proteins that are not meiosis-specific but are required for proper meiotic progression in AFP (Malik *et al.*, 2008; Villeneuve and Hillers, 2001) (Fig. 3.5). Orthologues of meiosis-specific genes are therefore only expected to be present in organisms that are sexual, or recently derived from sexual lineages.

What is known concerning meiosis and the mode of reproduction in *T. vaginalis*? Neither meiosis nor sex have been observed cytologically in trichomonads (Pereira-Neves *et al.*, 2003; Ribeiro *et al.*, 2000; Zubacova *et al.*, 2008), but sexual processes have been suggested in their hypermastigid relatives (Cleveland, 1947b, 1951, 1956). Meiosis could occur infrequently in *T. vaginalis* lab cultures, since there have been isolated reports of transiently-appearing quadrinucleated cells (Honigberg and Brugerolle, 1989) and the presence of low-levels (0.1%) of synapsed trophozoites (Drmota and Kral, 1997). On the one hand, the high copy number of DNA transposons in the *T. vaginalis* genome suggests that the parasite reproduces predominantly asexually, or that its lineage has recently lost sex (Arkhipova, 2005); however, the parasite's intact retroposons and reverse transcriptase homologues (Carlton *et al.*, 2007) may be relics of vertically transmitted elements typically maintained only in sexual lineages (Arkhipova and Meselson, 2005).

Completion of the *T. vaginalis* genome has enabled a genome-wide search for genes encoding DNA repair and recombination proteins in the organism (Carlton *et al.*, 2007; Malik *et al.*, 2008). Conserved DNA repair and recombination proteins from AFP and trypanosomes were used to search the *T. vaginalis* genome and EST databases. Evidence for expression of some of the repair genes comes from available ESTs. The parasite appears to contain genes encoding proteins required to carry out base excision repair (BER), nucleotide excision repair (NER), mismatch repair (MMR), DNA damage response (DDR), chromatin structure relevant to repair, repair involving DNA polymerase subunits, post-replication repair, homologous recombination (HR), meiosis, and nuclease activity for DNA editing and processing. Notably, the *T. vaginalis* genome lacks core non-homologous end-joining (NHEJ) machinery (Ku70, Ku80, Xrcc4, Lig4 and Artemis), indicating that DNA DSB repair must occur via HR and not by NHEJ, which opens a promising door for future homologous gene replacement investigations.

Eight of the ten homologues of meiosis-specific protein-coding genes (*Spo11–1, Hop1, Hop2,*

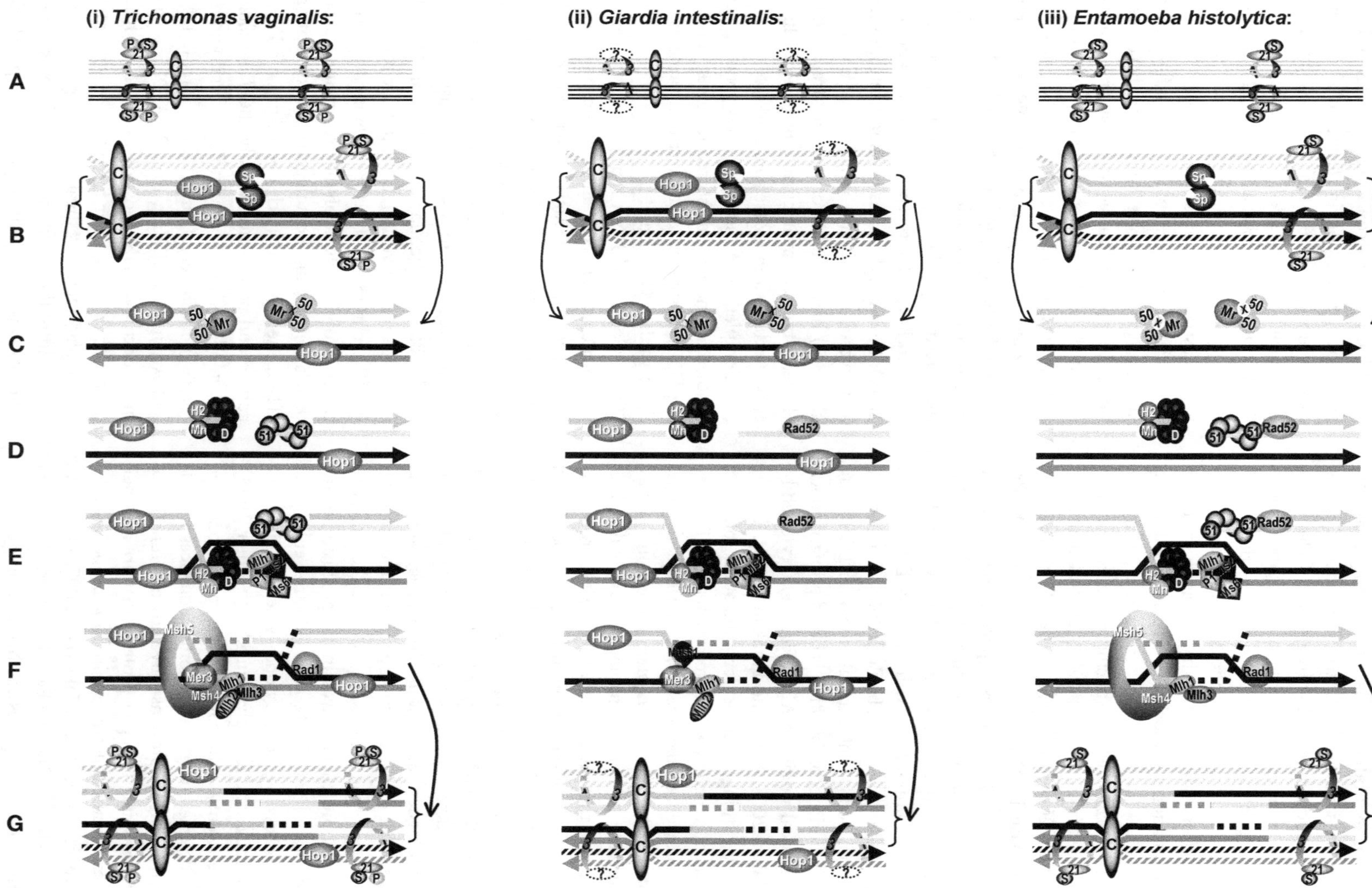

(i) *Trichomonas vaginalis*:
(ii) *Giardia intestinalis*:
(iii) *Entamoeba histolytica*:
A
B
C
D
E
F
G

Mnd1, Dmc1, Msh4, Msh5, and *Mer3*) were found (Fig. 3.5 and Fig. 3.4i). This suggests that *T. vaginalis* 'core meiotic gene' homologues are either evolutionary relics or that they are functional in meiotic recombination in a sexual cycle that has not yet been described. If meiotic recombination occurs in *T. vaginalis,* it may use these eight meiosis-specific proteins to form a synaptonemal complex and double-strand DNA breaks that are repaired by interhomologue recombination and crossover resolution with interference. Two *T. vaginalis* meiosis-specific gene homologues match ESTs, however the limited number of publicly available ESTs probably does not sample the transcriptome to saturation and excludes genes expressed at low levels, thus precluding further conclusions. Another ten out of 28 'core meiotic genes' usually present as single copies in other organisms are found in multiple copies in the *T. vaginalis* genome. Six genes are uniquely duplicated in *T. vaginalis,* while the other four are also duplicated in plants or fish; many of these appear to be rapidly evolving and could be either pseudogenes or adapting to new functions (Malik *et al.*, 2008).

Comparative analysis with *G. lamblia* and *E. histolytica*

Sexual reproduction has been observed directly or inferred by population genetic studies in AFP and some protists, but these methods are unable to distinguish rare sex from complete asexuality and several lineages, particularly highly diverged parasitic protists such as *G. lamblia* and *E. histolytica,* are often considered asexual for a lack of detectable genetic recombination (Tibayrenc and Ayala, 2002; Tibayrenc *et al.*, 1990). Compared to *T. vaginalis,* current investigations into DNA repair in *G. lamblia* and *E. histolytica* focus mainly on components of HR and meiosis, and range from phylogenomic identification of HR and meiotic protein homologues in their genomes (Malik, 2007; Malik *et al.*, 2008; Morrison *et al.*, 2007; Ramesh *et al.*, 2005; Smith and Nicolas, 1998) to functional analyses (Lopez-Casamichana *et al.*, 2008; Melo *et al.*, 2008; Poxleitner *et al.*, 2008) (Figs 3.4 and 3.5). Investigations of gene expression (Melo *et al.*, 2008) and cell biology (Poxleitner *et al.*, 2008) in *G. lamblia* indicate some meiosis-specific gene homologue activity in HR in cysts (Fig. 3.4ii). Gene expression analyses of HR genes in *E. histolytica* also indicate near-typical eukaryotic HR activity (Lopez-Casamichana *et al.*, 2008)(Fig. 3.4iii). An implication of these findings is that all of these meiosis, DNA recombination and repair genes evolved very early during eukaryotic evolution, since homologues are found in representatives of five highly divergent eukaryotic supergroups. These findings open promising doors to genetic analyses in these parasites.

Since *T. vaginalis* contains most of the conserved suite of 'core meiotic genes' whose homologues mediate meiosis in AFP (Malik *et al.*, 2008), and since trichomonads have not been observed to undergo meiosis (Dimota and Kral, 1997; Honigberg and Brugerolle, 1989; Zubacova *et al.*, 2008) or to recombine (Hampl *et al.*, 2001; Tibayrenc *et al.*, 1990), a major question of interest to studies on this parasite is whether *T. vaginalis* persists without sex. It is necessary to determine when in the *T. vaginalis* life cycle meiosis could occur and to discern whether the parasite uses standard two-step meiosis or

Figure 3.4 (opposite) Double-strand DNA break repair models of meiotic recombination hypothesized for (i) *Trichomonas vaginalis*, (ii) *Giardia lamblia,* and (iii) *Entamoeba histolytica.* Conserved proteins comprising the 'core meiotic machinery' are indicated, with meiosis-specific proteins highlighted as white text. Meiotic DSB repair may or may not result in crossovers. (A) Cohesins bind to sister chromatids of a homologous chromosome pair. (B) Spo11–1 makes a double-strand DNA break in a sister chromatid. Where present, Hop1 forms axial elements of the synaptonemal complex. (C) Resection of DNA ends by Mre11-Rad50-(Xrs2) complex, releasing Spo11 from DNA. The expected Xrs2 homologue (dotted lines in 'B', X in 'C') has not yet been found in *T. vaginalis* or *G. lamblia*. (D) Attachment of interhomologue recombination proteins Rad51, Rad52, and meiosis-specific Dmc1, Hop2 and Mnd1 proteins (some accessory Rad52 epistasis group proteins not shown). (E) Strand invasion by interhomologue recombination proteins, DNA synthesis, and mismatch repair by Msh2, Msh6, Mlh1, and Pms1. (F) Holliday junction resolution by crossing over using meiosis-specific proteins (G), or without crossovers (not shown) using general 'single strand annealing' proteins.

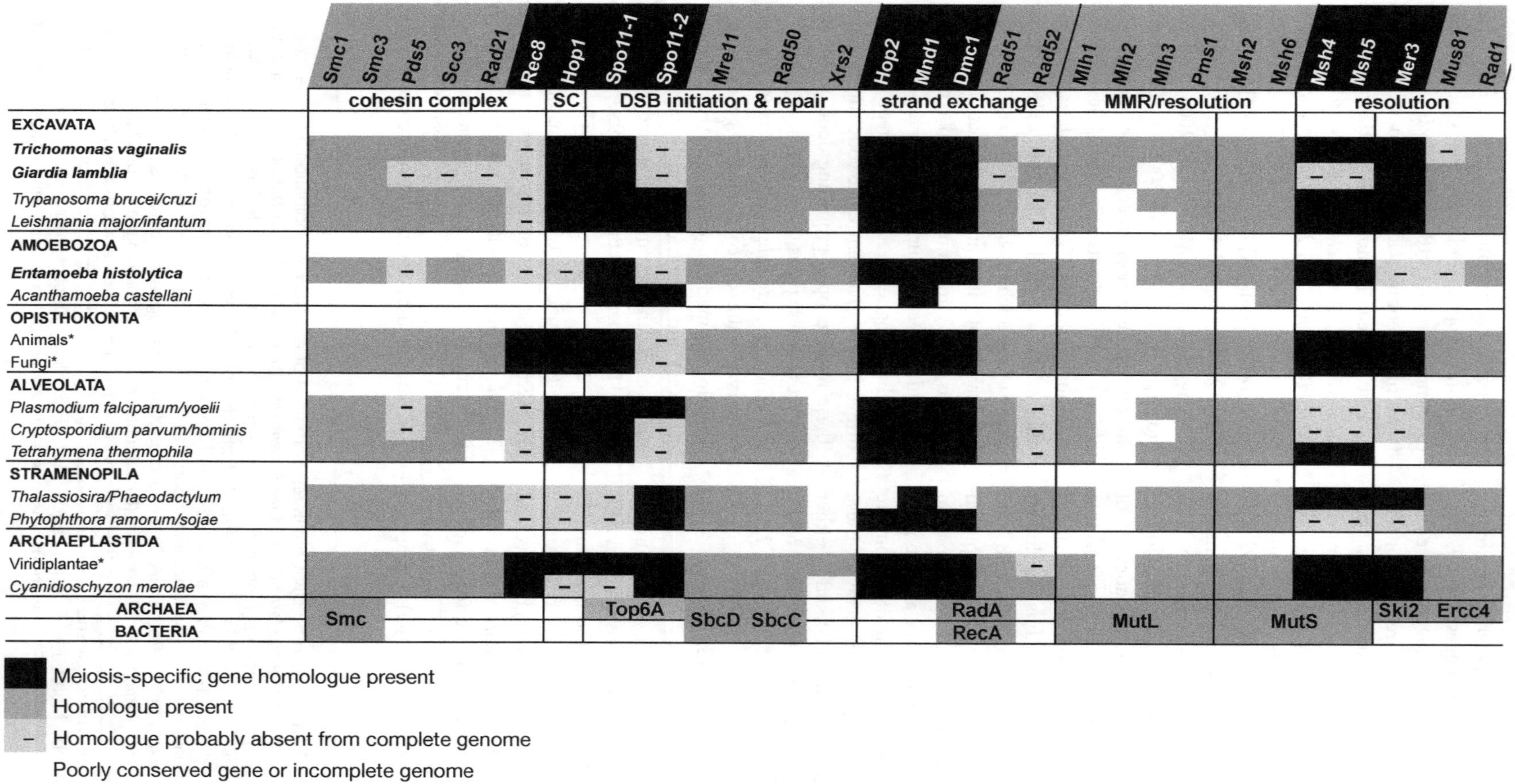

* consensus inferred for common ancestors, though some independent losses within group detailed by Malik *et al* 2008.

Figure 3.5 Core meiotic proteins conserved across eukaryotic supergroups. BLASTP searches using animal, fungal and plant protein sequence queries against genome databases were used to identify predicted homologous proteins in protists; these were then verified by phylogenetic analyses (Malik *et al.*, 2008). Abbreviations are as follows: SC, synaptonemal complex; DSB, DNA double-strand break repair; MMR, mismatch repair. A key to the shading is presented under the table; meiosis-specific gene homologues are highlighted in dark grey boxes. Data were collated from the following references: Klingbeil *et al.*, 2007; Malik *et al.*, 2007, 2008; Watkins and Gray, 2008. Meiotic and meiosis-specific gene homologues were lost independently in sexual AFP several times, thus the absence of some genes does not preclude the possibility of sex and meiosis.

has meiosis resembling that in hypermastigote parabasalids (Cavalier-Smith, 2002; Cleveland, 1947a, b, 1951, 1956; Haig, 1993). Since meiosis in hypermastigotes is induced in response to ecdysone (Cleveland, 1951), meiosis in *T. vaginalis* might be similarly induced in response to hormones of its animal hosts. Ultimately, it may be possible to exploit HR and perhaps perform lab-based genetic crosses, an extremely powerful tool for dissecting the genetics of these organisms.

Membrane trafficking and *T. vaginalis* pathobiology

Membrane trafficking, which mediates selective cargo transport through vesicular and tubular structures between internal membranes and between these and the cell surface, is one of the defining features of eukaryotes. It orchestrates numerous aspects of nutrient uptake, cellular differentiation, communication with the extracellular milieu via secretions, and the transmission of environmental cues (Alberts *et al.*, 2008). The initial steps occurring on the surface of a donor organelle depend on the action of a small GTPase and cargo adaptor proteins concentrating cargo and coat forming protein complexes inducing the formation of the transport intermediate. When vesicles reach their destinations, they fuse with their target membranes with the aid of the vesicle fusion machinery, which includes syntaxins, Sec1/Munc18 (SM) proteins and Rab GTPases (Bonifacino and Glick, 2004; McMahon and Mills, 2004). Membrane trafficking can be divided into two major interconnected, highly integrated and interdependent routes: the biosynthetic-secretory (hereafter the 'secretory pathway') and the endocytic (internalization) pathways. The secretory pathway involves the movement of material from the endoplasmic reticulum (ER) to the Golgi (and the recycling of resident ER components back to the ER) and from the *trans*-Golgi network (TGN) to various membranes including endosomes, lysosomes and the plasma membrane. The endocytic pathways involve movement of material from the cell surface into the endosomes (endocytosis) or phagosomes (phagocytosis), depending on the size of the internalized material, via related but distinct mechanisms (Gruenberg and van der Goot, 2006; Kinchen and Ravichandran, 2008). Endocytosis brings cargo to early endosomes, where they are subsequently differentially sorted into either recycling endosomes (travelling back to the cell surface) or late endosomes called multi-vesicular bodies (MVB) that can interact with lysosomes (Gruenberg and van der Goot, 2006), the digestive organelles where cargos destined for degradation end their route (Luzio *et al.*, 2007). Similarly, material internalized into phagosomes can be digested in phagolysosomes (Luzio *et al.*, 2007). Lysosomes are also involved in another major cellular pathway, autophagy, which degrades long-lived cytosolic organelles and proteins by sequestering cargos destined for degradation into newly formed double-membrane structures, the autophagosomes. Mature autophagosomes eventually fuse with lysosomes to allow degradation and ultimate recycling of the autophagocytosed cellular material (Levine and Klionsky, 2004; Luzio *et al.*, 2007).

In host–parasite interactions, such as between the human urogenital mucosa and *T. vaginalis*, membrane trafficking is at the forefront in orchestrating the quality and length of the relationship. While membrane trafficking has been investigated intensively from the host perspective by immunologists and cell biologists, views from the parasite perspective have been particularly stimulated by parasite genome sequencing projects (e.g. Field and Carrington, 2004; Nozaki and Nakada-Tsukui, 2006; Sheiner and Soldati-Favre, 2008). Asymptomatic colonization by *T. vaginalis* is common but these infections can degenerate into full-blown cellular wars leading to acute and chronic inflammation causing high distress and discomfort, affecting reproduction and leading to other health complications (Johnston and Mabey, 2008; Nanda *et al.*, 2006). Membrane trafficking is intimately linked with *T. vaginalis* pathobiology, mediating key processes including cell surface expression of adhesins and membrane-bound proteases (Hirt *et al.*, 2007), endocytosis of host proteins such as lactoferrin (which is key for iron homeostasis) (Sutak *et al.*, 2008), secretion of cytotoxic factors such as proteases and pore-forming proteins (Fiori *et al.*, 1999), and phagocytosis of bacteria and host cells (Rendon-Maldonado *et al.*, 1998). Importantly, *T. vaginalis* infections are also linked with higher susceptibility to HIV and possibly other viral infections, including genital herpes simplex viruses (HSV), and human papilloma viruses (HPV) (Nanda *et al.*, 2006; Pindak *et al.*, 1989; Rendon-Maldonado *et al.*, 2003). It is thought that multiple aspects of host–pathogen interactions are involved in *T. vaginalis* facilitating virus transmissions between humans, but *T. vaginalis* membrane trafficking is directly involved in the internalization of human viruses (including HSV and HIV) by direct endocytosis or via phagocytosis of virus-infected human cells, as demonstrated *in vitro* (Pindak *et al.*, 1989; Rendon-Maldonado *et al.*, 2003). Following uptake, viruses accumulate in *T. vaginalis* MVB-like structures and remain infectious for days (2–9 days depending on the virus), but there is no evidence for human virus replication (Pindak *et al.*, 1989; Rendon-Maldonado *et al.*, 2003). Upon infection of a new host by virus-loaded *T. vaginalis*, internalized viruses could be released on contact with host tissue, hence contributing to virus transmission (Lal *et al.*, 2005). Similarly *T. vaginalis* could also contribute to the spread of bacterial pathogens such as *Mycoplasma* spp., which are also internalized by *T. vaginalis* and divide intracellularly (Dessi *et al.*, 2006). Hence *T. vaginalis* could represent a Trojan horse for various human infectious agents.

Compared to the intensively investigated AFP taxa (Bock *et al.*, 2001), little was known about the molecular and cellular characteristics of *T. vaginalis* membrane trafficking prior the genome sequencing project. The genome sequence data have provided a unique opportunity to investigate the molecular components mediating membrane trafficking in order to generate an initial sketch of *T. vaginalis* membrane pathways and guide future experimental characterization.

Trichomonas vaginalis membrane trafficking gene complement

Using a combination of BLASTP, RPS-BLAST, multiple sequence alignments and, in selected cases, phylogenetics for identification, a considerable diversity of membrane trafficking machinery proteins appears to be encoded by the *T. vaginalis* genome (Carlton *et al.*, 2007). Homologues of well-established pieces of the vesicle formation machinery, including vesicle coats and GTPases, were identified along with diverse elements of the vesicle fusion machinery, including the N-ethylmaleimide-sensitive fusion protein (NSF) SNAREs, Rab GTPases, and SM proteins (Table 3.5). Additional components known to be involved in membrane trafficking such as dynamin and pieces of the endosomal sorting complex required for transport (ESCRT, see below) machinery were also identified. Fig. 3.6 presents a schematic representation of the membrane trafficking pathways thought to be present in *T. vaginalis* based on protein families encoded by the genome and on ultrastructural data (Benchimol, 2004).

Small GTPases of the Ras superfamily

There are five major subfamilies among the Ras-like small GTPases: Ras, Rho, Ran, Rab and ARF-Sar (hereafter simply ARF) (Colicelli, 2004). The Rab and ARF subfamilies are known to be key players in orchestrating different aspects of membrane trafficking (vesicle formation, motility, docking and fusion) (Behnia and Munro, 2005). *Trichomonas vaginalis* possesses orthologues of each of these five subfamilies (Table 3.5). The striking feature here is the massive amplification of the TvRab subfamily (293 members versus 70 in human), expanding on the original observation made from cDNA and genome analysis which characterized 65 Rabs (Lal *et al.*, 2005). This staggering number establishes a new record of Rab gene copies (Lal *et al.*, 2005) and also represents more then twice the number found in *E. histolytica* (91 Rabs; Saito-Nakano *et al.*, 2005), another human mucosal protozoan pathogen with an unexpectedly large gene copy number. This is a rather unexpected observation, in particular for a microbial eukaryote, since complexity of the Rab repertoire was initially thought to be correlated with the expanded endomembrane complexity linked to multicellularity and tissue diversity (Pereira-Leal and Seabra, 2001). ARF homologues are also present in large numbers (36 versus 30 in humans), whereas the sizes of the other Ras subfamilies are more in line with other taxa (Table 3.5). The Ran small GTPase subfamily, for instance, which regulates nucleocytoplasmic transport of RNA and proteins (see earlier sections), consists of only two paralogues in *T. vaginalis*, indicating that substantial gene duplication has taken place in only a subset of the small GTPases superfamily. The presence of conserved

Table 3.5 Protein family size for selected pieces of the membrane trafficking machinery

Protein family	*T. vaginalis*	*H. sapiens*	*A. thaliana*	*S. cerevisiae*
Vesicle formation: coat proteins				
Adaptin	73	24	23	13
Retromer	12	4	6	3
COPI	16	9	9	7
COPII	12	9	12	6
Vesicle fusion: syntaxin, syntaxin binding proteins				
Syntaxin	13	12	18	7
Syntaxin_E[a]	10	3	6	3
SM	18	7	8	4
SM_E[a]	16	3	2	2
Small GTPases				
Rab[b]	292	70	57	11
ARF[b]	36	30	21	7
Ran	2	1	4	2
Rho	12	23	11	6
Ras	14	35	0	4
ESCRT machinery – MVB, late endosomes				
ESCRT-I (4)	2	8	7	4
ESCRT-II (3)	4	3	3	3
ESCRT-III (10)	14	15	17	10

Modified from Carlton *et al.* (2007) and updated with more recent analyses. The values for adaptins are derived from Boehm and Bonifacino (2001). The values for COPI include all seven coatomer subunits (Bock *et al.*, 2001). For retromer, only Vps26, 29 and 35 were considered, as homologues for Vps5 and 17 proteins cannot be identified in diverse eukaryotes. For COPII, Sec 13, 23, 24 and 31 were counted (Bock *et al.*, 2001). The endocytic syntaxins (Syntaxin_E) are considered to be SynE homologues (i.e. Pep12, Vam3, Syntaxin 7, Syntaxin 12/13 and Syp21,22, 23) and TLG2 (Syntaxin 16) homologues. Vps33 and Vps45 homologues make up the endocytic SM (SM_E). The ESCRT machinery is made of different protein complexes three of which are found in most eukaryotes and these are listed here (complexes I, II and III), with the number of known subunits indicated in brackets (Leung *et al.*, 2008). The total number of subunits, counting paralogues when present, is given for each species. The ESCRT–III complex includes the complex III-associated proteins as defined by Leung *et al.* (2008) from which all ESCRT machinery values were derived.

a Values and categories in these rows denote a sub-total of the larger set in the row above.

b Small GTPases subfamilies involved in membrane trafficking. These numbers are contrasted with the size of the other Ras-like small GTPases subfamilies. Complement sizes for the genomes other than *T. vaginalis* were taken from the literature (Colicelli, 2004; Garcia-Ranea and Valencia, 1998; Vernoud *et al.*, 2003).

sequence features in over 90% of the TvRab proteins – conserved G-boxes, G1 to G5, and characteristic double cysteine-containing C-terminus motifs (Colicelli, 2004) – and the fact that a large number of the TvRab genes have EST support suggest that most members of this gene family are

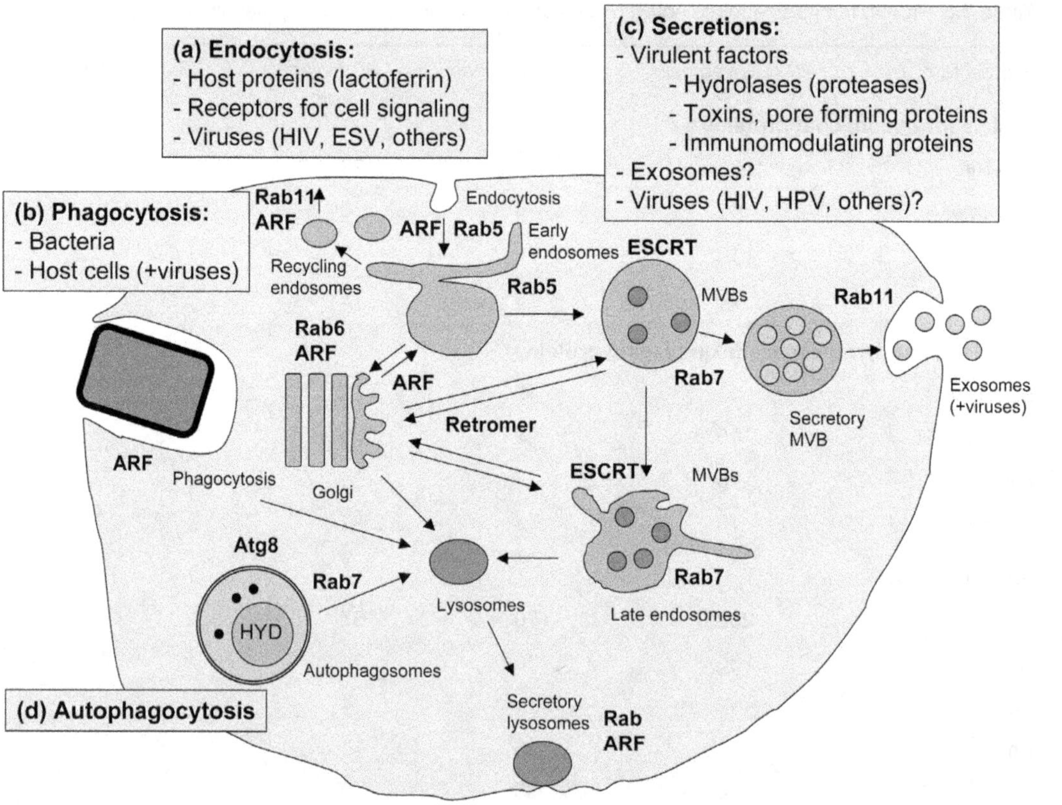

Figure 3.6 Schematic representation of membrane trafficking thought to take place in *T. vaginalis*. The illustrated pathways focus on (a) endocytosis; (b) phagocytosis, illustrated with a bacterial cell but this route also mediates host cell uptake (see text); (c) secretions, and (d) autophagocytosis of cytosolic proteins and organelles. An autophagosome with a double membrane is illustrated with a sequestered hydrogenosome (HYD) and protein complexes (black dots). Examples of processes involved in *T. vaginalis* pathobiology are indicated in the boxes. The ER to Golgi trafficking is not illustrated for simplicity. Select key markers identified by bioinformatic analyses (Tables 3.6 and 3.7) are indicated in bold. Exosomes are indicated but there is currently no experimental evidence (indicated by the '?') for their existence in *T. vaginalis*. If present, exosomes could contribute to the secretion of cytotoxic compounds by *T. vaginalis* and mediate toxicity at a distance to human cells, including immune cells, or modulate the overall immune status of the mucosal surface, since human exosomes appear to modulate the immune system (Schorey and Bhatnagar, 2008). Similarly, secretions derived from MVB could release internalized human viruses – see text and cited reference. A secretory lysosome poised to fuse with the plasma membrane is also illustrated as suggested by biochemical data (Scott *et al.*, 1995).

functional and unlikely to represent pseudogenes. Members of the Rab subfamilies known in other taxa to regulate the secretory (Rab1, Rab6 and Rab11) and the endocytic (Rab5 and Rab7) pathways are all present (Lal *et al.*, 2005).

Coat proteins

Four types of coat protein are known to mediate vesicle formation. These are the COPI, COPII, clathrin and the retromer coats (Bonifacino and Glick, 2004; McMahon and Mills, 2004; Seaman, 2005). The set of coat proteins involved in vesicle transport in the early secretory system is slightly expanded in *T. vaginalis* compared to the yeast *Saccharomyces cerevisiae* and similar to the gene complement of multicellular systems (human and *Arabidopsis*) (Table 3.5). The *T. vaginalis* complement of COPII genes is also similar to that seen in other model systems. Components of the coatomer complex (COPI) also showed moderate expansion, similar to *S. cerevisiae*, human and *Arabidopsis thaliana*.

In sharp contrast to genes for coat proteins from the early secretory system, the families encoding vesicle coat proteins responsible for post TGN and endosomal trafficking are remarkably expanded, even relative to their multicellular counterparts. The retromer complex is responsible for recycling of lysosomal hydrolase receptors from the endosome back to the TGN as well as recycling other components, including proteins on the cell surface (Bonifacino and Hurley, 2008). In *T. vaginalis*, there are 12 genes encoding the three core components of the retromer complex (Vps26, 29 and 35) (Table 3.5); this is twice the number found in *A. thaliana*, while *S. cerevisiae* and humans have three and four genes encoding these components, respectively. The adaptin complexes work as cargo adaptors with the clathrin coats to bind cargo at the TGN, endosomes and the plasma membrane, and they initiate endocytosis in the latter (Bonifacino and Glick, 2004). There are four complexes (named AP1, 2, 3 and 4) and four subunits per complex, producing a base number of 16 possible proteins. Both *A. thaliana* and humans have expanded their adaptin complement with 23 and 24 homologues respectively compared with 13 homologues in *S. cerevisiae*, which has lost the AP4 complex (Boehm and Bonifacino, 2001). In contrast, *T. vaginalis* has a massively expanded adaptin complement, comprising 73 paralogues with all four adaptin complexes represented. The unexpected expansion of adaptin and retromer paralogues is far greater than previously observed for any eukaryote, let alone a unicellular one.

Membrane fusion

Examination of several paralogous gene families involved in membrane fusion further confirmed the expansion of the endocytic system. Three key players were considered: Rabs, syntaxins and SM proteins. In the cells of well studied systems, paralogous proteins perform a similar function but at different locations and in different pathways. Within each of the protein families, some are involved in endocytosis whereas others are involved in the secretory pathway (Behnia and Munro, 2005; Chen and Scheller, 2001). In all three cases, *T. vaginalis* has significantly expanded the gene complement for the endocytic paralogues, both when compared to the secretory paralogues in its own genome and when compared to the endocytic complement in the genomes of three model eukaryotes (Table 3.5). In contrast NFS, the ATPase that promotes the disassembly of the fusion complex (Bock *et al.*, 2001), is only encoded by a single gene in all compared genomes.

Syntaxins are SNARE proteins residing at the surface of acceptor organelles, allowing vesicle fusion and potentially encoding targeting information (Bonifacino and Glick, 2004). We identified a total of 13 syntaxin homologues in *T. vaginalis*. This compares to 18, 12 and seven syntaxins in *A. thaliana*, human and yeast respectively (Bock *et al.*, 2001; Sanderfoot *et al.*, 2000). Based on the BLASTP and phylogeny-derived annotation, 10 out of the 13 *T. vaginalis* syntaxins are thought to be endocytic paralogues. This compares to only six out of 18 in *A. thaliana*, three out of 12 in human and three out of seven in yeast (Table 3.5).

SM proteins bind syntaxins and regulate their pairing with other SNARE proteins (Bonifacino and Glick, 2004). Eight, seven, and four SM protein homologues are known in *A. thaliana*, humans and *S. cerevisiae*, respectively, in comparison to 18 SM protein homologues identified in *T. vaginalis*. Strikingly, those SM proteins involved in fusion reactions in *S. cerevisiae* and mammalian endocytic pathways (Vps33 and Vps45) account for 16 of the *T. vaginalis* homologues, compared with only three endocytic homologues in humans and two in *A. thaliana* and *S. cerevisiae*.

Consistent with the other observed endocytic protein family expansions, *T. vaginalis* also possesses an expanded repertoire of two well-established endocytic Rab functional classes: Rab5, Rab7 and related Rabs (Pereira-Leal and Seabra, 2001). *T. vaginalis* has at least 43 'endocytic'-like Rabs (identified with a PHI-BLASTP search using TvRab5 and the TIG[A/S]XF motif, a characteristic of endocytic Rabs that is part of the G2 box and switch I region (Lal *et al.*, 2005)), with four Rab5s and six Rab7s and a collection of additional endocytic-like Rabs. This is a far larger number than in the three model systems: humans have nine (six Rab5 and Rab5-like and three Rab7/9, of which seven were identified with a PHI-BLASTP search using HsRab5 and the TIG[A/S]XF pattern), *A. thaliana* has 11 and *S. cerevisiae* has four (Pereira-Leal and Seabra, 2001). The 47 *T. vaginalis* 'endocytic-like' Rabs contrast with the 27 *T. vaginalis* 'secretory-like' Rabs (identified with a PHI-BLASTP search using TvRab11 and the TIG[V/I]XF pattern), which include the three TvRab1 and two TvRab11 paralogues (Lal *et al.*, 2005).

Multivesicular bodies and autophagosomes

The MVBs are late endosomal compartments that receive material from early endosomes and the TGN and can fuse to lysosomes (Luzio *et al.*, 2007; Williams and Urbe, 2007). They are characterized by internal vesicles derived from invaginations of the MVB membrane (Gruenberg and van der Goot, 2006), a process mediated by the ESCRT machinery required for the sorting of ubiquitinated endocytic cargo transmembrane proteins and other proteins into MVB endo-vesicles. The contents of MVBs are destined for either degradation in lysosomes (Luzio *et al.*, 2007) or for secretion; in the latter case, secreted endo-vesicles are called exosomes and promote intercellular communication in animals (Schorey and Bhatnagar, 2008). Some viruses, including HIV, have hijacked this cellular route and are secreted along the same pathway as exosomes (Schorey and Bhatnagar, 2008); *T. vaginalis* likely accumulates endocytosed human viruses (e.g. HIV and ESV) in such structures (Pindak *et al.*, 1989; Rendon-Maldonado *et al.*, 2003). The *T. vaginalis* genome encodes most components of ESCRT complexes I, II and III (Table 3.5) but, like most eukaryotes, does not possess elements of the ESCRT–0 complex found in animals and fungi (Leung *et al.*, 2008). MVB-like structures were also identified in *T. vaginalis* when investigating the cellular distribution of a G-protein alpha subunit, which could be involved in regulating secretions upon binding of *T. vaginalis* to extracellular matrix (ECM) proteins (Lal *et al.*, 2006). Together these data strongly suggest that MVBs are present in *T. vaginalis* and that investigating their properties could lead to important and interesting insight into *T. vaginalis* pathobiology.

Autophagy is a process shared by most, if not all, eukaryotes. It mediates rapid recycling of long-lived cytosolic material including organelles (e.g. mitochondria or hydrogenosomes) and proteins, as well as intracellular bacteria and viruses, through their sequestration into a double membrane sealed organelle and eventual degradation upon fusion with lysosomes (Levine and Klionsky, 2004; Luzio *et al.*, 2007). Autophagy is essential in AFP for response to starvation and other stresses, cellular differentiation, and development (Levine and Klionsky, 2004). The formation and fate of autophagosomes is dependent on the concerted activity of a collection of specific proteins (called Atg proteins) and proteins shared with other aspects of membrane trafficking (e.g. Rab GTPases, SNAREs) as autophagosomes interact with endosomes and ultimately fuse with lysosomes. Among the 12 Atg proteins that are shared among yeast, animals and plants (Levine and Klionsky, 2004), a total of nine were identified in *T. vaginalis* (Table 3.6). These include Atg8, a key autophagosome marker associated with the mature organelle (Levine and Klionsky, 2004). These molecular data strongly suggest that autophagy takes place in *T. vaginalis*, a process observed at the cellular level in the related species *T. foetus* (Benchimol, 1999). Autophagy is likely to represent an essential process underlying *T. vaginalis* response to starvation and other stresses, and could also be involved in cellular differentiation during transformation into amoeboid forms upon contact with host mucosal cells and ECM proteins.

Based on these genome-wide analyses, there is a coherent pattern emerging regarding various components of the membrane trafficking system, indicating that *T. vaginalis* has selectively generated and maintained a complex repertoire of endocytic markers. The differential amplification and magnitude of the amplifications of specific subfamilies are consistent with lineage-specific and relatively recent gene duplications, also called 'conservative' gene duplications, which are rationalized as contributing to lineage-specific adaptations (Vogel and Chothia, 2006). These observations suggest that *T. vaginalis* has developed a highly complex endocytic endomembrane system when compared to other aspects of its membrane trafficking machinery and as compared to several model organisms. This can be rationalized as either a requirement for more finely-tuned regulation of the endocytotic pathways (Bock *et al.*, 2001) to accommodate, for example, a large diversity of cargos, or as part of a broader diversification of endosomal subcompartments with distinct roles, possibly required during the different

Table 3.6 *T. vaginalis* protein homologues potentially involved in autophagy

Protein	Number of proteins	BLASTP *E*-values	
		S. cerevisiae	*H. sapiens*
Regulation of induction			
Atg1	1	2e-44	4e-42
Vesicle nucleation			
Atg6	1	2e-06	6e-05
Vesicle expansion and completion			
Atg3	1	5e-29	7e-36
Atg4	5	8e-18	1e-18
Atg7	1	2e-90	1e-100
Atg8	2	2e-27	6e-27
Atg10	1	–	5e-07
Recycling			
Atg9 (with 6 TMD)	1	5e-54	3e-27
Atg18	1	1e-13	5e-11

There are 30 Atg proteins in yeast, of which 12 are conserved across a broad range of eukaryotes (Levine and Klionsky, 2004) and nine are identifiable in *T. vaginalis*. Atg4 and Atg8 are characterized by more then one paralogue in *T. vaginalis*. The Atg9 proteins are characterized by six transmembrane domains (TMD). The *E*-values of the best BLASTP hit between *T. vaginalis* and yeast and human proteins are shown, with only the best hit listed when there is more then one entry. Dashes (–) indicate the absence of a hit.

stages of infections – male versus female urogenital tracts, mucosal surface versus intra-mucosal – and/or in response to different host innate and adaptive defence mechanisms. Hence the endocytic pathway is likely to play an even more important role in the survival and pathobiology of *T. vaginalis* than was thought initially. It could therefore represent a cellular system to be targeted for pharmacological control of the parasite (Stein *et al.*, 2003), an interesting alternative to classic antimicrobial therapy targets (Clatworthy *et al.*, 2007). Identification of molecular markers for MVB, phagosomes and autophagosomes should also provides new opportunities to study the subtleties of host–pathogen interactions, including the possible role of *T. vaginalis* as carrier of other infectious agents such as bacteria and viruses in MVB-like organelles. Additional bioinformatic analyses and future experimental cell biology work is required to address the functional significance of the extraordinarily large repertoire of proteins regulating membrane trafficking in *T. vaginalis*. Thus, *T. vaginalis* represents a fascinating challenge to our current understanding of protein family diversity regulating membrane trafficking, and as such could become an invaluable model system for comparative molecular cell biology of membrane trafficking.

Web resources: the *T. vaginalis* genome database, TrichDB

One of the crucial tasks of genome projects is to provide public access to, and comprehensive searching of, the genomic sequence data. The NIAID-funded EupathDB family of websites (http://eupathdb.org) provides this service for sequence data from a variety of human pathogens. Within the EuPathDB family, the TrichDB website (http://trichdb.org) specifically serves as a portal to genome-scale trichomonad datasets (Aurrecoechea *et al.*, 2009). Underlying TrichDB are multiple sequence datasets derived from direct sequencing and subsequent gene modelling (currently only from *T. vaginalis*, but with more species to be added): ~65,000 genomic DNA sequences (scaffolds and contigs); ~564,000 open reading frames of >50 nucleotides long, and their conceptual translations; and ~60,000 predicted protein-encoding DNA sequences and their cognate mRNA and amino acid sequences. Additionally, the 38,000 'repeat genes' sequences – virus-like, transposon-like, retrotransposon-like, and unclassified repeated genes, all with high copy number and low polymorphism as described earlier – are represented as separate datasets of DNA, transcript, and protein sequences. An EST dataset, containing (as of Oct 2008) ~92,000 ESTs from eleven libraries, is also incorporated into TrichDB. Complementing these sequence datasets are sequence annotations, which were manually curated as part of the *T. vaginalis* G3 genome sequencing project.

While the complete annotation and sequence datasets are all downloadable from TrichDB for local use, the primary intent of the site is to provide a web-based work environment for trichomonad genome data mining. To this end the site offers an extensive selection of gene, genomic DNA, EST, and ORF identification and retrieval tools, with future provision for microarray, SNP and SAGE data mining (Fig. 3.7). TrichDB also includes a graphical genome browser that can display annotations, BLAST matches, EST assemblies, and other features mapped to the genome.

The gene identification tools illustrate the range of TrichDB utilities available for querying the various sequence datasets. Users can search by 'Gene attribute' to select only protein-coding genes, various types of RNA-coding genes (e.g. rRNA, tRNA, small nuclear RNA), or repeat genes, or any combination of these. Exon/intron structure, gene ID, species, and product description keywords are other searchable gene attributes. Genes can also be retrieved based on their chromosomal location (i.e. scaffold/contig location in the current *T. vaginalis* assembly), whether they have EST evidence associated with them, features of their predicted protein products (isoelectric point, MW, secondary structure, cellular location, GO function, EC number), and hits to protein motif, domain, and BLAST similarity searches. In the latter case the user can either input a protein motif as a Perl-style regular expression and amino acid property code (e.g., 'CC6+RK' for 'two cysteines followed by one or more hydrophobic amino acids, followed by arginine, then lysine'), or input a nucleotide or protein sequence as a BLAST query. The tools differ somewhat for different datasets – the EST set adds a query by extent of gene overlap, for example. Queries can be combined with Boolean logic and given informative names by the user via the 'Query History' page, and saved to a free user account for future reference. Users can invoke help at each query option and can watch video tutorials available on-site.

A query typically returns a table of matching sequence IDs and user-configurable descriptive data. Each sequence ID in the table is also a link to a dedicated page for that sequence, which for a protein coding gene shows its DNA, mRNA and protein sequences, its protein features (domain database and GenBank BLAST hits), and a graphic display of the gene in genomic context, with the target gene and its neighbours shown as icons along with their EST data and GenBank hits. A link from this graphic takes the user to a highly user-configurable genome browser view (Fig. 3.8) where users can zoom in to or out of the genomic region, add or subtract various 'tracks' of data beyond the default set (e.g., ORFs, 3-frame translations, graphs of G+C content or gene density), and specify display properties and regions of interest; these track configurations can be saved. Importantly, users can also upload their own (or others') annotations for the region, a capability that may prove crucial in the future for keeping genome projects up-to-date.

In addition to data from the *T. vaginalis* G3 genome sequencing project, TrichDB relies on other, more general public databases for some of its data, including GenBank and dbEST, as well as TvXpress, a web-searchable database of trichomonad ESTs hosted at the Chang Gung Bioinformatics Centre (http://tvxpress.cgu.edu.tw). TrichDB solicits new datasets from the research community at regular intervals, and as a result will soon be providing new genomic, microarray, proteomic, and microRNA data, not only for *T. vaginalis* but for other trichomonad species too.

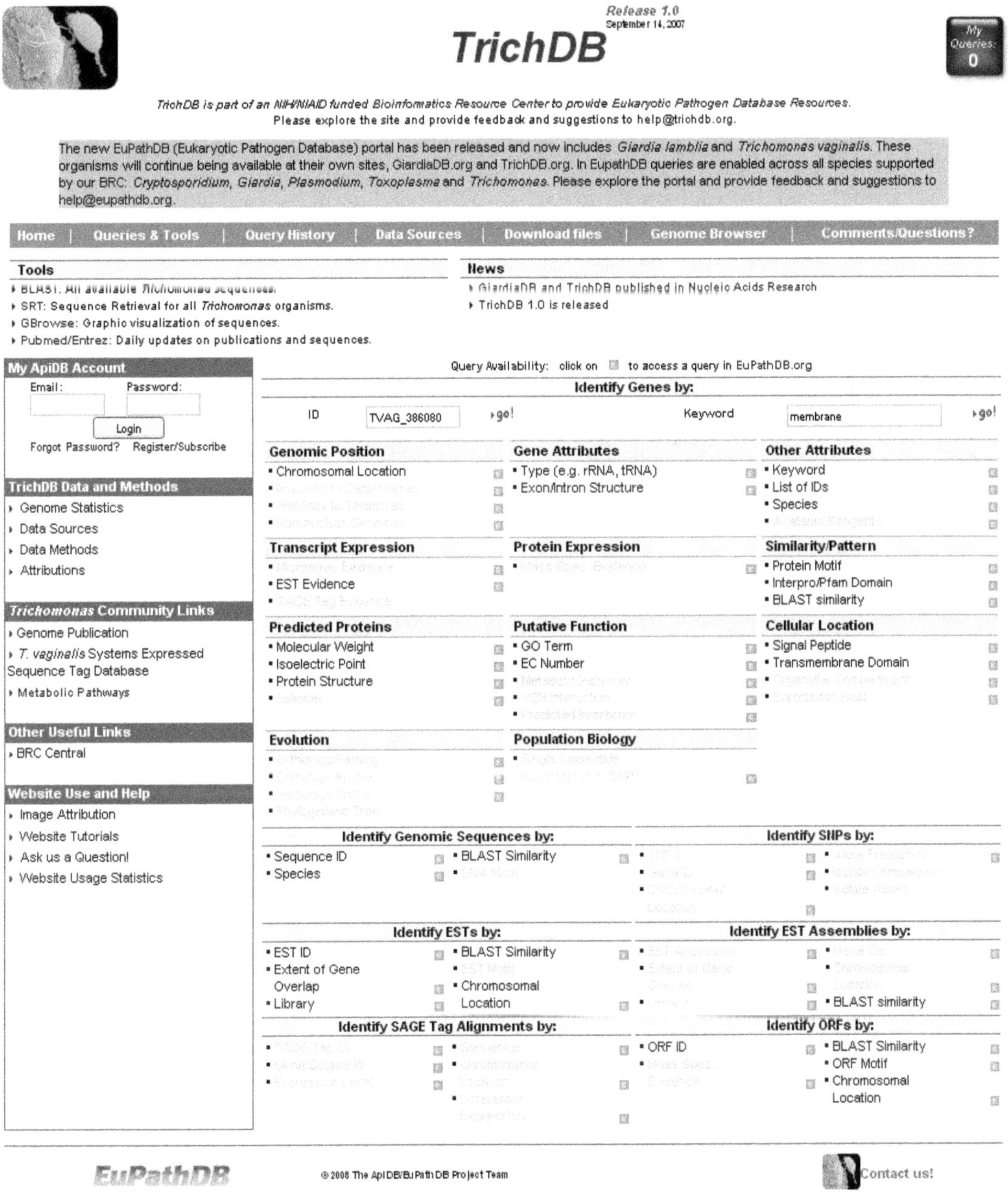

Figure 3.7 TrichDB.org front page. The front page aggregates links to news, information about the genome and the database, related websites, help pages, and access to the entire sequence query toolkit. Each available search type (currently unavailable searches are greyed out) takes the user to a query page with options customized for that search. The tabs at the top additionally include a link to a 'Download files' page where users can retrieve the complete annotation and comprehensive sets of genomic, transcript, CDS (coding sequence), protein and EST sequences.

What is next for *T. vaginalis* genomics?

The first *T. vaginalis* genome sequence was undoubtedly a success. Not only were fascinating insights into the biology of the organism obtained, but the research community united behind a common goal, ultimately resulting in a satisfying high-profile publication and widespread publicity (see for example: http://news.bbc.co.uk/2/hi/health/6248165.stm and http://www.usatoday.com/tech/science/2007-01-12-parasite-mapped_x.htm). What's next in store

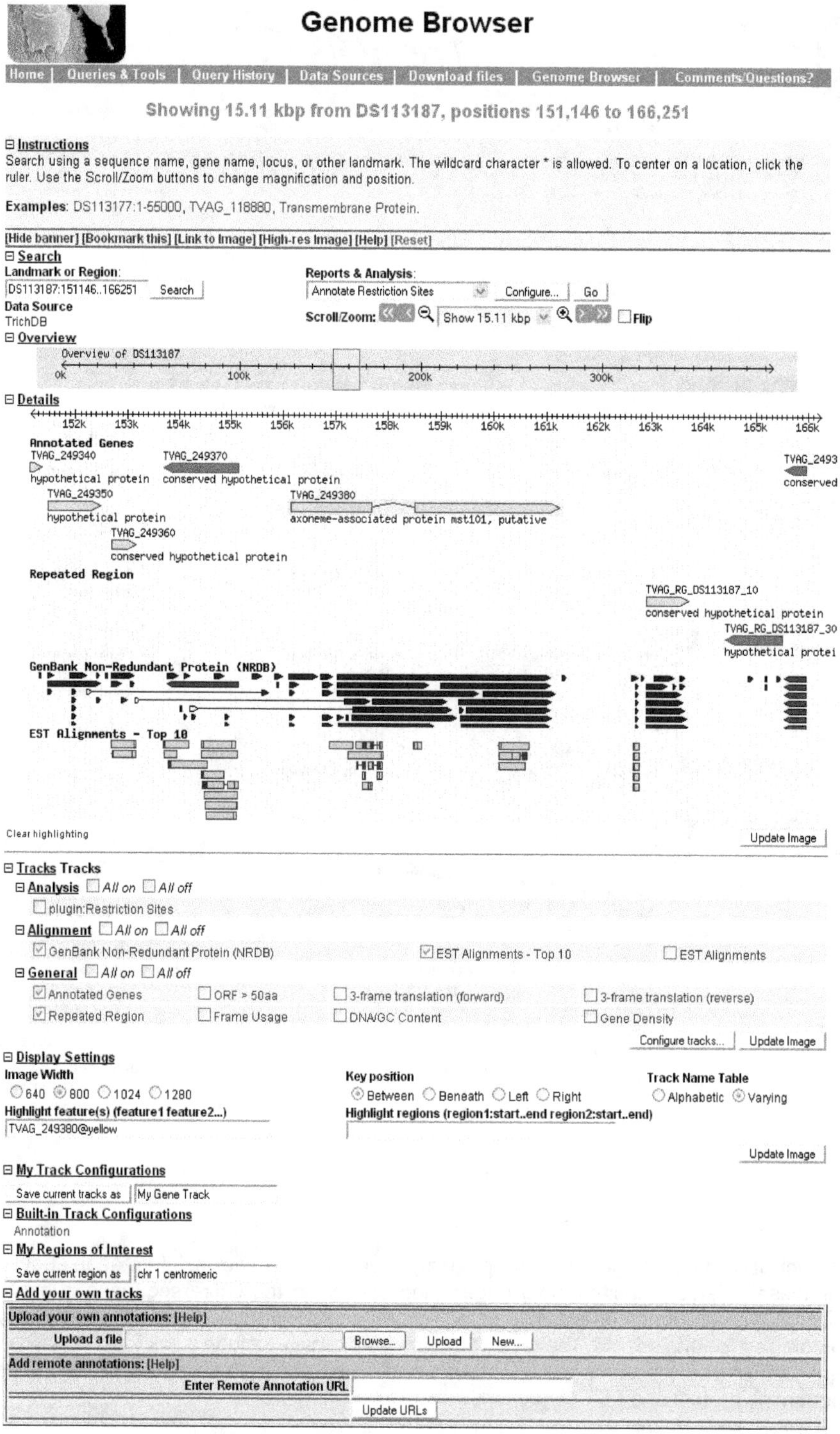

Figure 3.8 TrichDB genome browser view. The genome browser can be used to view any user-delineated section of the genome, or a particular gene can be used to 'anchor' the view, as here (showing one of the rare intron-containing genes in the *T. vaginalis* genome). The result is much like the genome context graphic at the top of a Gene Page view, but with vastly more user configurability. Users can zoom in and out from the region, and add or subtract various 'tracks' of graphic data to the default view (e.g., all ORFs > 50 amino acids, GC content, sequence alignments). Display width, text, layout and colour features can be altered, and configurations can be saved – all of which facilitate creation of publishable figures. Also notable is the 'Add your own tracks' feature, which allows users to add annotation to the region.

for *T. vaginalis* genomics? Most importantly, a reannotation of the current G3 sequence is under way; in particular all of the 59,681 genes are being re-evaluated in light of the identification of the *Maverick* gene family and other transposable element families. By depreciating those genes found to be within transposable elements, it will be possible to refine the total protein-coding gene count of the organism. In an interesting twist, it may be possible to use the *Maverick* elements to match and stitch together scaffolds by exploiting the combination of target site duplication and terminal inverted repeat sequences that uniquely identifies each *Maverick* element. Curation of the *T. vaginalis* genes also continues through searching the literature and manual editing of coding regions predicted by automated gene callers. All updates are being made to TrichDB and GenBank in order to facilitate ease of access.

In terms of functional genomics, the size and complexity of the *T. vaginalis* genome presents significant challenges to whole genome expression studies, comparative genomic hybridization experiments, and proteomic studies. With the completion of the first genome reannotation in sight, the revised gene count should ease this burden considerably, and it may be possible to provide the research community with, for example, microarrays through the NIAID-funded Pathogen Functional Genomics Resource Centre (http://pfgrc.jcvi.org/) which provides such resources for many microbial pathogens.

Finally, a second genome, that of the TO16 isolate, is currently being completed at Chang Gung University, with emphasis on finishing a large contiguous section of the genome. This will provide invaluable information concerning *T. vaginalis* genome architecture and structure. With the advent of the next generation of genome sequencing methods, it will no longer be such a time-consuming or expensive process to shotgun sequence a trichomonad genome, and so it is hoped that the genomes of other closely-related trichomonads such as *T. tenax, P. hominis,* and, more distantly, *T. foetus,* will be available in the future for comparative studies. Trichomonad genomics is finally coming of age.

References

Adl, S.M., Simpson, A.G., Farmer, M.A., Andersen, R.A., Anderson, O.R., Barta, J.R., Bowser, S.S., Brugerolle, G., Fensome, R.A., Fredericq, S., *et al.* (2005). The new higher level classification of eukaryotes with emphasis on the taxonomy of protists. J. Eukaryot. Microbiol. 52, 399–451.

Alberts, B., Johnson, A., Lewis, J., Raff, M., Roberts, K., and Walter, P. (2008). Molecular Biology of the Cell, 5th edn (New York, Garland Publisher, Inc.).

Arisue, N., Hasegawa, M., and Hashimoto, T. (2005). Root of the Eukaryota tree as inferred from combined maximum likelihood analyses of multiple molecular sequence data. Mol. Biol. Evol. *22*, 409–420.

Arkhipova, I., and Meselson, M. (2005). Deleterious transposable elements and the extinction of asexuals. Bioessays 27, 76–85.

Arkhipova, I.R. (2005). Mobile genetic elements and sexual reproduction. Cytogenet. Genome Res. *110*, 372–382.

Aronica, L., Bednenko, J., Noto, T., DeSouza, L.V., Siu, K.W., Loidl, J., Pearlman, R.E., Gorovsky, M.A., and Mochizuki, K. (2008). Study of an RNA helicase implicates small RNA-noncoding RNA interactions in programmed DNA elimination in *Tetrahymena*. Genes Dev. 22, 2228–2241.

Aurrecoechea, C., Brestelli, J., Brunk, B.P., Carlton, J.M., Dommer, J., Fischer, S., Gajria, B., Gao, X., Gingle, A., Grant, G., *et al.* (2009). *Giardia*DB and TrichDB: integrated genomic resources for the eukaryotic protist pathogens *Giardia lamblia* and *Trichomonas vaginalis*. Nucleic Acids Res. 37, D526–530.

Bastida-Corcuera, F.D., Okumura, C.Y., Colocoussi, A., and Johnson, P.J. (2005). *Trichomonas vaginalis* lipophosphoglycan mutants have reduced adherence and cytotoxicity to human ectocervical cells. Eukaryot. Cell 4, 1951–1958.

Behnia, R., and Munro, S. (2005). Organelle identity and the signposts for membrane traffic. Nature *438*, 597–604.

Benchimol, M. (1999). Hydrogenosome autophagy: an ultrastructural and cytochemical study. Biol. Cell *91*, 165–174.

Benchimol, M. (2004). Trichomonads under microscopy. Microsc. Microanal. *10*, 528–550.

Bock, J.B., Matern, H.T., Peden, A.A., and Scheller, R.H. (2001). A genomic perspective on membrane compartment organization. Nature *409*, 839–841.

Boehm, M., and Bonifacino, J.S. (2001). Adaptins: the final recount. Mol. Biol. Cell *12*, 2907–2920.

Bonifacino, J.S., and Glick, B.S. (2004). The mechanisms of vesicle budding and fusion. Cell *116*, 153–166.

Bonifacino, J.S., and Hurley, J.H. (2008). Retromer. Curr. Opin. Cell Biol. *20*, 427–436.

Bowden, F.J., and Garnett, G.P. (1999). Why is *Trichomonas vaginalis* ignored? Sex. Transm. Infect. 75, 372–374.

Bowler, C., Allen, A.E., Badger, J.H., Grimwood, J., Jabbari, K., Kuo, A., Maheswari, U., Martens, C., Maumus, F., Otillar, R.P., *et al.* (2008). The *Phaeodactylum* genome reveals the evolutionary history of diatom genomes. Nature *456*, 239–244.

Bradley, P.J., Lahti, C.J., Plümper, E., and Johnson, P.J. (1997). Targeting and translocation of proteins into the hydrogenosome of the protist *Trichomonas*: similarities with mitochondrial protein import. EMBO J. *16*, 3484–3493.

Brown, M.T., Goldstone, H.M., Bastida-Corcuera, F., Delgadillo-Correa, M.G., McArthur, A.G., and Johnson, P.J. (2007). A functionally divergent hydrogenosomal peptidase with protomitochondrial ancestry. Mol. Microbiol. *64*, 1154–1163.

Burki, F., and Pawlowski, J. (2006). Monophyly of Rhizaria and multigene phylogeny of unicellular bikonts. Mol. Biol. Evol. *23*, 1922–1930.

Carlton, J.M., Hirt, R.P., Silva, J.C., Delcher, A.L., Schatz, M., Zhao, Q., Wortman, J.R., Bidwell, S.L., Alsmark, U.C., Besteiro, S., *et al.* (2007). Draft genome sequence of the sexually transmitted pathogen *Trichomonas vaginalis*. Science *315*, 207–212.

Carpenter, K.J., and Keeling, P.J. (2007). Morphology and phylogenetic position of *Eucomonympha imla* (Parabasalia: Hypermastigida). J. Eukaryot. Microbiol. *54*, 325–332.

Cavalier-Smith, T. (2002). Origins of the machinery of recombination and sex. Heredity *88*, 125–141.

Cepicka, I., Kutisova, K., Tachezy, J., Kulda, J., and Flegr, J. (2005). Cryptic species within the *Tetratrichomonas gallinarum* species complex revealed by molecular polymorphism. Vet. Parasitol. *128*, 11–21.

Cerutti, H., and Casas-Mollano, J.A. (2006). On the origin and functions of RNA-mediated silencing: from protists to man. Curr. Genet. *50*, 81–99.

Chen, Y.A., and Scheller, R.H. (2001). SNARE-mediated membrane fusion. Nat. Rev. Mol. Cell Biol. *2*, 98–106.

Clatworthy, A.E., Pierson, E., and Hung, D.T. (2007). Targetting virulence: a new paradigm for antimicrobial therapy. Nat. Chem. Biol. *3*, 541–548.

Cleveland, L.R. (1947a). Sex produced in the protozoa of *Cryptocercus* by molting. Science *105*, 16–17.

Cleveland, L.R. (1947b). The origin and evolution of meiosis. Science *105*, 287–280.

Cleveland, L.R. (1951). Hormone-induced sexual cycles of flagellates. J. Morphol. *88*, 385–439.

Cleveland, L.R. (1956). Brief accounts of the sexual cycles of the flagellates of *Cryptocercus*. J. Protozool. *3*, 161–180.

Cohen, M.S., Hoffman, I.F., Royce, R.A., Kazembe, P., Dyer, J.R., Daly, C.C., Zimba, D., Vernazza, P.L., Maida, M., Fiscus, S.A., *et al.* (1997). Reduction of concentration of HIV-1 in semen after treatment of urethritis: implications for prevention of sexual transmission of HIV-1. AIDSCAP Malawi Research Group. Lancet *349*, 1868–1873.

Colicelli, J. (2004). Human RAS superfamily proteins and related GTPases. Sci. STKE *2004*, RE13.

Coombs, G.H. (1982). Proteinases of *Leishmania mexicana* and other flagellate protozoa. Parasitology *84*, 149–155.

Dacks, J.B., Walker, G., and Field, M.C. (2008). Implications of the new eukaryotic systematics for parasitologists. Parasitol. Int. *57*, 97–104.

DeRose-Wilson, L.J., and Gaut, B.S. (2007). Transcription-related mutations and GC content drive variation in nucleotide substitution rates across the genomes of *Arabidopsis thaliana* and *Arabidopsis lyrata*. BMC Evol. Biol. *7*, 66.

Dessi, D., Rappelli, P., Diaz, N., Cappuccinelli, P., and Fiori, P.L. (2006). *Mycoplasma hominis* and *Trichomonas vaginalis*: a unique case of symbiotic relationship between two obligate human parasites. Front. Biosci. *11*, 2028–2034.

Drmota, T., and Kral, J. (1997). Karyotype of *Trichomonas vaginalis*. Eur. J. Protistol. *33*, 131–135.

Du, T.T., and Zamore, P.D. (2005). microPrimer: The biogenesis and function of microRNA. Development *132*, 4645–4652.

Duboucher, C., Pierce, R.J., Dei-Cas, E., Capron, M., and Viscogliosi, E. (2008). Recent advances in pulmonary trichomonosis. Trends Parasitol. *24*, 201–202.

Edgcomb, V., Viscogliosi, E., Simpson, A.G.B., Delgado-Viscogliosi, P., Roger, A.J., and Sogin, M.L. (1998). New insights into the phylogeny of trichomonads inferred from small subunit rRNA sequences. Protist *149*, 359–366.

Egel, R. (2008). Meiotic crossing-over and disjunction: overt and hidden layers of description and control. In Genome Dynamics and Instability, Volume 2. Recombination and Meiosis: Crossing Over and Non-Disjunction, R. Egel and D.-H. Lankenau, eds. (Berlin: Springer Verlag), pp. 1–30.

Ehmsen, K.T., and Heyer, W.-D. (2008). Biochemistry of meiotic recombination: formation, processing, and resolution of recombination intermediates. In Genome Dynamics and Instability, Volume 3. Recombination and Meiosis: Models, Means, and Evolution, D.-H. Lankenau, ed. (Berlin: Springer Verlag), pp. 91–164.

El-Sayed, N.M., Myler, P.J., Bartholomeu, D.C., Nilsson, D., Aggarwal, G., Tran, A.N., Ghedin, E., Worthey, E.A., Delcher, A.L., Blandin, G., *et al.* (2005). The genome sequence of *Trypanosoma cruzi*, etiologic agent of Chagas disease. Science *309*, 409–415.

Feschotte, C., and Pritham, E.J. (2007). DNA transposons and the evolution of eukaryotic genomes. Annu. Rev. Genet. *41*, 331–368.

Field, M.C., and Carrington, M. (2004). Intracellular membrane transport systems in *Trypanosoma brucei*. Traffic *5*, 905–913.

Fiori, P.L., Rappelli, P., and Addis, M.F. (1999). The flagellated parasite *Trichomonas vaginalis*: new insights into cytopathogenicity mechanisms. Microbes Infect. *1*, 149–156.

Fiori, P.L., Rappelli, P., Addis, M.F., Sechi, A., and Cappuccinelli, P. (1996). *Trichomonas vaginalis* haemolysis: pH regulates a contact-independent mechanism based on pore-forming proteins. Microb. Pathog. *20*, 109–118.

Fiori, P.L., Rappelli, P., Rocchigiani, A.M., and Cappuccinelli, P. (1993). *Trichomonas vaginalis* haemolysis: evidence of functional pores formation on red cell membranes. FEMS Microbiol. Lett. *109*, 13–18.

Fire, A., Xu, S., Montgomery, M.K., Kostas, S.A., Driver, S.E., and Mello, C.C. (1998). Potent and specific genetic interference by double-stranded RNA in *Caenorhabditis elegans*. Nature *391*, 806–811.

Flynt, A.S., and Lai, E.C. (2008). Biological principles of microRNA-mediated regulation: shared themes amid diversity. Nat. Rev. Genet. *9*, 831–842.

Galvin, S.R., and Cohen, M.S. (2004). The role of sexually transmitted diseases in HIV transmission. Nat. Rev. Microbiol. *2*, 33–42.

Garcia-Ranea, J.A., and Valencia, A. (1998). Distribution and functional diversification of the ras superfamily in *Saccharomyces cerevisiae*. FEBS Lett. *434*, 219–225.

Gerbod, D., Sanders, E., Moriya, S., Noël, C., Takasu, H., Fast, N.M., Delgado-Viscogliosi, P., Ohkuma, M., Kudo, T., Capron, M., *et al.* (2004). Molecular phylogenies of Parabasalia inferred from four protein genes and comparison with rRNA trees. Mol. Phylogenet. Evol. *31*, 572–580.

Gerhold, R.W., Yabsley, M.J., Smith, A.J., Ostergaard, E., Mannan, W., Cann, J.D., and Fischer, J.R. (2008). Molecular characterization of the *Trichomonas gallinae* morphologic complex in the United States. J. Parasitol. *94*, 1335–1341.

Gilbert, R.O., Elia, G., Beach, D.H., Klaessig, S., and Singh, B.N. (2000). Cytopathogenic effect of *Trichomonas vaginalis* on human vaginal epithelial cells cultured *in vitro*. Infect. Immun. *68*, 4200–4206.

Gookin, J.L., Birkenheuer, A.J., St John, V., Spector, M., and Levy, M.G. (2005). Molecular characterization of trichomonads from feces of dogs with diarrhea. J. Parasitol. *91*, 939–943.

Gruenberg, J., and van der Goot, F.G. (2006). Mechanisms of pathogen entry through the endosomal compartments. Nat. Rev. Mol. Cell Biol. *7*, 495–504.

Haber, J. (2008). Evolution of models of homologous recombination. In Recombination and Meiosis: Models, Means and Evolution, R. Egel and D.-H. Lankenau, eds. (Berlin: Springer Verlag), pp. 1–64.

Haig, D. (1993). Alternatives to meiosis: the unusual genetics of red algae, microsporidia, and others. J. Theor. Biol. *163*, 15–31.

Hampl, V., Cepicka, I., Flegr, J., Tachezy, J., and Kulda, J. (2004). Critical analysis of the topology and rooting of the parabasalian 16S rRNA tree. Mol. Phylogenet. Evol. *32*, 711–723.

Hampl, V., Horner, D.S., Dyal, P., Kulda, J., Flegr, J., Foster, P.G., and Embley, T.M. (2005). Inference of the phylogenetic position of oxymonads based on nine genes: support for Metamonada and Excavata. Mol. Biol. Evol. *22*, 2508–2518.

Hampl, V., Vanacova, S., Kulda, J., and Flegr, J. (2001). Concordance between genetic relatedness and phenotypic similarities of *Trichomonas vaginalis* strains. BMC Evol. Biol. *1*, 11.

Hampl, V., Vrlik, M., Cepicka, I., Pecka, Z., Kulda, J., and Tachezy, J. (2006). Affiliation of *Cochlosoma* to trichomonads confirmed by phylogenetic analysis of the small-subunit rRNA gene and a new family concept of the order Trichomonadida. Int. J. Syst. Evol. Microbiol. *56*, 305–312.

Hirt, R.P., Lal, K., Pinxteren, J., Warwicker, J., Healy, B., Coombs, G.H., Field, M.C., and Embley, T.M. (2003). Biochemical and genetic evidence for a family of heterotrimeric G-proteins in *Trichomonas vaginalis*. Mol. Biochem. Parasitol. *129*, 179–189.

Hirt, R.P., Noël, C.J., Sicheritz-Ponten, T., Tachezy, J., and Fiori, P.L. (2007). *Trichomonas vaginalis* surface proteins: a view from the genome. Trends Parasitol. *23*, 540–547.

Honigberg, B.M., and Brugerolle, G. (1989). Structure. In Trichomonads Parasitic in Humans, B.M. Honigberg, ed. (New York: Springer Verlag), pp. 5–35.

Howard-Till, R.A., and Yao, M.C. (2006). Induction of gene silencing by hairpin RNA expression in *Tetrahymena thermophila* reveals a second small RNA pathway. Mol. Cell Biol. *26*, 8731–8742.

Hrdy, I., Hirt, R.P., Dolezal, P., Bardonová, L., Foster, P.G., Tachezy, J., and Embley, T.M. (2004). *Trichomonas* hydrogenosomes contain the NADH dehydrogenase module of mitochondrial complex I. Nature *432*, 618–622.

Ikegami, A., Honma, K., Sharma, A., and Kuramitsu, H.K. (2004). Multiple functions of the leucine-rich repeat protein LrrA of *Treponema denticola*. Infect. Immun. *72*, 4619–4627.

Johnson, E.H., Windsor, J.J., and Clark, C.G. (2004). Emerging from obscurity: biological, clinical, and diagnostic aspects of *Dientamoeba fragilis*. Clin. Microbiol. Rev. *17*, 553–570.

Johnston, V.J., and Mabey, D.C. (2008). Global epidemiology and control of *Trichomonas vaginalis*. Curr. Opin. Infect. Dis. *21*, 56–64.

Jones, G., and Franklin, F. (2008). Meiosis in *Arabidopis thaliana*: Recombination, chromosome organization and meiotic progression. In Recombination and Meiosis: Crossing-over and Disjunction, R. Egel and D.-H. Lankenau, eds. (Berlin, Springer Verlag), pp. 279–306.

Kinchen, J.M., and Ravichandran, K.S. (2008). Phagosome maturation: going through the acid test. Nat. Rev. Mol. Cell Biol. *9*, 781–795.

Kissinger, P., Amedee, A., Clark, R.A., Dumestre, J., Theall, K.P., Myers, L., Hagensee, M.E., Farley, T.A., and Martin, D.H. (2009). *Trichomonas vaginalis* treatment reduces vaginal HIV-1 shedding. Sex. Transm. Dis. 36, 11–16.

Kleina, P., Bettim-Bandinelli, J., Bonatto, S.L., Benchimol, M., and Bogo, M.R. (2004). Molecular phylogeny of Trichomonadidae family inferred from ITS-1, 5.8S rRNA and ITS-2 sequences. Int. J. Parasitol. *34*, 963–970.

Klemba, M., and Goldberg, D.E. (2002). Biological roles of proteases in parasitic protozoa. Annu. Rev. Biochem. *71*, 275–305.

Klingbeil, M.M., Burton, P., Barnes, R., and McCulloch, R. (2007). The three Rs of the trypanosomatid genomes: Replication, recombination and repair. In Trypanosomes: After the Genome, D. Barry, R. McCulloch, J.C. Mottram, and A. Acosta-Serrano, eds. (Norfolk UK, Horizon Scientific Press), pp. 133–175.

Lal, K., Field, M.C., Carlton, J.M., Warwicker, J., and Hirt, R.P. (2005). Identification of a very large Rab GTPase family in the parasitic protozoan *Trichomonas vaginalis*. Mol. Biochem. Parasitol. *143*, 226–235.

Lal, K., Noël, C.J., Field, M.C., Goulding, D., and Hirt, R.P. (2006). Dramatic reorganisation of *Trichomonas* endomembranes during amoebal transformation: a possible role for G-proteins. Mol. Biochem. Parasitol. *148*, 99–102.

Lander, E.S., and Waterman, M.S. (1988). Genomic mapping by fingerprinting random clones: a mathematical analysis. Genomics *2*, 231–239.

Lehker, M.W., and Alderete, J.F. (1999). Resolution of six chromosomes of *Trichomonas vaginalis* and conservation of size and number among isolates. J. Parasitol. *85*, 976–979.

Leippe, M., and Herbst, R. (2004). Ancient weapons for attack and defense: the pore-forming polypeptides of pathogenic enteric and free-living amoeboid protozoa. J. Eukaryot. Microbiol. *51*, 516–521.

Leiros, H.K., Kozielski-Stuhrmann, S., Kapp, U., Terradot, L., Leonard, G.A., and McSweeney, S.M. (2004). Structural basis of 5-nitroimidazole antibiotic resistance: the crystal structure of NimA from *Deinococcus radiodurans*. J. Biol. Chem. *279*, 55840–55849.

Leung, K.F., Dacks, J.B., and Field, M.C. (2008). Evolution of the multivesicular body ESCRT machinery; retention across the eukaryotic lineage. Traffic *9*, 1698–1716.

Levine, B., and Klionsky, D.J. (2004). Development by self-digestion: molecular mechanisms and biological functions of autophagy. Dev. Cell *6*, 463–477.

Lin, W.C., Li, S.C., Shin, J.W., Hu, S.N., Yu, X.M., Huang, T.Y., Chen, S.C., Chen, H.C., Chen, S.J., Huang, P.J., Gan, R.R., Chiu, C.H., and Tang, P. (2009). Identification of microRNA in the protist *Trichomonas vaginalis*. Genomics 93, 487–493.

Lin, Z., Kong, H., Nei, M., and Ma, H. (2006). Origins and evolution of the *recA/RAD51* gene family: evidence for ancient gene duplication and endosymbiotic gene transfer. Proc. Natl. Acad. Sci. U.S.A. *103*, 10328–10333.

Liston, D.R., and Johnson, P.J. (1999). Analysis of a ubiquitous promoter element in a primitive eukaryote: early evolution of the initiator element. Mol. Cell Biol. *19*, 2380–2388.

Liston, D.R., Lau, A.O.T., Ortiz, D., Smale, S.T., and Johnson, P.J. (2001). Initiator recognition in a primitive eukaryote: IBP39, an initiator-binding protein from *Trichomonas vaginalis*. Mol. Cell Biol. *21*, 7872–7882.

Loftus, B., Anderson, I., Davies, R., Alsmark, U.C., Samuelson, J., Amedeo, P., Roncaglia, P., Berriman, M., Hirt, R.P., Mann, B.J., *et al.* (2005). The genome of the protist parasite *Entamoeba histolytica*. Nature *433*, 865–868.

Lopez-Casamichana, M., Orozco, E., Marchat, L.A., and Lopez-Camarillo, C. (2008). Transcriptional profile of the homologous recombination machinery and characterization of the EhRAD51 recombinase in response to DNA damage in *Entamoeba histolytica*. BMC Mol. Biol. *9*, 35.

Lun, Z.R., Chen, X.G., Zhu, X.Q., Li, X.R., and Xie, M.Q. (2005). Are *Tritrichomonas foetus* and *Tritrichomonas suis* synonyms? Trends Parasitol. *21*, 122–125.

Luzio, J.P., Pryor, P.R., and Bright, N.A. (2007). Lysosomes: fusion and function. Nat. Rev. Mol. Cell Biol. *8*, 622–632.

Lyons, E.J., and Carlton, J.M. (2004). Mind the gap: bridging the divide between clinical and molecular studies of the trichomonads. Trends Parasitol. *20*, 204–207.

Macrae, I.J., Zhou, K., Li, F., Repic, A., Brooks, A.N., Cande, W.Z., Adams, P.D., and Doudna, J.A. (2006). Structural basis for double-stranded RNA processing by Dicer. Science *311*, 195–198.

Malik, S.-B. (2007). The early evolution of meiotic genes. PhD thesis, Department of Biology, University of Iowa, 238 pages.

Malik, S.-B., Ramesh, M.A., Hulstrand, A.M., and Logsdon, J.M., Jr. (2007). Protist homologs of the meiotic *Spo11* gene and topoisomerase VI reveal an evolutionary history of gene duplication and lineage-specific loss. Mol. Biol. Evol. *24*, 2827–2841.

Malik, S.-B., Pightling, A.W., Stefaniak, L.M., Schurko, A.M., and Logsdon, J.M., Jr. (2008). An expanded inventory of conserved meiotic genes provides evidence for sex in *Trichomonas vaginalis*. PLoS ONE 3, e2879.

Matsuzaki, M., Misumi, O., Shin, I.T., Maruyama, S., Takahara, M., Miyagishima, S.Y., Mori, T., Nishida, K., Yagisawa, F., Yoshida, Y., *et al.* (2004). Genome sequence of the ultrasmall unicellular red alga *Cyanidioschyzon merolae* 10D. Nature *428*, 653–657.

Matzke, M.A., and Birchler, J.A. (2005). RNAi-mediated pathways in the nucleus. Nat. Rev. Genet. *6*, 24–35.

McClelland, R.S. (2008). *Trichomonas vaginalis* infection: can we afford to do nothing? J. Infect. Dis. *197*, 487–489.

McMahon, H.T., and Mills, I.G. (2004). COP and clathrin-coated vesicle budding: different pathways, common approache. Curr. Opin. Cell Biol. *16*, 379–391.

Mello, C.C., and Conte, D. (2004). Revealing the world of RNA interference. Nature *431*, 338–342.

Melo, S.P., Gomez, V., Castellanos, I.C., Alvarado, M.E., Hernandez, P.C., Gallego, A., and Wasserman, M. (2008). Transcription of meiotic-like-pathway genes in *Giardia intestinalis*. Mem. Inst. Oswaldo Cruz *103*, 347–350.

Miller, M., Liao, Y., Gomez, A.M., Gaydos, C.A., and D'Mellow, D. (2008). Factors associated with the prevalence and incidence of *Trichomonas vaginalis* infection among African American women in New York City who use drugs. J. Infect. Dis. *197*, 503–509.

Moreira, D., von der Heyden, S., Bass, D., Lopez-Garcia, P., Chao, E., and Cavalier-Smith, T. (2007). Global eukaryote phylogeny: Combined small- and large-subunit ribosomal DNA trees support monophyly of Rhizaria, Retaria and Excavata. Mol. Phylogenet. Evol. *44*, 255–266.

Morrison, H.G., McArthur, A.G., Gillin, F.D., Aley, S.B., Adam, R.D., Olsen, G.J., Best, A.A., Cande, W.Z., Chen, F., Cipriano, M.J., *et al.* (2007). Genomic minimalism in the early diverging intestinal parasite *Giardia lamblia*. Science *317*, 1921–1926.

Mukherjee, M., Brown, M.T., McArthur, A.G., and Johnson, P.J. (2006a) Proteins of the glycine decarboxylase complex in the hydrogenosome of *Trichomonas vaginalis*. Eukaryot. Cell *5*, 2062–2071.

Mukherjee, M., Sievers, S.A., Brown, M.T., and P.J. Johnson (2006b) Identification and biochemical characterization of serine hydroxymethyl transferase in the hydrogenosome of *Trichomonas vaginalis*. Eukaryot. Cell 5, 2072–2078.

Müller, M. (1993). The hydrogenosome. J. Gen. Microbiol. *139*, 2879–2889.

Murphy, D., Dancis, B., and Brown, J.R. (2008). The evolution of core proteins involved in microRNA biogenesis. BMC Evol. Biol. *8*, 92.

Musatovova, O., and Alderete, J.F. (1999). The *Trichomonas vaginalis* phenotypically varying P270 immunogen is highly conserved except for numbers of repeated elements. Microb. Pathog. *27*, 93–104.

Nakayashiki, H., Kadotani, N., and Mayama, S. (2006). Evolution and diversification of RNA silencing proteins in fungi. J. Mol. Evol. *63*, 127–135.

Nanda, N., Michel, R.G., Kurdgelashvili, G., and Wendel, K.A. (2006). Trichomoniasis and its treatment. Expert Rev. Anti. Infect. Ther. *4*, 125–135.

Noël, C., Noda, S., Mantini, C., Dolan, M.F., Moriya, S., Delgado-Viscogliosi, P., Kudo, T., Capron, M., Pierce, R.J., Ohkuma, M., *et al.* (2007). Molecular phylogenetic position of the genera *Stephanonympha* and *Caduceia* (Parabasalia) inferred from nuclear small subunit rRNA gene sequences. J. Eukaryot. Microbiol. *54*, 93–99.

Nozaki, T., and Nakada-Tsukui, K. (2006). Membrane trafficking as a virulence mechanism of the enteric protozoan parasite *Entamoeba histolytica*. Parasitol. Res. *98*, 179–183.

Okumura, C.Y., Baum, L.G., and Johnson, P.J. (2008). Galectin-1 on cervical epithelial cells is a receptor for the sexually transmitted human parasite *Trichomonas vaginalis*. Cell. Microbiol. *10*, 2078–2090.

Pal, D., Banerjee, S., Cui, J., Schwartz, A., Ghosh, S.K., and Samuelson, J. (2009). *Giardia*, *Entamoeba*, and *Trichomonas* enzymes activate metronidazole (nitroreductases) and inactivate metronidazole (nitroimidazole reductases). Antimicrob. Agents Chemother. *53*, 458–64

Pereira-Leal, J.B., and Seabra, M.C. (2001). Evolution of the Rab family of small GTP-binding proteins. J. Mol. Biol. *313*, 889–901.

Pereira-Neves, A., and Benchimol, M. (2007). Phagocytosis by *Trichomonas vaginalis*: new insights. Biol. Cell *99*, 87–101.

Pereira-Neves, A., Ribeiro, K.C., and Benchimol, M. (2003). Pseudocysts in trichomonads – new insights. Protist *154*, 313–329.

Pindak, F.F., Mora de Pindak, M., Hyde, B.M., and Gardner, W.A., Jr. (1989). Acquisition and retention of viruses by *Trichomonas vaginalis*. Genitourin. Med. *65*, 366–371.

Poxleitner, M.K., Carpenter, M.L., Mancuso, J.J., Wang, C.J., Dawson, S.C., and Cande, W.Z. (2008). Evidence for karyogamy and exchange of genetic material in the binucleate intestinal parasite *Giardia intestinalis*. Science *319*, 1530–1533.

Pritham, E.J., Putliwala, T., and Feschotte, C. (2007). *Mavericks*, a novel class of giant transposable elements widespread in eukaryotes and related to DNA viruses. Gene *390*, 3–17.

Prucca, C.G., Slavin, I., Quiroga, R., Elias, E.V., Rivero, F.D., Saura, A., Carranza, P.G., and Lujan, H.D. (2008). Antigenic variation in *Giardia lamblia* is regulated by RNA interference. Nature *456*, 750–754.

Quon, D.V.K., Delgadillo, M.G., Khachi, A., Smale, S.T., and Johnson, P.J. (1994). Similarity between a ubiquitous promoter in an ancient eukarytote and mammalian initiator elements. Proc. Natl. Acad. Sci. U.S.A. *91*, 4579–4583.

Ramesh, M.A., Malik, S.-B., and Logsdon, J.M., Jr. (2005). A phylogenomic inventory of meiotic genes: Evidence for sex in *Giardia* and an early eukaryotic origin of meiosis. Curr. Biol. *15*, 185–191.

Rendon-Maldonado, J., Espinosa-Cantellano, M., Soler, C., Torres, J.V., and Martinez-Palomo, A. (2003). *Trichomonas vaginalis*: *in vitro* attachment and internalization of HIV-1 and HIV-1-infected lymphocytes. J. Eukaryot. Microbiol. *50*, 43–48.

Rendon-Maldonado, J.G., Espinosa-Cantellano, M., Gonzalez-Robles, A., and Martinez-Palomo, A. (1998). *Trichomonas vaginalis*: *in vitro* phagocytosis of lactobacilli, vaginal epithelial cells, leukocytes, and erythrocytes. Exp. Parasitol. *89*, 241–250.

Ribeiro, K.C., Arnholdt, A.C., and Benchimol, M. (2002). *Tritrichomonas foetus*: induced division synchrony by hydroxyurea. Parasitol. Res. *88*, 627–631.

Ribeiro, K.C., Monteiro-Leal, L.H., and Benchimol, M. (2000). Contributions of the axostyle and flagella to closed mitosis in the protists *Tritrichomonas foetus* and *Trichomonas vaginalis*. J. Eukaryot. Microbiol. *47*, 481–492.

Ronquist, F., and Huelsenbeck, J.P. (2003). MrBayes 3: Bayesian phylogenetic inference under mixed models. Bioinformatics *19*, 1572–1574.

Ruvkun, G. (2001). Molecular biology – Glimpses of a tiny RNA world. Science *294*, 797–799.

Saito-Nakano, Y., Loftus, B.J., Hall, N., and Nozaki, T. (2005). The diversity of Rab GTPases in *Entamoeba histolytica*. Exp. Parasitol. *110*, 244–252.

Sanderfoot, A.A., Assaad, F.F., and Raikhel, N.V. (2000). The *Arabidopsis* genome. An abundance of soluble N-ethylmaleimide-sensitive factor adaptor protein receptors. Plant Physiol. *124*, 1558–1569.

Schorey, J.S., and Bhatnagar, S. (2008). Exosome function: from tumor immunology to pathogen biology. Traffic *9*, 871–881.

Schumacher, M.A., Lau, A.O.T and Johnson, P.J. (2003). Structural basis of core promoter recognition in a primitive eukaryote. Cell *115*, 413–424.

Schwebke, J.R., and Barrientes, F.J. (2006). Prevalence of *Trichomonas vaginalis* isolates with resistance to metronidazole and tinidazole. Antimicrob. Agents Chemother. *50*, 4209–4210.

Schwebke, J.R., and Burgess, D. (2004). Trichomoniasis. Clin. Microbiol. Rev. *17*, 794–803.

Scott, D.A., North, M.J., and Coombs, G.H. (1995). The pathway of secretion of proteinases in *Trichomonas vaginalis*. Int. J. Parasitol. *25*, 657–666.

Seaman, M.N. (2005). Recycle your receptors with retromer. Trends Cell Biol. *15*, 68–75.

Shabalina, S.A., and Koonin, E.V. (2008). Origins and evolution of eukaryotic RNA interference. Trends Ecol. Evol. 23, 578–587.

Sheiner, L., and Soldati-Favre, D. (2008). Protein trafficking inside *Toxoplasma gondii*. Traffic *9*, 636–646.

Shi, H., Chamond, N., Tschudi, C., and Ullu, E. (2004). Selection and characterization of RNA interference-deficient trypanosomes impaired in target mRNA degradation. Eukaryot. Cell 3, 1445–1453.

Shuter, B.J., Thomas, J.E., Taylor, W.D., and Zimmerman, A.M. (1983). Phenotypic correlates of genomic DNA content in unicellular eukaryotes and other cells. Am. Nat. *122*, 26–44.

Silva, J.C., Bastida, F., Bidwell, S.L., Johnson, P.J., and Carlton, J.M. (2005). A potentially functional *mariner* transposable element in the protist *Trichomonas vaginalis*. Mol. Biol. Evol. *22*, 126–134.

Simpson, A.G.B. (2003). Cytoskeletal organization, phylogenetic affinities and systematics in the contentious taxon Excavata (Eukaryota). Int. J. Syst. Evol. Microbiol. *53*, 1759–1777.

Simpson, A.G.B., and Patterson, D.J. (1999). The ultrastructure of *Carpediemonas membranifera* (Eukaryota) with reference to the 'Excavate hypothesis'. Eur. J. Protistol. *35*, 353–370.

Smid, O., Matuskova, A., Harris, S.R., Kucera, T., Novotny, M., Horvathova, L., Hrdy, I., Kutejova, E., Hirt, R.P., Embley, T.M., *et al.* (2008). Reductive evolution of the mitochondrial processing peptidases of the unicellular parasites *Trichomonas vaginalis* and *Giardia intestinalis*. PLoS Pathog. *4*, e1000243.

Smith, K.N., and Nicolas, A. (1998). Recombination at work for meiosis. Curr. Opin. Genet. Dev. *8*, 200–211.

Sorvillo, F., and Kerndt, P. (1998). *Trichomonas vaginalis* and amplification of HIV-1 transmission. Lancet *351*, 213–214.

Stark, D.J., Beebe, N., Marriott, D., Ellis, J.T., and Harkness, J. (2006). Dientamoebiasis: clinical importance and recent advances. Trends Parasitol. *22*, 92–96.

Stein, M.P., Dong, J., and Wandinger-Ness, A. (2003). Rab proteins and endocytic trafficking: potential targets for therapeutic intervention. Adv. Drug Deliv. Rev. *55*, 1421–1437.

Stockdale, H.D., Givens, M.D., Dykstra, C.C., and Blagburn, B.L. (2009). *Tritrichomonas foetus* infections in surveyed pet cats. Vet. Parasitol. *160*, 13–17.

Sutak, R., Lesuisse, E., Tachezy, J., and Richardson, D.R. (2008). Crusade for iron: iron uptake in unicellular eukaryotes and its significance for virulence. Trends Microbiol. *16*, 261–268.

Sutcliffe, S., Giovannucci, E., Alderete, J.F., Chang, T.H., Gaydos, C.A., Zenilman, J.M., De Marzo, A.M., Willett, W.C., and Platz, E.A. (2006). Plasma antibodies against *Trichomonas vaginalis* and subsequent risk of prostate cancer. Cancer Epidemiol. Biomarkers Prev. *15*, 939–945.

Tibayrenc, M., and Ayala, F.J. (2002). The clonal theory of parasitic protozoa: 12 years on. Trends Parasitol. *18*, 405–410.

Tibayrenc, M., Kjellberg, F., and Ayala, F.J. (1990). A clonal theory of parasitic protozoa: the population structures of *Entamoeba, Giardia, Leishmania, Naegleria, Plasmodium, Trichomonas*, and *Trypanosoma* and their medical and taxonomical consequences. Proc. Natl. Acad. Sci. U.S.A. *87*, 2414–2418.

Ullu, E., Tschudi, C., and Chakraborty, T. (2004). RNA interference in protozoan parasites. Cell. Microbiol. *6*, 509–519.

Vacquier, V.D. (1998). Evolution of gamete recognition proteins. Science *281*, 1995–1998.

Van der Pol, B. (2007). *Trichomonas vaginalis* infection: the most prevalent nonviral sexually transmitted infection receives the least public health attention. Clin. Infect. Dis. *44*, 23–25.

Van Der Pol, B., Kwok, C., Pierre-Louis, B., Rinaldi, A., Salata, R.A., Chen, P.L., van de Wijgert, J., Mmiro, F., Mugerwa, R., Chipato, T., *et al.* (2008). *Trichomonas vaginalis* infection and human immunodeficiency virus acquisition in African women. J. Infect. Dis. *197*, 548–554.

Vanacova, S., Yan, W., Carlton, J.M., and Johnson, P.J. (2005). Spliceosomal introns in the deep-branching eukaryote *Trichomonas vaginalis*. Proc. Natl. Acad. Sci. U.S.A. *102*, 4430–4435.

Vernoud, V., Horton, A.C., Yang, Z., and Nielsen, E. (2003). Analysis of the small GTPase gene superfamily of *Arabidopsis*. Plant Physiol. *131*, 1191–1208.

Villeneuve, A.M., and Hillers, K.J. (2001). Whence meiosis? Cell *106*, 647–650.

Vinogradov, A.E. (2005). Dualism of gene GC content and CpG pattern in regard to expression in the human genome: magnitude versus breadth. Trends Genet. *21*, 639–643.

Vogel, C., and Chothia, C. (2006). Protein family expansions and biological complexity. PLoS Comput. Biol. *2*, e48.

Wang, A.L., and Wang, C.C. (1985). Isolation and characterization of DNA from *Tritrichomonas foetus* and *Trichomonas vaginalis*. Mol. Biochem. Parasitol. *14*, 323–335.

Watkins, R.F., and Gray, M.W. (2008). Sampling gene diversity across the supergroup Amoebozoa: large EST data sets from *Acanthamoeba castellanii, Hartmannella vermiformis, Physarum polycephalum, Hyperamoeba dachnaya* and *Hyperamoeba* sp. Protist *159*, 269–281.

Whelan, S., and Goldman, N. (2001). A general empirical model of protein evolution derived from multiple protein families using a maximum-likelihood approach. Mol. Biol. Evol. *18*, 691–699.

Williams, R.L., and Urbe, S. (2007). The emerging shape of the ESCRT machinery. Nat. Rev. Mol. Cell Biol. *8*, 355–368.

World Health Organization (2001). Global prevalence and incidence of selected curable sexually transmitted infections: overview and estimates (Geneva: World Health Organization).

Yao, C., Donelson, J.E., and Wilson, M.E. (2003). The major surface protease (MSP or GP63) of *Leishmania* sp. Biosynthesis, regulation of expression, and function. Mol. Biochem. Parasitol. *132*, 1–16.

Yuh, Y.S., Liu, J.Y., and Shaio, M.F. (1997). Chromosome number of *Trichomonas vaginalis*. J. Parasitol. *83*, 551–553.

Zubacova, Z., Cimburek, Z., and Tachezy, J. (2008). Comparative analysis of trichomonad genome sizes and karyotypes. Mol. Biochem. Parasitol. *161*, 49–54.

Dissecting the Actin Cytoskeleton of *Entamoeba histolytica* from a Genomic Perspective

4

Chung Chau Hon, Kumiko Nakada-Tsukui, Tomoyoshi Nozaki and Nancy Guillén

Abstract

Host tissue invasion by *E. histolytica* is driven by motility and phagocytosis, which are both regulated primarily by reorganization of the actin cytoskeleton. In this chapter, the structural and signalling components of the actin cytoskeleton from available genomes of three *Entamoeba* species, *E. histolytica, E.dispar* and *E. invadens,* are annotated manually and compared systematically. These protein families include the actin superfamily, actin nucleators, calponin-related actin-binding proteins (ABPs), gelsolin-related ABPs, myosins, small GTPases and their regulators, as well as phospholipid signalling regulators. Recent findings on these proteins in *E. histolytica* are also reviewed. With an emphasis on the reliability of manual annotations, we aim to give an overview of the existence and diversity of the actin cytoskeleton and its related proteins in *E. histolytica.* We also hope this chapter will provide useful information for researchers wishing to investigate the many uncharacterized actin-cytoskeletal components and their regulators in *E. histolytica,* which should shed light on the molecular mechanisms of the parasite's motility and its phagocytosis-driven tissue invasion.

Introduction

Entamoeba histolytica is a highly motile unicellular parasite that is able to dramatically alter its cell shape based on a versatile cytoskeleton (Meza *et al.,* 2006). In eukaryotic cells, the cytoskeleton consists of three major elements, namely microfilaments (i.e. actin filaments), intermediate filaments and microtubules. Analysis of the genome of *E. histolytica* suggests the absence of intermediate filaments (Clark *et al.,* 2007). Microtubules in *E. histolytica,* as in other eukaryotes, are built from heterodimers of α and β tubulin. These genes are present in *E. histolytica* genome although they encode highly divergent proteins (Roy and Lohia, 2004). In addition, γ tubulin, which participates in microtubule assembly, was also identified in *E. histolytica* (Gómez-Conde *et al.,* 2000). The microtubules of *E. histolytica* were demonstrated to be intra-nuclear (Vayssié *et al.,* 2004), but the regulation of their assembly and disassembly is largely unknown. In *E. histolytica,* research on microtubules is rather in its infancy when compared with that on the actin cytoskeleton.

The actin cytoskeleton is the major skeletal component in *E. histolytica,* responsible for controlling the cell shape as well as many pivotal cellular processes, including motility, phagocytosis, cytokinesis and amoeba–substrate interactions. Assembly, disassembly and spatial arrangements of actin filaments are dynamic events that are tightly regulated by a large repertoire of actin-binding proteins (ABPs). Although a preliminary scan of the pipeline-annotated genome of *E. histolytica* suggested the presence of most essential genes for an actin cytoskeleton and its regulators, systematic analyses of these gene families based on manual annotations have not yet been performed. Recently, the draft genome sequences of two other *Entamoeba* species, *E. dispar* and *E. invadens,* have been released, allowing us to compare and contrast these genes among the three Entamoebae. To this end, we have scanned for the existence of the structural and signalling components of the actin cytoskeleton among the three genomes. These protein families include the actin superfamily, actin nucleators, calponin-related ABPs, gelsolin-related ABPs, myosins, small GTPases and their regulators, as well as phospholipid signalling regulators. Orthologues of these proteins in the three Entamoebae were then annotated manually using InterProScan implemented in BLAST2GO (Götz *et al.*, 2008) and compared systematically using phylogenetic analyses. Recent findings on these proteins in *E. histolytica* are also reviewed in this chapter. With an emphasis on the reliability of manual annotations, we aim to provide readers with an overview of the existence and diversity of the actin cytoskeleton and its related proteins in *E. histolytica*. This chapter is expected to help researchers to utilize efficiently the genome information to formulate testable hypotheses regarding the actin cytoskeleton of *E. histolytica*.

Structural components of the actin cytoskeleton

Conventional actin

Actin is one of the most conserved and ubiquitous proteins in eukaryotes being found from *Giardia* to vertebrates. It is an ATP-binding protein that exists in two forms: monomeric globular (G-) actin and polymeric filamentous (F-) actin. The actin filament is the major component of the eukaryotic cytoskeleton. Its polymerization and depolymerization are extremely dynamic and tightly regulated processes. It has been noted that purified actin binds tightly to DNase I and inhibits its endonuclease activity (Podolski and Steck, 1988). Although the physiological significance of this tight association is unclear, this property has been employed for crystallization of actin (Kabsch *et al.*, 1990) and measurement of the G-actin content (Fox *et al.*, 1981). Many species express several isoforms of actin, for example at least five, six, ten, fifteen and eighteen actin isoforms have been identified in *Caenorhabditis elegans* (MacQueen *et al.*, 2005), mammals (Vandekerckhove and Weber, 1978), *Arabidopsis thaliana* (McKinney and Meagher, 1998), *Dictyostelium discoideum* (Joseph *et al.*, 2008) and *Paramecium tetraurelia* (Sehring *et al.*, 2007), respectively. In contrast, some unicellular eukaryotes possess only one actin isoform, including some apicomplexan species (Gordon and Sibley, 2005), *Saccharomyces cerevisiae* (Shortle *et al.*, 1982), *Leishmania* species (Kapoor *et al.*, 2008), *Trypanosoma cruzi* (De Melo *et al.*, 2008), and *Giardia* (Drouin *et al.*, 1995). These findings suggest the number of actin isoforms can vary greatly among species, but seems to be independent of the complexity of organisms.

Conventional actin in Entamoeba

The first evidence for the existence of actin in *E. histolytica* was provided by immuno-staining using anti-human actin antibodies (Kettis *et al.*, 1977). The actin protein was then purified and shown to bind phalloidin but not DNase I (Gadasi, 1982; Meza *et al.*, 1983). Later, multiple distinct mRNA transcripts encoding a single actin isoform were identified in *E. histolytica* (Edman *et al.*, 1987). The presence of multiple actin gene copies in the *E. histolytica* genome was further demonstrated using Southern blotting (Huber *et al.*, 1987). Our genomic searches support the presence of multiple gene copies for an identical actin isoform in each of the three *Entamoeba* genomes. In both *E. histolytica* and *E. dispar*, eight actin gene copies encoding a single actin isoform were found, except for one gene copy in *E. dispar* (EDI251490) that showed a single nucleotide non-synonymous mutation. In *E. invadens*, 11 and 5 genes were identified that encode two distinct actin isoforms differing by only two amino acids (Table 4.1). It should be noted that some of these *Entamoeba* actin gene sequences appeared to be truncated, but careful examination of their genomic sequences revealed those truncations are likely to be sequencing artefacts, e.g. frame-shifts or incomplete coverage. The diversity of conventional actin isoforms in *Entamoeba* is extremely limited when compared with its amoebozoan relative *D. discoideum* (Fig. 4.1). Nonetheless, the finding of an evolutionarily luxurious redundant number of gene copies encoding a single protein is common to both *Entamoeba* and *D. discoideum*, as the isoform ACT1 in *D. discoideum* is encoded by 17 genes (Joseph *et al.*, 2008). It is noteworthy that such gene redundancy for a single actin isoform was also observed in *Trichomonas vaginalis* (Bricheux and Brugerolle, 1997). The selective advantage for maintenance of redundant actin genes in these eukaryotes is unclear but it has been

Table 4.1 List of genes for selected structural components of the actin cytoskeleton in three *Entamoeba* species

Category	Gene name/ Group[a]	*Entamoeba* species		
		E. histolytica	*E. dispar*	*E. invadens*
Actin superfamily				
Actin isoforms	Actin	EHI_182900	EDI_325560	EIN_026070[b]
		EHI_163750	EDI 251490	EIN_034700
		EHI_159150	EDI_237510	EIN_072180
		EHI_142730	EDI_192970[b]	EIN_089410
		EHI_140120	EDI_125480	EIN_147500[b]
		EHI_126190	EDI_043510[b]	EIN_182410
		EHI_107290	EDI_012660	EIN_218770
		EHI_043640[b]	EDI_004490	EIN_219020
		n/a	n/a	EIN_219670
		n/a	n/a	EIN_220530
		n/a	n/a	EIN_222210
		n/a	n/a	EIN_222280
		n/a	n/a	EIN_222660
		n/a	n/a	EIN_299810[b]
Conserved ARPs	ARP2	EHI_111050	EDI_053710	EIN_212990
	ARP3	EHI_198930	EDI_286940	EIN_056670
Orphan ARPs	EtmbARPi	EHI_163580	EDI_209470	EIN_166290
	EtmbARPii	EHI_197810	EDI_241590	EIN_109520
	EtmbARPiii	EHI_008780	EDI_287230	EIN_295910
	EtmbARPvi	EHI_039070	EDI_071000	EIN_026850
Orphan ARPs with fused domains	EtmbARPv1	EHI_131230	EDI_309310	EIN_092760
	EtmbARPv2	EHI_048630	EDI_230520	EIN_093130
	EtmbARPvi	EHI_148100	EDI_335070	EIN_156330
Actin nucleation-related proteins				
ARP2/3 complex subunits	ARPC1	EHI_045000	EDI_145840[b]	n/a
		EHI_094120	EDI_163950[b]	n/a
	ARPC2	EHI_091250	EDI_015750	EIN_222900
		EHI_199690	n/a	n/a
	ARPC3	EHI_172960	EDI_063860	EIN_135030
		EHI_174910	EDI_078800	n/a
	ARPC4	EHI_030820	EDI_082410	EIN_192890
		EHI_152660	EDI_092890	n/a
	ARPC5	EHI_103440	EDI_330860	EIN_134860
		EHI_103700	EDI_346050	n/a

Table 4.1 Continued

Category	Gene name/ Group[a]	*Entamoeba* species		
		E. histolytica	*E. dispar*	*E. invadens*
WASp-related proteins (containing a conserved VCA domain)	WASH	EHI_048670	EDI_203910	EIN_235690
		EHI_050810	EDI_230560	n/a
		EHI_167840	n/a	n/a
	MIM	EHI_007000	EDI_061520	EIN_020440
		EHI_008450	EDI_062410	EIN_181260
		EHI_016130	EDI_276470	EIN_294590
Formins	Formin-1	EHI_125300	EDI_115990	EIN_127950
	Formin-2	EHI_118260	EDI_322460	EIN_235800
	Formin-3	EHI_192460	EDI_195010	EIN_235510
	Formin-4	EHI_197550	EDI_002900	EIN_235730
	Formin-5	EHI_126070	EDI_012340	EIN_108020
	Formin-6	EHI_024650	EDI_342220	EIN_236320
	Formin-7	EHI_197230	EDI_045890	EIN_288940
	Formin-8	EHI_090250	EDI_222570	EIN_219120
Proteins containing multiple WH2 domains	n/a	EHI_059870	EDI_188070	EIN_181240
Calponin-related ABPs				
Filamins	FLN-A	EHI_104630	EDI_145990	EIN_165800
		EHI_094060[b]	EDI_343370[b]	EIN_248380
		n/a	n/a	EIN_165810[b]
α-Actinins	α-Actinin 1	EHI_164430	EDI_207850	EIN_037840
		EHI_164440	n/a	n/a
	α-Actinin 2	EHI_199000	EDI_033890	n/a
		n/a	EDI_263690	n/a
Proteins containing Sel-1 repeats and two CH domains	n/a	EHI_038800	EDI_070760	EIN_289010
Proteins containing only two CH domains	Group A[a]	EHI_083140	EDI_077810	EIN_133170
		EHI_103430	EDI_330870	EIN_184760
		EHI_104560	EDI_345940	EIN_262340
	Group B[a]	EHI_164920	EDI_282600	EIN_015820
	Group C[a]	EHI_155290	EDI_237860	EIN_289070
	Group D[a]	EHI_093850	EDI_263610	EIN_037940
		EHI_191900	EDI_341030	EIN_101390
		n/a	EDI_217330[b]	n/a
	Group E[a]	EHI_154330	EDI_336560	EIN_104420
		n/a	n/a	EIN_249390

Category	Gene name/ Group[a]	*Entamoeba* species		
		E. histolytica	*E. dispar*	*E. invadens*
Gelsolin-related ABPs				
Amoebozoa-specific gelsolin-related ABPs	Proflightless	EHI_021270	EDI_201430	EIN_147290
		EHI_150430	EDI_291740	EIN_252790
		n/a	n/a	EIN_252990
	GRP125	EHI_009570	EDI_025780	EIN_110860[b]
	ABPH/Villidin	EHI_122800	EDI_154560	EIN_136070
Proteins containing multiple Gel domains and a villin head	Four Gel domains[a]	EHI_021260	EDI_201440	EIN_017500
	Three Gel domains[a]	EHI_007950	EDI_031340	EIN_242000
	One Gel domain[a]	EHI_007480	EDI_152310	EIN_218750
Myosin-related proteins				
Myosin heavy chain	Myosin IB	EHI_110810	EDI_340740	EIN_220600
	Myosin II	EHI_110180	EDI_175070	EIN_238710
		n/a	n/a	EIN_048380[b]
		n/a	n/a	EIN_078360[b]
		n/a	n/a	EIN_121880[b]
		n/a	n/a	EIN_140020[b]
		n/a	n/a	EIN_159520[b]
		n/a	n/a	EIN_193280[b]
		n/a	n/a	EIN_225650[b]
		n/a	n/a	EIN_262420[b]
Myosin regulatory light chain	RLC	EHI_023500	EDI_225770	EIN_196240
Unconventional myosin light chain kinases A	MLCKA	EHI_137120	EDI_162900	EIN_136660
		EHI_137980	EDI_134090	EIN_059640
		n/a	n/a	EIN_148410

a No orthologues can be identified in other eukaryotes and groups were assigned based on phylogenetic analyses of the conserved domains.

b Gene appears to be abnormally truncated, inserted or deleted. These may be sequencing artefacts or pseudogenes.

n/a Not applicable. Orthologues may be truly absent from a particular *Entamoeba* species or missing due to incompleteness of the genomic sequences.

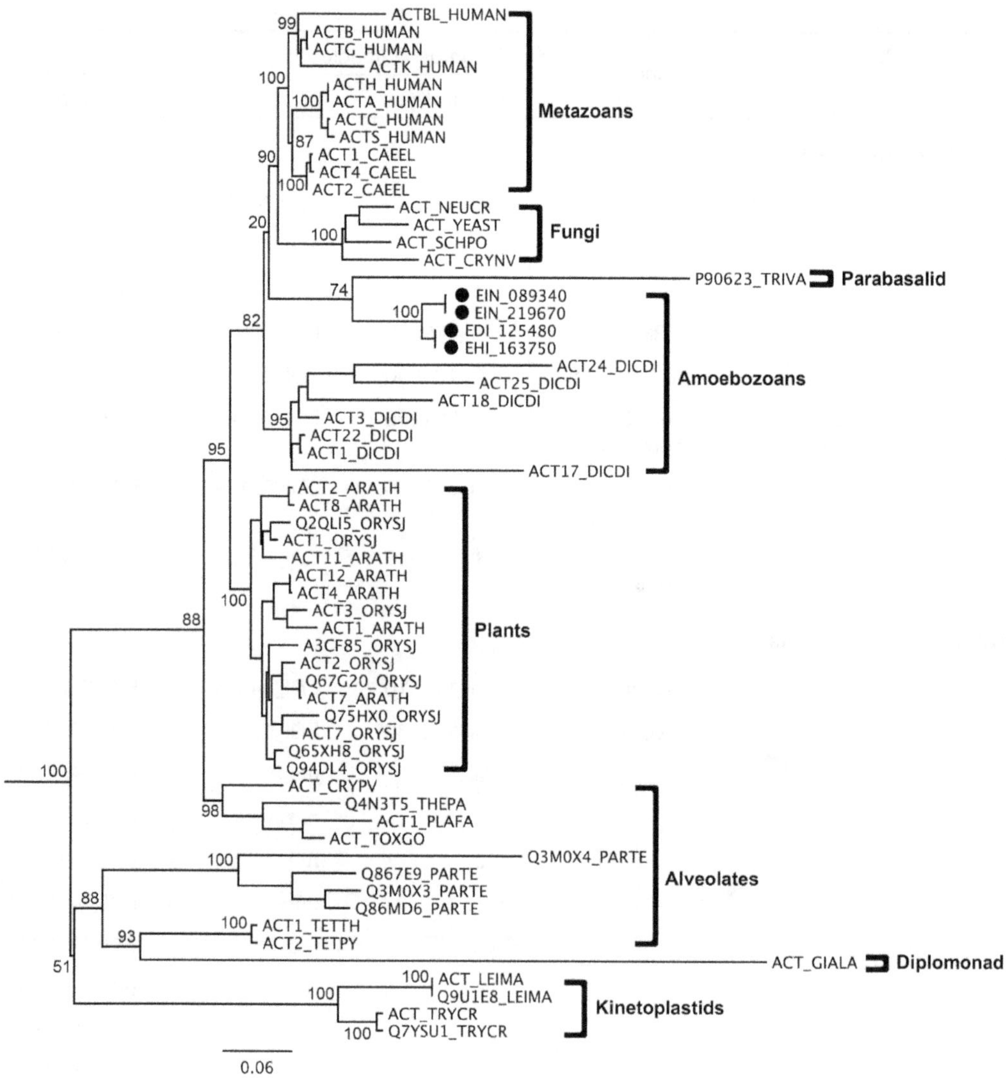

Figure 4.1 Bayesian phylogeny of actin isoforms in various species. The actin sequences were retrieved from UniProt (http://www.uniprot.org/) or Pathema (http://pathema.jcvi.org/) and aligned using MUSCLE with default parameters followed by careful manual adjustments (Edgar, 2004). The tree is a Bayesian phylogeny constructed using BEAST (Drummond and Rambaut, 2007) under a Lognormal Uncorrelated Relaxed Clock model (Drummond *et al.*, 2006). The number at the nodes indicates the Bayesian posterior probability support (as percentages) summarized from trees sampled at every 1000th step of a Bayesian Markov Chain Monte Carlo (BMCMC chain) of 10,000,000 steps. The tree topology is from the tree with maximum clade credibility and branch lengths are represented as the median of all sampled trees. The root position of the tree was jointly estimated in the BMCMC chain. The actins of *Entamoeba* species are highlighted with black dots. The scale bar represents the number of amino acid substitutions per site. The sequence names are presented as the original identifiers from UniPro with the abbreviated species names in the last part of the tag: HUMAN, *Homo sapiens*; CAEEL, *Caenorhabditis elegans*; NEUCR, *Neurospora crassa*; YEAST, *Saccharomyces cerevisiae*; SCHPO, *Schizosaccharomyces pombe*; CRYNV, *Cryptococcus neoformans* var. *grubii*; TRIVA, *Trichomonas vaginalis*; DICDI, *Dictyostelium discoideum*; ARATH, *Arabidopsis thaliana*; ORYSJ, *Oryza sativa* ssp. *japonica*; CRYPV, *Cryptosporidium parvum*; THEPA, *Theileria parva*; PLAFA, *Plasmodium falciparum*; TOXGO, *Toxoplasma gondii*; TRYCR, *Trypanosoma cruzi*; LEIMA, *Leishmania major*; PARTE, *Paramecium tetraurelia*; TETPY, *Tetrahymena pyriformis*; TETTH, *Tetrahymena thermophila;* GIALA, *Giardia lamblia*; EIN, *Entamoeba invadens*; EDI, *Entamoeba dispar*; EHI, *Entamoeba histolytica*.

proposed to be related to co- or post-translational modifications of actin that occur only at a specific time-point of their life cycle or in particular sub-cellular locations (Joseph *et al.*, 2008). Comparison of the *Entamoeba* actin isoforms shows that several of the major differences (4 of 20 amino acid substitutions) are located immediately adjacent to the N-termini (Fig. 4.2). Important modifications, such as arginylation, occur in the N-terminal region of actin (Kashina, 2006) and these modifications have been demonstrated to be essential for the biological functions of some actin isoforms in more complex eukaryotes (Karakozova *et al.*, 2006). The potential roles of the N-terminal differences between *Entamoeba* actin isoforms remain to be investigated.

Actin-related proteins

Despite the evolutionary conservativeness of conventional actin among eukaryotes, a diverse group of proteins bearing considerable sequence similarity to actin, named Actin-Related Proteins (ARPs), is present in most if not all eukaryotes (Muller *et al.*, 2005). At least eleven clusters of ARPs (ARP1 to 11) were found to be conserved from plants to vertebrates, suggesting these ARPs are evolutionarily ancient (Goodson and Hawse, 2002). While some of these ARPs have

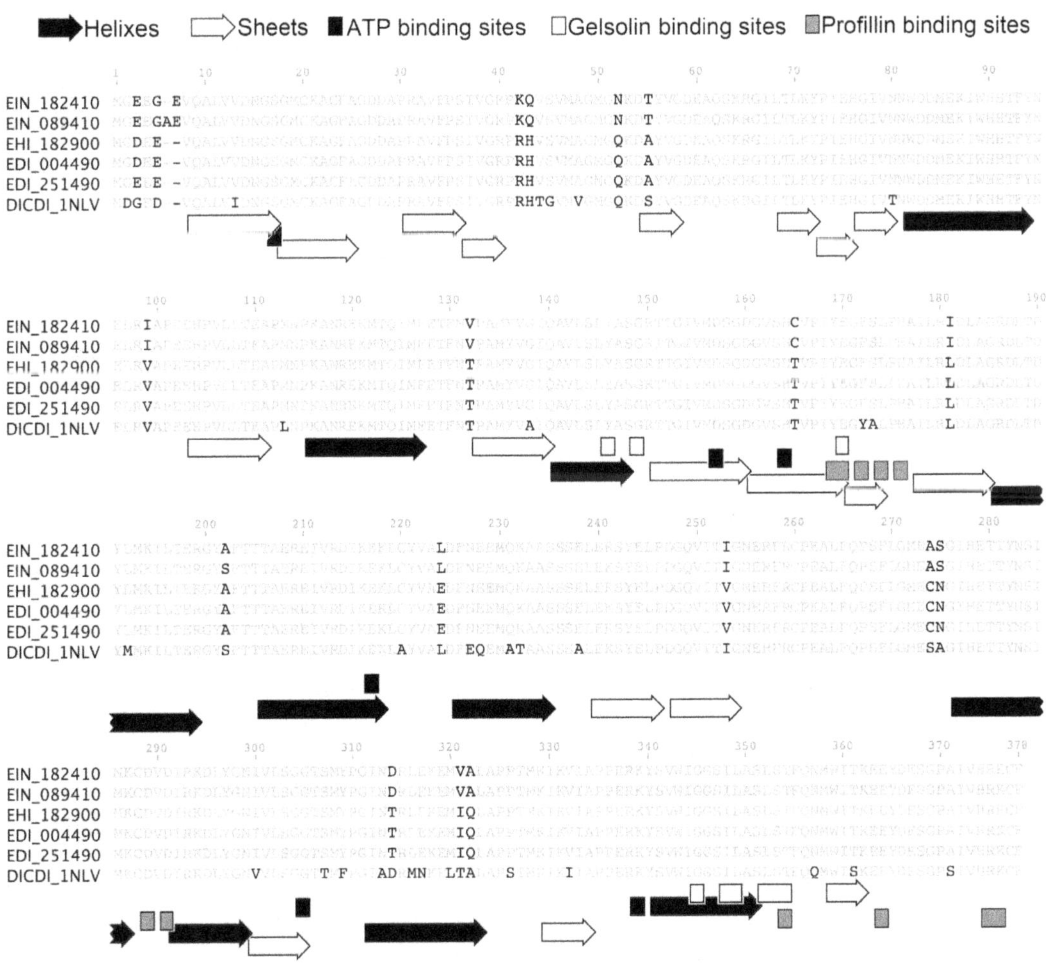

Figure 4.2 Alignment of *Entamoeba* actin isoforms. The alignment was generated using CLUSTAL with default parameters. The primary sequence of the crystallized ACT1 protein of *D. discoideum* (PDB entry: 1NLV) was also aligned and its secondary structural features as well as binding sites for ATP, gelsolin and profilin are annotated at the bottom of the alignment. Variable residues are highlighted in black.

well-defined roles in regulation of the cytoskeleton, e.g. ARP2/3 complex in actin nucleation (Goley and Welch, 2006), recent findings suggest an emerging role in chromatin remodelling, e.g. ARP4 in the SWI2/SNF2 complex (Olave *et al.*, 2002). Although a conserved actin-like domain (InterPro entry: IPR004000) is present in all members of this actin superfamily, phylogenetic analyses failed to delineate the evolutionary relationships between them due to their relatively high inter-cluster divergence (Muller *et al.*, 2005). In addition, a number of 'orphan' ARPs, which cannot be classified into any of the conserved ARP clusters, are present in many eukaryotes, e.g. actin-like 7a in humans (Olave *et al.*, 2002). We cannot exclude the possibility that some of these orphan ARPs may share the same ancestry but such phylogenetic signals are too weak to be detected (Muller *et al.*, 2005). Numerous orphan ARPs have also been identified from the three *Entamoeba* genomes, in addition to the conventional actin and two conserved ARPs (ARPs 2 and 3).

Cytoplasmic actin-related proteins

Cytoplasmic ARPs, i.e. ARP 1 to 3, 10 and 11, exist as subunits of two multi-protein complexes, namely dynactin and ARP2/3 complex. The dynactin complex regulates Dynein and Kinesin-2, which are both cytoplasmic motor proteins that transport cellular cargo along microtubules (Berezuk and Schroer, 2007; Gross, 2003; Hirokawa, 1998). The dynactin complex consists of eight subunits, including ARP1 and ARP10 (in yeast) or ARP11 (in animals) (Schroer, 2004). No orthologues of these three ARPs can be identified from the three *Entamoeba* genomes. Moreover, other subunits of the dynactin complex, including p62, p25, p27 and p50, are all absent from the *Entamoeba* genomes, suggesting the dynactin complex may not be present in *Entamoeba*. In support of this conclusion, orthologues of the dynein heavy and light chains (InterPro entries: IPR004273 and IPR001372) are also absent from the *Entamoeba* genomes. Additionally, although a conserved group of kinesin motor domain-containing proteins is present in *E. histolytica* (Dastidar *et al.*, 2007), these kinesins are not orthologues of kinesin-2 in other eukaryotes. Together these findings strongly suggest that the dynactin-regulated dynein/kinesin-2 pathway of cargo transport along microtubules found in other eukaryotes is absent in *Entamoeba*.

The ARP2/3 complex, which is composed of two ARPs (ARP2 and 3) and five additional subunits, is a major actin nucleator and has been found to be present in the *Entamoeba* genomes. This complex, as well as its activators, the WASP-related proteins, will be discussed later in this chapter.

Nuclear actin-related proteins

Nuclear ARPs, i.e. ARPs 4 to 9, exist as subunits of SWI2/SNF2 ATP-dependent chromatin–remodelling complexes such as SWI2/SNF2, IN80 and SWR1 (Blessing *et al.*, 2004). These complexes all contain an ATPase subunit belonging to the SWI2/SNF2 ATPase family. Phylogenetic analysis of all *Entamoeba* ARPs suggests that nuclear ARPs are absent from the three *Entamoeba* genomes. In addition, analysis of all proteins containing a SNF2 domain (InterPro entry: IPR000330) suggests the absence of SWI2/SNF2 and SWR1 subfamilies of ATPases, which are characterized by the presence of a bromodomain (InterPro entry: IPR001487) and a HSA (InterPro entry: IPR006562) domain, respectively (Mizuguchi *et al.*, 2004; Roberts and Orkin, 2004). The absence of these ATPases and the related nuclear ARPs strongly suggests these chromatin–remodelling complexes are not present in *Entamoeba*. However, potential orthologues of the INO80 ATPase subfamily, which are characterized by the presence of a split ATPase domain in the core ATPase subunit (Bao and Shen, 2007), are present in the *Entamoeba* genomes. Nonetheless, these proteins are very divergent from their potential orthologues in other eukaryotes and a more thorough phylogenetic analysis is needed to confirm their evolutionary origins. In any case, conserved INO80 complexes are not expected to be present in *Entamoeba* since the related nuclear ARPs (ARPs 4, 5 and 8) are absent. Nevertheless, we cannot exclude the possibility that orphan ARPs may have replaced the functions of these conserved nuclear ARPs during evolution.

Orphan actin-related proteins

Orphan ARPs refer to ARPs that share no apparent phylogenetic relationship with the conserved subfamilies ARP1 to 11 (Muller *et al.*, 2005). Seven orphan ARPs were identified and found to be conserved among the three Entamoebae (Fig. 4.3). To avoid confusion with the naming of ARP1 to 11, these *Entamoeba* orphan APRs were designated EtmbARPi, ii, iii, iv, v1, v2 and vi, and were numbered in Roman numerals according to their similarity to conventional actin in descending order, i.e. EtmbARPi shares the highest similarity with conventional actin. For instance, two of them share a very close phylogenetic relationship and are thought to result

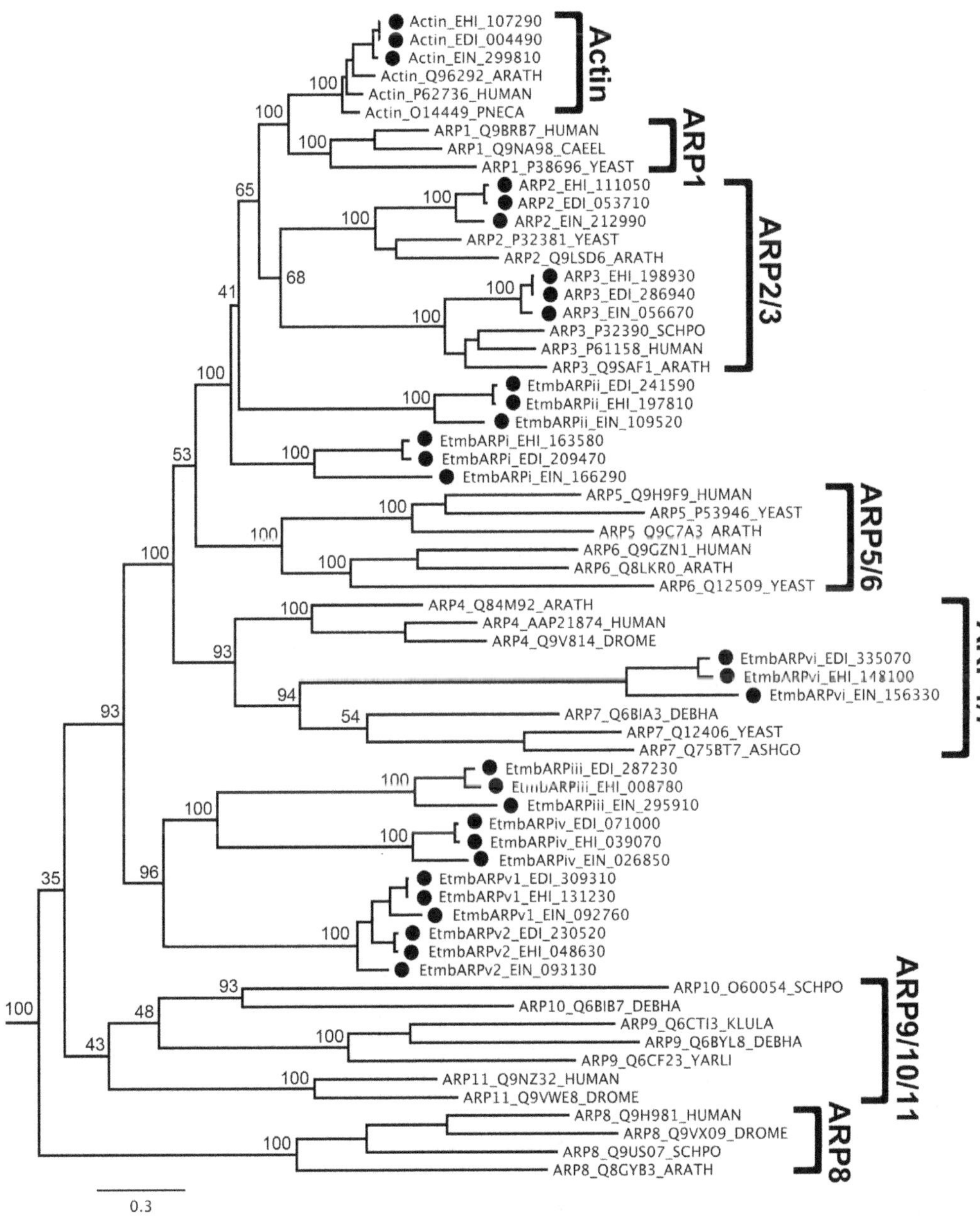

Figure 4.3 Bayesian phylogeny of ARPs of *Entamoeba* species. The ARP sequences were retrieved, aligned and a Bayesian phylogeny constructed as described in the legend to Fig. 4.1. The ARPs of *Entamoeba* species are highlighted using black dots. The taxon names are as indicated in legend to Fig. 4.1 with the following additions: DROME, *Drosophila melanogaster*; DEBHA, *Debaryomyces hansenii*; ASHGO, *Ashbya gossypii*; KLULA, *Kluyveromyces lactis*; YARLI, *Yarrowia lipolytica*.

from a recent duplication, and thus were named EtmbARPv1 and v2 (Fig. 4.3). It is notable that EtmbARPv1, v2 and vi possess extra domains at their N-termini and these fusion ARPs will be discussed in the next section. EtmbARPi and ii share a highly supported monophyletic clade with conventional actin and ARPs 1 to 3, suggesting that these two EtmbARPs may originate from ancient duplications of the common ancestor of these cytoplasmic ARPs and actin. EtmbARPiii, vi and v were clustered into a relatively deep monophyletic clade that shows no specific relationship to other ARPs. The actin-like domain of EtmbARPvi is phylogenetically related to ARP4 and APR7, which are involved in the SWI2/SNF2 chromatin remodelling complexes (Blessing *et al.*, 2004). However, EtmbARPvi cannot be classified as ARP4 or APR7 because of the presence of extra filamin domains in its N-terminal region.

Despite the conservation of these orphan EtmbARPs among the *Entamoeba* genomes, their physiological roles remain to be investigated. Interestingly, numerous additional orphan ARPs were identified that are unique to *E. invadens* (n=31) and are phylogenetically unrelated to the EtmbARPs mentioned above. Many of these *E. invadens* orphan ARPs seem to result from recent duplications and appear to be truncated to varying extents. A more thorough analysis is needed to delineate the evolutionary significance and origins of these orphan ARPs in *E. invadens.*

Actin-related proteins fused with other domains

Orphan ARPs EtmbARPv1, v2 and vi were found to possess extra domains at the N-terminal end of their actin-like domain. EtmbARPv1 and v2 are characterized by the presence of a loosely conserved coiled-coil region and pericardin-like repeats (InterPro entry: IPR009765) at their N-termini and are probably the result of duplication of a recent common ancestor (Fig. 4.3). Despite their similar domain architecture, the sequences of their extra N-terminal domains are poorly conserved on both the paralogous and orthologous level. The function of these proteins is currently unknown and no orthologues can be identified. Based on the poor conservation of their N-terminal domains, we suspect their original physiological functions may have degenerated. EtmbARPvi is characterized by the presence of two filamin domains (InterPro entry: IPR001298) and is believed to be an orthologue of filactin in *D. discoideum,* based on their similar domain architecture (Joseph *et al.*, 2008). The function of filactin is currently unknown but its evolutionary conservation among *Entamoeba* species and *D. discoideum* implies that it fulfils an important role in amoebozoa. Detailed comparisons of the filamin domain from filactins and other members of the filamin family, e.g. ABP120, may shed light on the evolutionary origin of this conserved fusion ARP.

Actin nucleators

Spontaneous polymerization of actin monomers is unfavourable due to the instability of actin dimers, in addition to inhibition by sequestering proteins such as thymosin β4 and profilin. Nucleation of actin monomers is thus the major rate-limiting step for actin polymerization (Le Clainche and Carlier, 2008). Three main classes of actin nucleator have been identified in eukaryotes, including the ARP2/3 complex, formins and spire (Goley and Welch, 2006). While orthologues of the ARP2/3 complex subunits and formins are all present in the three *Entamoeba* genomes (Majumder and Lohia, 2008), no conserved spire orthologue can be identified.

ARP2/3 complex and WASP-related proteins

The ARP2/3 complex mimics an actin dimer to initiate the formation of a new actin filament that branches off from an existing filament, thus forming networks with y-shaped branches (Goley and Welch, 2006). It is composed of ARP2 and 3, in addition to five other subunits (ARPC1 to 5). ARP2 and 3 are highly conserved among eukaryotes, despite their apparent absence in some unicellular eukaryotes such as apicomplexa, algae and microsporidia (Muller *et al.*, 2005). Orthologues of all the subunits of the ARP2/3 complex were identified in the three *Entamoeba* genomes except for ARPC1 of *E. invadens* (Table 4.1), which may possibly be due to the low coverage of its draft genome. Additional preliminary analyses suggest that ARP2 and 3 orthologues are also present in *T. vaginalis* (TVAG354020 and TVAG371880) but are absent in *G. lamblia.* Although the ARP2/3 complex was shown to be essential for the viability of numerous organisms

(Goley and Welch, 2006), its absence in some eukaryotes suggests that other actin nucleators may take on its roles in actin nucleation. The function of the ARP2/3 complex in *Entamoeba* has not been investigated systematically. Nonetheless, a previous study demonstrated the presence of ARP2/3 subunits in phagosomes of *E. histolytica* (Marion *et al.*, 2005). In addition, Zaki and co-workers demonstrated that the function of ARP2 in *D. discoideum* could be complemented genetically by ARP2 of *E. histolytica* (Zaki *et al.*, 2007). These findings support the likely role of the ARP2/3 complex in actin nucleation in *Entamoeba*.

The ARP2/3 complex nucleates actin monomers upon activation by various nucleation promoting factors (Goley and Welch, 2006). In particular, members of the Wiskott-Aldrich Syndrome Protein (WASP) and the Suppressor of cAMP Receptor (SCAR) family have been widely characterized (Vartiainen and Machesky, 2004). Five members of the WASP/SCAR family have been identified in mammals. In contrast, *D. discoideum*, *C. elegans* and *Drosophila melanogaster* each contain one WASP and one SCAR protein, while *S. cerevisiae* and *Schizosaccharomyces pombe* encode only a single WASP/SCAR protein (Vartiainen and Machesky, 2004). BLASTP searches using human and *D. discoideum* WASP/SCAR orthologues against the proteomes of the three Entamoebae failed to identify any protein that contains WASP-homology domain 1 (WH1) or SCAR-homology domain (SHD), which are signatures in the N-terminal regulatory regions of WASP and SCAR proteins respectively, suggesting the absence of WASP/SCAR orthologues in *Entamoeba*. Recently, a new family of WASP/SCAR related proteins, named WASH proteins, was identified in the human genome and has been found to be conserved from metazoa to *E. histolytica* (Linardopoulou *et al.*, 2007).

Despite the divergence of the N-terminal regulatory region among members of the WASP, SCAR and WASH families, their C-terminal 'bridge' region, i.e. verprolin (or WH2) homology, cofilin homology and acidic motif (collectively named the VCA domain), is quite conserved (Linardopoulou *et al.*, 2007). The WH2 and acidic motifs of the VCA domain bind to G-actin monomers and the ARP2/3 complex, respectively, forming a bridge between them. Functional characterization of WASH and N-WASP (neural WASP) demonstrated that its VCA domain alone is sufficient for activation of ARP2/3-dependent actin nucleation (Linardopoulou *et al.*, 2007). By searching for the WH2 domain (InterPro entry: IPR003124) in the *Entamoeba* genomes, we identified two families of orthologues that contain the VCA domain, including the recently identified WASH and an additional family named 'Missing In Metastasis' (MIM) (Machesky and Johnston, 2007). These MIM orthologues contain an N-terminal IRSp53/MIM homology domain (InterPro entry: IPR013606) and a C-terminal VCA domain, which are conserved among the three Entamoebae despite the existence of several N-terminally truncated isoforms (Table 4.1). The sequence homology of VCA domains between the WASP/SCAR/WASH and MIM proteins was not systematically compared in previous studies, except for their WH2 region (Lee *et al.*, 2007). Comparing the critical residues within the VCA domain of the WASP/SCAR/WASH family in metazoa with WASH and MIM orthologues in *Entamoeba* suggests that both families may be able to activate the ARP2/3 complex, based on the fact that the VCA domain itself is functionally independent (Lee *et al.*, 2007). These findings suggest that although conserved orthologues of WASP/SCAR proteins are absent in Entamoebae, closely related genes, identified by the presence of a conserved VCA domain, may complement the role of WASP/SCAR proteins in ARP2/3-dependent actin nucleation.

Formins

Unlike the ARP2/3 complex, formins are members of a multi-domain protein family that nucleates actin for assembly of unbranched filaments (Wallar and Alberts, 2003). Formin Homology (FH) 2 domain (InterPro entry: IPR015425), which is conserved among all formins, binds directly to actin and is sufficient to nucleate actin monomers (Pruyne *et al.*, 2002). The FH2 domain initiates filament assembly and remains associated with the barbed end, enabling rapid insertion of actin subunits and protecting the filament end from capping proteins. Besides the conserved FH2 domain, a number of other regulatory domains are present in some but not all groups of formins, notably FH1 domain, GTPase binding domain (GBD) and the loosely defined FH3 domain (Wallar and Alberts, 2003). Phylogenetic analyses suggest that formin isoforms are present in all eukaryotic genomes characterized so far, except *G. lamblia* (Grunt *et al.*, 2008; Higgs and Peterson, 2005). Exhaustive searches among 36 complete or nearly complete eukaryotic genomes suggest that most of these genomes encode at least two FH2-containing proteins and, strikingly, that the unicellular protists *Naegleria gruberi* and *P. tetraurelia* possess 15 and 14 FH2-containing proteins respectively (Grunt *et al.*, 2008). Majumder and coworkers identified eight formin isoforms in the *E. histolytica* genome (Majumder and Lohia, 2008). Our InterProScan results suggest the orthologues of these eight formin isoforms are also present in both *E. dispar* and *E. invadens,* and no additional FH2-containing proteins were identified.

The eight isoforms of *E. histolytica* formin (Ehformin) can be classified into three groups based on domain architecture and phylogenetic analysis of their FH2 domain, designated as: group I, Ehformin-1 to -3; group II, Ehformin-4; and group III, Ehformin-5 to -8. Ehformins-1 to-3 possess all three FH domains, GBD, a coiled-coil region and a protein kinase C1 domain, sharing a similar domain architecture with some of the Diaphanous-related formins (DRF) in *D. discoideum.* Ehformin-4 possesses all three FH domains and is phylogenetically related to the Cappuccino group of formins. Ehformin-5 to -8 possess only the FH2 domain and a coiled-coil region, and are phylogenetically distinct from all other known formins. Majumder and coworkers showed that Ehformin-1 and -2 are associated with F-actin in pseudopodia, and in pinocytic and phagocytic vesicles (Majumder and Lohia, 2008). Moreover, previous proteomic analyses showed the presence of Ehformin-1 and -2 (Marion *et al.*, 2005) and Ehformin-3 (Okada *et al.*, 2006) in phagosomes. These findings suggest possible roles for Ehformin-1 to -3 in regulation of actin cytoskeleton during phagocytosis and motility, although their ability to nucleate actin remains to be investigated. Interestingly, Ehformins-1 and -2 colocalize with the microtubular assembly in the nucleus and over-expression of these proteins delays the process of cell division, implying their possible involvement in genome segregation and thus mitosis and cytokinesis in *E. histolytica* (Majumder and Lohia, 2008). Notably, the FH1 domain is absent from Ehformin-5 to -8 and from ForC of *D. discoideum,* which was previously demonstrated to be involved in the local actin cytoskeletal reorganization mediating cell-cell adhesion (Kitayama and Uyeda, 2003).

In summary, the presence of numerous unique but conserved formin isoforms in the three *Entamoeba* genomes and the accumulated evidence on their involvement in both the nucleoskeleton and cytoskeleton provides hints to their possible roles in regulation of actin filament dynamics. Further investigation of the divergent Ehformin-5 to -8 may also open avenues to discover novel cellular functions of formins.

Proteins containing multiple WH2 domains

Spire proteins are members of a new class of actin nucleator that contains four tandemly repeated WH2 domains that bind consecutive actin monomers, forming a nucleus for barbed end actin filament elongation (Baum and Kunda, 2005). Although conserved orthologues of spire are absent from the *Entamoeba* genomes, proteins containing four (or three) tandem repeats of the WH2 domain were identified from all three Entamoebae (Table 4.1). These WH2 tandem repeat proteins are different from the spire proteins characterized in other eukaryotes. First, the regulatory domains, i.e. KIND, FYVE and Spir-box, are absent. Second, the loop region between the WH2 domains is extremely large [~200 amino acids (aa)]. No orthologue of similar domain architecture can be found in UniPro and the possible roles of these *Entamoeba* WH2 tandem repeat proteins deserve further investigations.

Calponin-related and gelsolin-related actin binding proteins

ABPs are principally defined as proteins that are able to bind actin monomers, or polymers, or both. This diverse group of proteins regulates various aspects of actin dynamics through distinct mechanisms, such as controlling the filament length by severing activity (e.g. villin), producing bundles of microfilaments (e.g. filamin), cross-linking actin filaments to form a meshwork (e.g. α-actinin) and controlling the G-actin pool by preventing the polymerization of actin (e.g.

thymosin β4). Despite the existence of more than 160 distinct ABPs (dos Remedios *et al.*, 2003), the vast majority of them can be grouped into a small number of conserved families based on the presence of particular functional domains. Among these domains, the best characterized are the WH2 domain (as discussed above), gelsolin-homology (Gel) domain, calponin-homology (CH) domain and myosin motor domain (Dominguez, 2004). In this section, the diversity and recent findings on *Entamoeba* proteins containing the CH or Gel domain will be discussed.

Calponin-related actin-binding proteins

The CH domain (InterPro entry: IPR016146) was first identified in calponin, which is an ABP that plays a major regulatory role in muscle contraction. This domain exists in a variety of functionally distinct proteins as a single copy (e.g. calponin) or tandem repeats ranging from two (e.g. α-actinin and filamin) to four copies (e.g. fimbrin). Because a fully functional actin binding domain requires a tandem pair of CH domains (Korenbaum and Rivero, 2002), we will confine our discussion to proteins containing at least two CH domains. According to the domain architecture suggested by our InterProScan results, there are at least four categories of *Entamoeba* proteins containing two CH domains, including the previously characterized filamins (Vargas *et al.*, 1996) and α-actinins (Virel and Backman, 2006), as well as uncharacterized Sel-1 repeat containing proteins (InterPro entry: IPR006597) and a group of proteins containing two CH domains only.

Filamins cross-link cortical actin to stabilize the dynamic three-dimensional web of actin filaments. Members of the filamin family contain a N-terminal actin binding region of two CH domains, followed by a rod region consisting of a variable number of filamin repeats (Popowicz *et al.*, 2006). We previously identified the first filamin in *E. histolytica*, EhFLN-A, formerly named gelation factor ABP-120 (Vargas *et al.*, 1996). EhFLN-A carries two N-terminal CH domains with consensus sites for actin binding, and a rod region composed of five filamin repeats. The first three filamin domains are quasi-repetitive units of about 72 aa while the fourth domain is interrupted by a stretch of polar residues rich in lysines and glutamic acid (designated d100). The fifth domain, designated the END domain, is composed of 143 aa and is homologous to sixth filamin repeat of *D. discoideum* FLN-A (Vargas *et al.*, 1996). FLN-A of *D. discoideum* was demonstrated to associate with both phosphatidic acid and PI(3)P *in vitro*, which are new lipid targets for filamins (Díaz-Valencia *et al.*, 2007). In addition, the integrity of the fourth repeat of EhFLN-A is essential for its interaction with PI(3)P. In fact, transfected trophozoites over-expressing the d100 domain showed enhanced motility and chemotactic responses. On the other hand, the END domain was demonstrated to be necessary and sufficient for *in vitro* dimerization of EhFLN-A and for the interaction between EhFLN-A and sulfatide (SO(3)–3galactosylceramide), a glycosphingolipid component of the *E. histolytica* plasma membrane (Díaz-Valencia *et al.*, 2005). These results suggest the potential role of EhFLN-A in regulation of the cortical cytoskeleton. Detailed investigation suggests that orthologues of EhFLN-A are conserved among the three *Entamoeba* genomes and, in addition, a number of isoforms were identified. These isoforms possess conserved and intact tandem CH domains but their rod domains are truncated to different extents. Briefly, three FLN-A isoforms exist in *E. invadens*. Two of them share the same domain architecture as EhFLN-A and are 74% identical, while the other isoform has the first two filamin repeats deleted. One extra isoform is present in both *E. histolytica* and *E. dispar*. This isoform is C-terminally truncated with only an incomplete filamin repeat following the two intact CH domains. It is worth mentioning that proteins containing various number of tandem filamin repeats (two to five copies) but lacking the N-terminal CH domains were found to be present in the three *Entamoeba* genomes (EDI146020, EDI219930, EDI309790, EDI346060, EHI104390, EHI094030, EHI103450, EIN248390 and EIN274360). We suspect these filamin repeat-containing proteins arose from C-terminal truncation of the same ancestral gene, based on their relatively conserved first two filamin repeats. Their expression and biological roles remain to be investigated.

α-Actinin dimerizes and cross-links actin filaments into bundles. Members of the α-actinin family are defined by the presence of three distinct domains, a N-terminal actin binding domain of two CH domains, a central rod domain with a varying number of spectrin repeats (SR) (InterPro entry: IPR002017), and a C-terminal calcium-binding domain with at least two EF-hand motifs (InterPro entry: IPR011992). Virel and co-workers identified and characterized two isoforms of α-actinin in *E. histolytica* (Eh-α-actinin 1 and 2) (Virel *et al.*, 2007; Virel and Backman, 2006). Despite the presence of the typical CH domains and EH-hand motifs in both isoforms, the rod domain of Eh-α-actinin 2 probably forms two SR, while the rod domain of Eh-α-actinin 1 is significantly shorter and it is uncertain whether it forms a typical SR or a coiled-coil structure. Their analyses implied that Eh-α-actinin 1 and α-actinin in *Encephalitozoon cuniculi* are the most ancestral form of all α-actinins (Virel and Backman, 2007). Both Eh-α-actinins were demonstrated to bind actin *in vitro* and to organize filaments in bundles (Virel *et al.*, 2007; Virel and Backman, 2006). In addition, proteomic analysis of phagosomes has shown that Eh-α-actinin 2 is enriched with myosin IB and that both Eh-α-actinins may be functionally related to surface receptors (Blazquez *et al.*, 2008; Seigneur *et al.*, 2005). Orthologues of α-actinin 1 are present in all three Entamoebae, while two orthologues of α-actinin 2 were found in *E. dispar* but none in *E. invadens* (Table 4.1).

InterProScan revealed that all three *Entamoeba* genomes contain a conserved orthologue possessing two CH domains and Sel-1 like repeats (EDI070760, EHI038800 and EIN289010). Searching in UniPro for proteins with similar domain architecture yielded a hit from *T. vaginalis* (UniPro ID: A2DKT4_TRIVA). This protein family may represent a new class of CH-related-ABPs that is absent in other eukaryotes.

Phylogenetic analysis of the *Entamoeba* proteins containing only two CH domains suggests the existence of at least five phylogenetically distinct families of potential ABPs that are conserved among the three Entamoebae (Table 4.1). BLASTP searches using the regions of these proteins outside of the CH domains did not yield any significant hits, suggesting that these proteins may represent a diverse group of novel ABPs with biological functions conserved among Entamoebae. Further investigation of their actin binding abilities and biological activities may shed light on their possible roles in regulation of the cytoskeleton.

Gelsolin-related actin-binding proteins

The gelsolin superfamily encodes a number of ABPs that regulate actin dynamics by severing, capping and nucleating filaments, e.g. gelsolin, adseverin and villin (McGough *et al.*, 2003). These proteins contain a triple repeats of the Gel domain (InterPro entry: IPR007122) and in many cases an extra actin binding domain, the villin headpiece (InterPro entry: IPR003128), is present at their C-termini (Archer *et al.*, 2005). EhABPH is the only *Entamoeba* gelsolin-related ABP characterized so far (Ebert *et al.*, 2000) and is an orthologue of villidin, which was later characterized in *D. discoideum* (Gloss *et al.*, 2003). Both possess an extra coronin domain (InterPro entry: IPR015505). Although their physiological functions are unclear, orthologues of EhABPH are conserved among the three Entamoebae (Table 4.1) and since EhABP1 was shown to be enriched in the uroid during capping (Ebert *et al.*, 2000) this suggests a possible role in regulation of the cytoskeleton.

Orthologues of *D. discoideum* GRP125 (Stocker *et al.*, 1999), which are characterized by the presence of five gel domains and several coiled-coil regions, were also found in the three *Entamoeba* genomes (Table 4.1). GRP125 has been demonstrated to be essential for phototaxis of *D. discoideum* but its ability to bind and regulate actin remains to be investigated (Bandala-Sanchez *et al.*, 2006). In addition, orthologues of metazoan flightless I, specifically proflightless (Archer *et al.*, 2005), are present in Entamoebae as at least two isoforms (Table 4.1), although their functions are unknown. Indeed, detailed examination of all *Entamoeba* proteins containing Gel domains revealed the presence of several, potentially novel, gelsolin-related ABPs with domain architectures that are distinct from those in other eukaryotes but conserved among Entamoebae, e.g. proteins containing a villin headpiece plus one, three or four Gel domains (Table 4.1). Intriguingly, in many cases, knockout phenotypes of gelsolin-related ABPs are not lethal, which has been interpreted as indicating possible compensatory mechanisms involving other members of the same family (Silacci *et al.*, 2004). Phylogenetic and biochemical characterization of this diverse but conserved group of gelsolin-related ABPs in *Entamoeba* may offer a better understanding of the function of gelsolin-related ABPs in unicellular eukaryotes.

Myosins

Myosins comprise a large family of motor proteins that hydrolyse ATP to generate mechanical force for driving the contraction of actin filaments. They are multimeric complexes composed of at least a heavy chain (Mhc) and a regulatory light chain. The Mhc of all myosin classes is characterized by a highly conserved motor domain containing the ATPase and actin binding domain that are essential for its motor function (Thompson and Langford, 2002). In addition, Mhc also possesses a neck domain containing IQ motifs for interaction with the light chain and N- or C-terminal extensions of varying length depending on the myosin classes. At least 35 distinct classes of myosins have been identified based on a phylogenetic analysis of 2269 motor domain sequences (Odronitz and Kollmar, 2007). That study inferred that the motor domain may represent the ancestor of all myosin classes and diverse protein domains have been incorporated subsequently during evolution. The motor domain first incorporated the IQ motif at its C-terminus, giving rise to the basic core for the myosin I family. Subsequent incorporation of a SH3 domain provided the scaffold for all the other myosins. Together, these domains formed the head region of Mhc, which over the course of evolution fused with the variety of tails that account for the functional specificity of myosins. Myosins are present in almost all characterized eukaryotic genomes with only three exceptions, the unicellular alga *Cyanidioschyzon merolae* as well as the protist parasites *G. lamblia* and *T. vaginalis* (Richards and Cavalier-Smith, 2005). Most eukaryotes possess multiple myosins and the highest number of myosin classes found in a single organism is at least 15, in *Phytophthora* species (Odronitz and Kollmar, 2007). In *D. discoideum*, 13 myosins of at least five classes have been identified (Kollmar, 2006). However, we identified only two classes of myosin (myosin II and IB) from the three *Entamoeba* genomes.

Conventional myosin II

Myosin II, the first myosin identified from skeletal muscles, has been referred to as the 'conventional' myosin. The tail domain of its Mhc consists of a coiled-coil domain enabling the formation of parallel dimers and in turn self-associating into bipolar filaments. These bipolar myosin II filaments operate as organized filamentous arrays to sustain cell contraction and drive motility. A functional myosin II is a double–headed complex composed of two sets of Mhc, an essential light chain and a regulatory light chain (RLC). Mhc of myosin II in *E. histolytica* has been identified as being encoded by a single-copy gene yielding a 6.4 kb mRNA transcript (Arhets *et al.*, 1992). In particular, the coiled-coil domain in its tail is interrupted by two proline residues, resembling myosins from smooth muscle and non-muscle cells and indicating that its tail domain folds over onto itself twice (Meza *et al.*, 2006). The Mhc gene of myosin II is also present in the genome of the other two Entamoebae, and an additional number of truncated copies were also found in *E. invadens* (Table 4.1). In smooth muscle and non-muscle cells of animals, the contractile activity of myosin II is regulated through phosphorylation of its RLC by Myosin Light Chain Kinases (MLCKs). Orthologues of *D. discoideum* RLC and unconventional MLCK, i.e. MLCKA, have been identified in *E. histolytica* (Labruyère and Guillén, 2006), as well as in the other two Entamoebae (Table 4.1), but no orthologue of the conventional MLCK of other eukaryotes was found. To date, the role of these kinases in phosphorylation of Mhc remains unclear. Nevertheless, the tail domain of *E. histolytica* possesses two potential sites for phosphorylation by protein kinase C (PKC) and casein kinase II, and drugs that inhibit kinases dramatically reduce the activity of myosin II (Arhets *et al.*, 1995), indicating the importance of phosphorylation in myosin II activation.

Unconventional myosin IB

Myosin I is an ancient member of the myosin superfamily that is present in both unicellular and multicellular eukaryotes but is not distributed as widely as myosin II (Thompson and Langford, 2002). It is a single-headed, non-filamentous molecule and its Mhc possesses several IQ motifs that bind to RLCs such as calmodulin. The tail domain of myosin IB in amoebozoa contains a polybasic region (tail homology I domain; TH1) that is responsible for membrane binding, a glycine, proline, alanine or glutamine-rich region (TH2) that contributes to the second actin-binding site, and a Src homology (SH3) domain (also known as TH3) that is present in many cytoskeletal proteins and may be involved in protein–protein interaction (Yumura and Uyeda, 2003). Myosin IB of *E. histolytica* was identified by PCR amplification of the SH3 domain from the genome (Vargas *et al.*, 1997a). Although it is orthologous to myosin IB of other amoebozoa, striking similarities with vertebrate unconventional myosins were also observed. The predicted actin- and ATP-binding sites are located in the head domain. The phosphorylation region of its Mhc is homologous to that of metazoan myosins I, with an acidic residue at the phosphorylation site. In the neck domain, an IQ motif is present, indicating its potential binding with calmodulin. In fact, myosin IB from *E. histolytica* has been a useful model for characterization of the myosin I subfamily because it is the only unconventional myosin present, thus simplifying interpretation, in contrast to the seven types of myosin I found in *D. discoideum* (Kollmar, 2006). A single copy of myosin IB was identified in both the *E. dispar* and the *E. invadens* genomes (Table 4.1).

The TH2 domain of *E. histolytica* myosin IB has been shown to be functional in F-actin binding *in vitro* and *in vivo*, implying that *Entamoeba* myosin IB may cross-link actin filaments through the two actin-binding sites within its Mhc (Marion *et al.*, 2005; Marion *et al.*, 2004b; Vargas *et al.*, 1997b). In fact, *E. histolytica* cells over-expressing the Mhc of myosin IB display exaggerated microfilament cross-linking activity. The increase in cytoskeleton rigidity upon cross-linking of microfilaments increases the cytoplasmic viscosity and reduces the phagocytic activity of the parasite (Marion *et al.*, 2004b).

In *D. discoideum*, the SH3 domains of myosin IB and IC interact with CARMIL, a protein that binds and activates the actin nucleator ARP2/3 complex, to promote *de novo* polymerization of actin filaments (Jung *et al.*, 2001). In *E. histolytica*, the SH3 domain of myosin IB was shown to be non-essential for modulating the density of actin network, and myosin IB overproduction may enhance the cross-linking of pre-existing actin polymers rather than inducing *de novo* actin polymerization (Marion *et al.*, 2004b). In contrast to the multiple myosin I copies in *D. discoideum*, we expect the single copy of myosin IB has broad and distinct physiological roles in Entamoebae.

Signalling components of the actin cytoskeleton

Small GTP-binding proteins

The small GTP-binding protein (G protein) superfamily is a group of monomeric G proteins that function as molecular switches, cycling between an inactive GDP-bound form and an active GTP-bound form. Small G proteins can be classified phylogenetically into five families, namely the Ras, Rho, Rab, Arf and Ran families (Takai *et al.*, 2001). While Ras proteins are mainly involved in cell proliferation, Rho proteins are primarily involved in cytoskeleton regulation. Of the other families, Rab and Arf proteins are mainly involved in membrane trafficking in the cytosol and Ran proteins in nuclear-cytosol transport (Takai *et al.*, 2001). Previous analysis of the *E. histolytica* genome suggested the presence of 20, 25, 91, 11 and 1 member(s) of the Ras, Rho, Rab, Arf and Ran families, respectively (Clark *et al.*, 2007; Loftus *et al.*, 2005; Saito-Nakano *et al.*, 2005). Based on our InterProScan results, the *E. histolytica* genome encodes 166 proteins containing an intact small G protein domain or Ras-like GTPase domain (InterPro entries: IPR005225 and IPR001806). Among them, 17, 22, 73, 12 and 1 proteins contain Ras-, Rho-, Rab-, Arf- and Ran-type G protein domains, respectively (InterPro entries: IPR003577, IPR003578, IPR003579, IPR006689 and IPR002041), while 41 do not have significant similarity to these five conserved small G protein domains. Although phylogenetic analysis can assign these 41 unclassified small G proteins into the five subtypes based on tree topology, the clustering is not statistically supported by significant bootstrap values. However, this still suggests that these 41 unclassified small G proteins may be divergent members of the known subtypes rather than representing novel small G protein subtypes. In contrast, if only the 125 proteins with significant domain hits are analysed, clustering into the five subtypes is supported by significant bootstrap values. Therefore, to be conservative, only proteins with significant hits to the conserved small G protein domains are discussed. In this section, the Rho proteins and their regulators in the *E. histolytica* genome are reannotated and their possible roles in cytoskeletal regulation are discussed. The diversity of *E. histolytica* small G proteins involved in membrane traffic has been reviewed elsewhere (Clark *et al.*, 2007; Saito-Nakano *et al.*, 2005).

Rho proteins in E. histolytica

The roles of Rho proteins in controlling the dynamics of the actin cytoskeleton are well established (Jaffe and Hall, 2005). Phylogenic analysis of Rho proteins from 20 eukaryotic genomes suggested Rho proteins of complex eukaryotes could be classified into eight subfamilies, namely Rho, Rac, Cdc42, RhoDF, Rnd, RhoBTB, RhoUV and RhoH (Boureux *et al.*, 2007). Among the 23 Rho proteins in the human genome, RhoA, Rac1 and Cdc42, of the Rho, Rac and Cdc42 subfamilies, respectively, are best studied. RhoA promotes the formation of stress fibres and focal adhesions, Rac1 induces the formation of lamellipodia and membrane ruffles, and Cdc42 subfamily stimulates formation of filopodia (Jaffe and Hall, 2005; Takai *et al.*, 2001). In addition, RhoD of the RhoDF subfamily forms a complex with hDia2C, which bridges actin filaments and endosomes and regulates endosome dynamics (Gasman *et al.*, 2003). Members of the human Rnd subfamily have been shown to participate in neurite extension in the brain and contraction in smooth muscles (Chardin, 2006). RhoBTB is involved in the tumour suppression of breast carcinoma, head and neck cancers (Beder *et al.*, 2006).

The genome of *E. histolytica* encodes 22 Rho family proteins (EhRho), which is similar to the number in *E. dispar* (n=21) but only just over half the number in *E. invadens* (n=37). Three of the unclassified small GTPases, EHI_153460, EHI_046630 and EHI_087390, may represent divergent Rho proteins of *E. histolytica* but they do not possess a Rho-type domain (InterPro entry: IPR003578) and cannot be reliably grouped with the 22 Rho proteins in our phylogenetic analysis. These three proteins are thus excluded from the following discussions. Six EhRho family proteins were identified prior to the availability of *E. histolytica* genome, namely EhRho1 (Lohia and Samuelson, 1993), EhRacA to D (Lohia and Samuelson, 1996) and EhRacG (Guillén *et al.*, 1998). Among the 22 proteins, EhRacA shows the highest identity (74%) to human Rac1 (Didsbury *et al.*, 1989). Except for the previously annotated EhRho proteins, we designate the rest of EhRho proteins in descending order of their overall percentage identity to EhRacA, i.e., EhRacH to T. Where EhRho proteins have a mutual identity of higher than 90%, they are considered to be isoforms and designated with an additional number (or letter), e.g. EhRacA1/2 or EhRho1A/B (Fig. 4.4). It should be noted that some differences have been found when compared with previously sequenced EhRho genes (Lohia and Samuelson, 1993, 1996), and thus the sequences used in this chapter were based on the latest release of *E. histolytica* genome sequence. Also, EhRacE and EhRacF were previously described as pseudogenes (Lohia and Samuelson, 1996) and therefore are excluded from the following discussion.

Phylogenetic analysis of the GTPase domain of Rho proteins from *E. histolytica* (EhRho), human (HsRho) and *D. discoideum* (DdRho) indicate that the EhRho proteins cannot be classified into one of the eight conserved subfamilies seen in other eukaryotes, with the exception of the Rac and RhoH subfamily (Fig. 4.4). Indeed, EhRacA1, A2, B, G and H cluster with human and *D. discoideum* Rac proteins with high Bayesian posterior probability support and are thus considered as orthologues of the Rac subfamily. This observation is consistent with the hypothesis that the Rac subfamily may be the origin of all Rho proteins and therefore its orthologues would be expected to be present in all eukaryotes (Boureux *et al.*, 2007). Notably, EhRacT clusters specifically with human RhoH, which is known to be involved in T cell signalling and to have Rac antagonizing activity (Aspenstrom *et al.*, 2007). It would be interesting to test if EhRacT is able to antagonize the activity of other EhRac proteins. In addition, several EhRho proteins shared a close phylogenetic relationship with DdRho proteins, including EhRacC/DdRacC, EhRacR/DdRacE and EhRacS/DdRacI (Fig. 4.4). DdRacC is involved in actin rearrangement

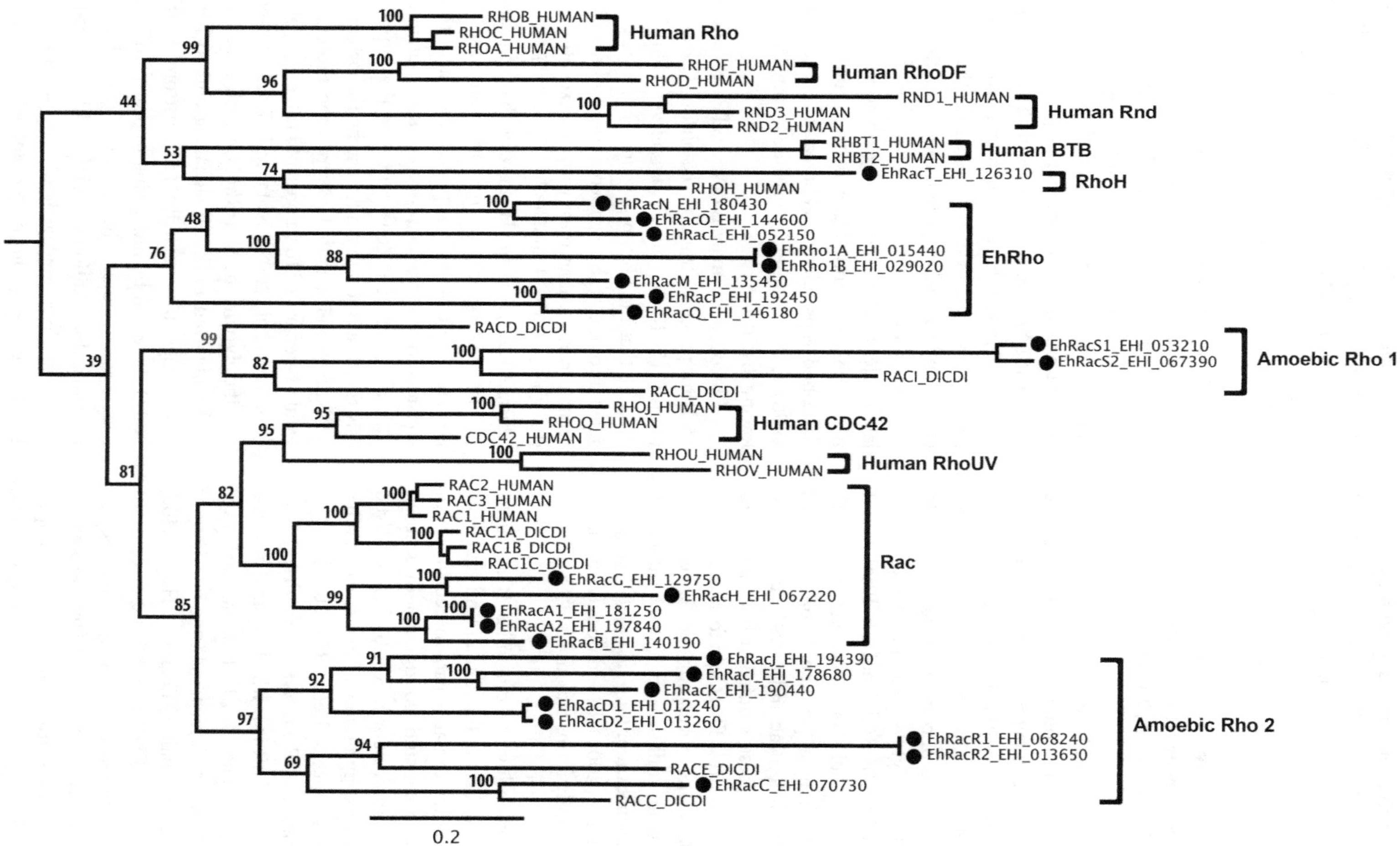

Figure 4.4 Bayesian phylogeny of Rho proteins from *E. histolytica, D. discoideum* and humans. The Rho proteins sequences (Rho domain only) were retrieved, aligned and a Bayesian phylogeny constructed as described in the legend of Fig. 4.1. The highly divergent Rho proteins of *D. discoideum* and humans that showed no apparent phylogenetic relationship with EhRho proteins were excluded from this alignment. The EhRho proteins are highlighted using black dots. The taxon names are as indicated in the legend to Fig. 4.1.

during chemotaxis by regulating WASP and PI3K (Han *et al.*, 2006). As *E. histolytica* does not possess a conserved orthologue of WASP, EhRacC may regulate the actin cytoskeleton through other effectors, which could possibly be other WASP-related proteins such as WASH and MIM as discussed earlier. DdRacE was found to be involved in cytokinesis (Larochelle *et al.*, 1996) and essential for organization of the cortical cytoskeleton to maintain proper cortical integrity (Gerald *et al.*, 1998). Despite the close phylogenetic relationship between DdRacC and DdRacE, chimeras of DdRacC/DdRacE failed to rescue DdRacE null mutants, suggesting that small GTPases engage several distinct protein interfaces to mediate signalling in various cellular processes (Agarwal *et al.*, 2004). Characterizing the functions of EhRacC and EhRacR should provide insights to the physiological roles of these phylogenetically closely related small GTPases. EhRacS clusters with DdRacD, I and L with high Bayesian posterior probability support (Amoebic Rho 1 clade in Fig. 4.4). Similarly, EhRacC, D, I, J, K and R, as well as DdRacE and C form a monophyletic group (Amoebic Rho 2 clade in Fig. 4.4). These two lineages may be specific to amoebozoa. Lastly, an *E. histolytica* specific lineage (EhRho in Fig. 4.4) was also identified and these EhRho proteins showed no significant phylogenetic relationship to any HsRho or DdRho proteins. It should be noted that *D. discoideum* also possesses divergent Rho proteins that are unique to this organism (Rivero *et al.*, 2001).

To gain better insight into the potential functional roles of the EhRho proteins, we examined their domain structures and potential modification sites. EhRho proteins are 182 to 217 aa in length and contain no additional discernible domains, such as the broad-complex, tramtrack, bric à brac (BTB) domain found in the RhoBTB subfamily. EhRacL, R, S, T and EhRho1 have a Rho-specific insertion shorter than other EhRho proteins. This Rho-specific insertion partially determines the specificity of function of Rac towards other small G proteins. *Clostridium botulinum* C3 exoenzyme inhibits Rho-mediated signalling pathways by ADP ribosylation of the asparagine residue of the FENY motif in Rho proteins (Sekine *et al.*, 1989). Only three EhRho proteins, EhRacG1/2 and EhRacH, possess this FENY motif. The closest variants of this motif (with only one amino acid substitution), FENF, FENS and FDNY, are found in EhRho1A/B and EhRacL (FENF), EhRacS1/2 (FENS), and EhRacA, B, C, D1/2, I, J, K, M, P and Q (FDNY). Other EhRho proteins apparently lack these motifs. The effects of C3 exoenzyme on recombinant EhRacG and EhRho1 were examined previously, and EhRacG, but not EhRho1, was ADP-ribosylated (Godbold *et al.*, 2002; Guillén *et al.*, 1998).

The C-termini of Rho proteins normally possess signals for attachment of geranylgeranyl or farnesyl moieties. These lipid modifications are involved in membrane association (Takai *et al.*, 2001). Most EhRho proteins have a conserved farnesylation signal, CXXX, while EhRacM and J possess CCXX and CCXXX geranylgeranylation motifs, respectively, which are usually found in Rab proteins (Pechlivanis and Kuhlmann, 2006). Five EhRho proteins, EhRacN, O, Q, R and T, do not possess any of these motifs. EhRho1 has a CXXXXXXXX motif at its carboxyl terminus but it is not clear if this motif works as a prenylation signal. Finally, EhRacA1/2, B, G, J and L possess a cysteine residue predicted to be potentially palmitoylated. There is no predicted N-terminal myristoylation site.

Regulators of Rho proteins

Like other families of small G proteins, the activity of Rho proteins is determined by the ratio of their GTP/GDP-bound forms (Takai *et al.*, 2001). This GTP-GDP cycle is tightly regulated by three families of proteins: guanine nucleotide exchange factors (GEFs), which enhance the exchange of bound GDP for GTP (Cerione and Zheng, 1996); GTPase-activating proteins (GAPs), which increase the intrinsic rate of hydrolysis of bound GTP (Tcherkezian and Lamarche-Vane, 2007); and GDP dissociation inhibitors (GDIs), which sequester G proteins in their GDP-bound inactive state and regulate their intracellular localization (Olofsson, 1999). In a quiescent state, a GDP-bound Rho protein binds GDI and remains in the cytosol. External or internal cues promote their release from the inhibitory complex and enable them to associate with the plasma membrane, where they are activated by GEFs and send a signal to the effector proteins. By contrast, GAPs inactivate G proteins by accelerating the intrinsic GTPase activity, leading to the GDP-bound form. Previous analysis suggested that the *E. histolytica* genome encodes 62 GEFs, 58 GAPs, and one GDI orthologue. The number of GEFs and GAPs is more than twice the number of Rho proteins in *E. histolytica,* which is similar to the situation in other organisms, e.g. human, yeast, *D. discoideum* and *Drosophila* (Tcherkezian and Lamarche-Vane, 2007). In this section, the diversity of GAP and GDI in *E. histolytica* will be discussed. For GEF, a recent review should be consulted (Vargas and Gonzalez-de la Rosa, 2007).

GAP

The majority of mammalian GAPs possess 1 to 7 extra functional domains, contributing to their diverse physiological roles and regulatory mechanisms (Tcherkezian and Lamarche-Vane, 2007). GAPs are not only involved in the activation of the GTPase activity of Rho proteins, but also mediate cross-talk between Rho proteins and other signalling pathways. Our InterProScan results suggest the genomes of *E. histolytica, E. dispar* and *E. invadens* encode 63, 60 and 92 proteins containing an intact GAP domain (InterPro entries: IPR000198 or IPR008936). Among the 63 GAPs in *E. histolytica,* 51 possess only a GAP domain while 12 harbour extra functional domains, including pleckstrin homology (PH), CDC25 subfamily of GEF (GEF_CDC25), inositol polyphosphate phosphatase, catalytic (IPPc), Dbl homology subfamily of GEF (GEF_DH), LIM-type Zinc finger (LIM), WD40 repeat, Diacylglycerol/Phorbol ester binding (DAPPE_bd) and IQ calmodulin-binding (IQ) domains (InterProScan entries: IPR001849, IPR001895, IPR000300, IPR000219, IPR001781, IPR001680, IPR002219 and IPR000048, respectively) (Fig. 4.5). These extra domains are generally conserved among the three Entamoebae. In *D. discoideum,* 46 GAPs were identified and while 21 of them possess only a GAP domain, 25 possess multiple domains and can be categorized into 17 groups based on domain architecture (Vlahou and Rivero, 2006). The level of complexity in domain architecture of GAPs in *D. discoideum* is therefore higher than in *E. histolytica.*

Among the *Entamoeba* GAPs without an extra functional domain, at least two well-supported large clusters, along with eight small clusters, can be identified (Fig. 4.5). These two large clusters are likely to have expanded by ancient gene duplications before the divergence of the *Entamoeba* species. Nonetheless, the underlying reasons and advantages of possessing such a large repertoire of simple GAPs are unknown. An interesting observation is that the GAPs with extra functional domains tend to cluster together. Notably, the three GAPs with N-terminal PH domains appear to originate from the same ancestral gene, and the same is true for those with N-terminal GEF_DH domains. The GAPs with IPPc domains, as well as the GAPs with WD40 and DAPPE_bd domains, appear to result from relatively recent duplications in Entamoebae.

GAPs with domain architectures similar to some of the EhGAPs have been functionally characterized in humans, such as the N-terminal PH domain in ArhGAP15 (Seoh *et al.,* 2003), the GEF_DH domain in Bcr and Abr (Chuang *et al.,* 1995), and the IPPc domain in OCRL-1 (Attree *et al.,* 1992). The PH domain is involved in the binding to phosphoinositides and is responsible, at least in part, for the membrane localization of RhoGAP (Krugmann *et al.,* 2002). The PH domain is required to localize ArhGAP15 to the cell periphery and over-expression of the full-length ArhGAP15, but not the PH domain-deleted mutant, resulted in an increase in actin stress fibres and cell contraction (Seoh *et al.,* 2003). The GEF_DH domain is present in GEFs of the Dbl family that are responsible for activation of Rho proteins. Bcr is an oncogene product that possesses GAP, GEF_DH, serine/threonine kinase, PH, proline rich and calcium-dependent lipid binding (C2) domains. Abr is structurally similar to Bcr, but lacks the kinase domain. In these two proteins, the GAP and GEF_DH domains bind to their targets in a non-competitive manner. It has been suggested that these two domains may interact simultaneously with two GTPases and may coordinately regulate Rho proteins (Chuang *et al.,* 1995). OCRL-1 is associated with a human X-linked developmental disorder characterized by a pleiotropic phenotype, including eye and renal abnormalities and severe mental retardation. The GAP domain of OCRL-1 does not possess the conserved arginine residue that is required for binding to GDP-bound Rac1. It is considered

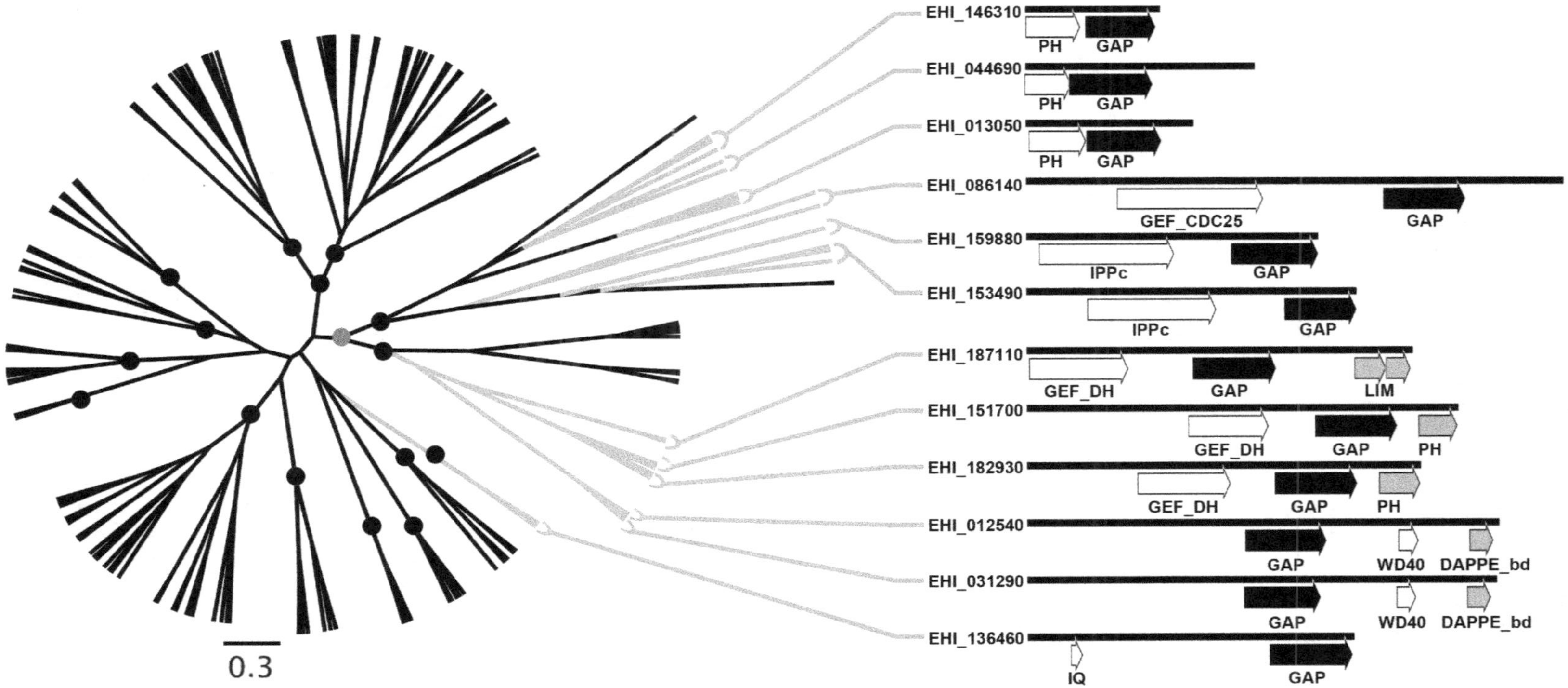

Figure 4.5 Unrooted Bayesian phylogeny and domain architecture of GAPs of the three *Entamoeba* species. The GAP sequences of the *Entamoeba* species (GAP domain only) were retrieved, aligned, and a Bayesian phylogeny constructed as described in the legend to Fig. 4.1. The grey and black dots indicate a Bayesian posterior probability support higher than 70% and 90% respectively. Grey and black branches represent GAPs with and without extra functional domains, respectively. The domain architecture of EhGAPs possessing extra functional domains is summarized in the right panel and linked to their phylogenetic positions on the tree using grey lines.

that this unusual property may contribute to a low turnover rate of substrate *in vitro*. The IPPc domain of OCRL-1 is catalytically active for phosphoinositide 5-phosphate hydrolysis both *in vitro* and *in vivo* (Faucherre *et al.*, 2003; Zhang *et al.*, 1995). OCRL-1 is a potent inhibitor of the formation of Rac-mediated membrane ruffles in Swiss 3T3 cells and is active as a phosphatidylinositol 4,5-bisphosphate (PtdIns(4,5)P2) 5-phosphatase in Rac-induced membrane ruffles (Faucherre *et al.*, 2003). It is well known that in *E. histolytica* the phosphatidylinositol 3-kinase (PI3K) inhibitor wortmannin inhibits endocytosis and phagocytosis (Ghosh and Samuelson, 1997), suggesting the importance of membrane phosphoinositides in cytoskeleton regulation. Since the IPPc domain of *Entamoeba* GAPs does not have the conserved arginine residue, it may act similarly to OCRL-1. To date, no functional characterization has been reported on GAPs of *E. histolytica*.

GDI

The genomes of *E. histolytica* and *E. dispar* each encode only one GDI (EHI_147570 and EDI_324570), while *E. invadens* possesses two (EIN_222930 and EIN_101600). In *D. discoideum*, one of two GDIs is expressed only in the late stage of development and its disruption does not affect normal development, suggesting that a single GDI is sufficient for the normal development of *D. discoideum* (Rivero *et al.*, 2002; Rivero and Somesh, 2002; Vlahou and Rivero, 2006). GDI binds to GDP-bound and lipid-modified Rho proteins and extracts them from the membrane. GDI displays three distinct biochemical activities. First, it blocks the dissociation of GDP from Rho and keeps it in the inactive GDP-bound form (Fukumoto *et al.*, 1990). Second, it binds to the GTP-bound active form of Rho protein and inhibits its activity (Del Pozo *et al.*, 2002; Hart *et al.*, 1992). Third, it extracts Rho proteins from membranes and affects their sub-cellular localization (Michaelson *et al.*, 2001). However, recent studies revealed that over-expression of a mutant form of CDC42 or Rac that is defective in GDI binding causes filopodia formation or membrane ruffling like wild-type CDC42 or Rac (Gandhi *et al.*, 2004; Gibson and Wilson-Delfosse, 2001). Therefore, GDI may not have an escort role for the delivery of nascent lipid modified Rho proteins. Furthermore, a constitutively active CDC42 mutant that also harbours a defect in GDI binding failed to localize to the plasma membrane and was distributed to the Golgi. In addition, this mutant showed a reduced transforming potential. When only a constitutively active mutation was introduced into CDC42, it localized to both the Golgi and the plasma membrane and still showed transforming activity. Altogether, these data suggest the ability of GDI to shuttle and target CDC42 to different sub-cellular compartments. Although *E. histolytica* does not possess a conserved orthologue of CDC42, investigating the interactions between EhGDI and other small G proteins should shed lights on the physiological roles EhGDI.

The N-terminus of GDI is responsible for binding to Rho proteins. Although EhGDI has a N-terminus shorter than mammalian GDI, it possesses conserved asparagine and serine residues that bind directly to threonine and valine residues of Rho proteins (Dovas and Couchman, 2005). Among the 22 EhRho proteins, 20 possess the conserved threonine and valine residues and thus are likely to bind to EhGDI. It was reported that deletion of 20 or 30 residues from the N-terminus caused loss of the GTPase inhibitory function of RhoGDIa on RhoA and Rac1 without affecting GDP dissociation inhibitory activity (Golovanov *et al.*, 2001). Interestingly, the N-terminus of EhGDI is 30 aa shorter than that of mammalian GDIs, suggesting that EhGDI may have unique properties in GTPase inhibition.

Phosphorylation of GDI appears to be one of the key regulatory mechanisms for GDI dissociation from Rho proteins. In fact, RhoGDI α is a substrate for protein kinase C α (PKC α) as well as p21 activated kinase (PAK) (DerMardirossian *et al.*, 2004). Indeed, EhGDI possesses several potential serine/threonine phosphorylation sites. A number of previous studies demonstrated that activation of PKC and PAK in *E. histolytica* stimulates cytoskeleton-related activities, i.e. induction of actin expression, enhancement of adherence to fibronectin, pseudopod formation and phagocytosis (Arias-Romero *et al.*, 2006; Dutta *et al.*, 2007; Labruyère *et al.*, 2003; Ortiz *et al.*, 2000; Santiago *et al.*, 1994). Therefore, it is logical to speculate that EhGDI may be involved in PKC- and PAK-induced cytoskeletal rearrangement by regulating the activity of Rho proteins. PAK kinases are discussed further below.

PAKs as the downstream effectors of Rho proteins

Rho proteins regulate a wide variety of cellular activities through interactions with multiple target proteins and each of these effectors in turn modulates multiple downstream signalling pathways (Jaffe and Hall, 2005). PAK is one of the effectors of Rho proteins and a key enzyme that translates Rho signalling into kinase activity. In mammals, three PAKs (i.e. PAKs 1 to 3) were found initially as CDC42 and Rac1 binding proteins and their kinase activities were stimulated with the GTP-bound but not with the GDP-bound form of small G proteins (Manser *et al.*, 1994). PAKs are serine/threonine kinases with significant homology to the catalytic domains of Ste20, a protein kinase from yeast (Manser *et al.*, 1994). In humans, six PAKs (i.e. PAKs 1 to 6) have been identified. Since the kinase activity of PAKs 4, 5 and 6 is not stimulated by adding the GTP-bound form of Rho proteins, they were named group B PAKs. In contrast, activity of PAKs 1 to 3 is stimulated by the GTP-bound form of Rho proteins, and these were named group A PAKs (Bokoch, 2003; Dan *et al.*, 2001; Jaffer and Chernoff, 2002). Group A PAKs are characterized by the presence of a proline-rich region at their N-termini, a PAK domain for Rho binding (PBD) and a C-terminal kinase domain. Group B PAKs contain a PBD domain at their N-termini followed by a C-terminal kinase domain.

The kinome of *E. histolytica* has been analysed systematically and 14 PAKs and two PAK-like proteins were identified (Anamika *et al.*, 2008). The kinase domains of the two PAK-like proteins, EHI_004390 and EHI_117680, are relatively divergent and have thus been excluded from our analysis. Only six PAKs possess a discernible PBD domain (InterPro entry: IPR000095). The PBD domain, also known as the CDC42/Rac interactive binding (CRIB) domain, is a small domain that binds CDC42 and Rho-like small GTPases. Binding to Rho proteins is one of the key characteristics of PAKs but whether all the EhPAKs function as Rho effectors remains unknown. So far, three putative PAKs (EhPAKs 1 to 3) have been characterized from *E. histolytica* (Arias-Romero *et al.*, 2006; Dutta *et al.*, 2007; Labruyère *et al.*, 2003). Among them, EhPAK1 was the first PAK characterized in *E. histolytica*; it possesses a kinase domain but lacks the PBD domain (Gangopadhyay *et al.*, 1997; Labruyère *et al.*, 2003). Comparison with PAKs from other organisms suggested that EhPAK1 possessed a relatively short N-terminus. However, we failed to recover a sequence identical to the reported EhPAK1 from the *E. histolytica* genome. Instead, the closest sequence is EHI_103610, which possesses a N-terminus that is 146 aa longer and includes 2 substitutions. This extra 146 aa region is also present in three other closely related EhPAKs (Fig. 4.6), strongly suggesting that the first annotation of EhPAK1 may have been incorrectly truncated (Gangopadhyay *et al.*, 1997). Nevertheless, it was demonstrated that the N-terminal region of this truncated EhPAK1 still interacted with human Rac1 (Labruyère *et al.*, 2003). This finding suggests that other EhPAKs without a PBD domain may also possess a cryptic domain for Rho protein binding and still function as Rho effectors. Phylogenetic analysis of the kinase domain of PAKs from humans, *D. discoideum* and *E. histolytica* suggests that all EhPAKs can be placed into one of three clusters, with one cluster sharing a relatively close phylogenetic relationship with human and *D. discoideum* PAKs (Fig. 4.6). Most of the EhPAKs within the other clusters contain PBD domains, including the previously identified EhPAK2 (EHI_148900) and EhPAK3 (EHI_148280), in contrast to EhPAK1. Another feature of this EhPAK cluster is the presence of a N-terminal PH domain in some members, resulting in a similar domain architecture to PAKC of *D. discoideum* and Cla4p of *S. cerevisiae*. Over-expression of EhPAK2 resulted in a 90% decrease in the parasite's ability to invade collagen matrix and severe defects in capping (Arias-Romero *et al.*, 2006). In the other two EhPAK clusters, some novel domain architecture was observed, including three tandem PH domains in EHI_010750, a C2 Peptidase domain (InterPro entry: IPR001300) in EHI_023610 and a Rab-type Ras protein domain in EHI_146320. Interestingly, the incorporation of a Rab-type Ras protein domain into a EhPAK seems to be

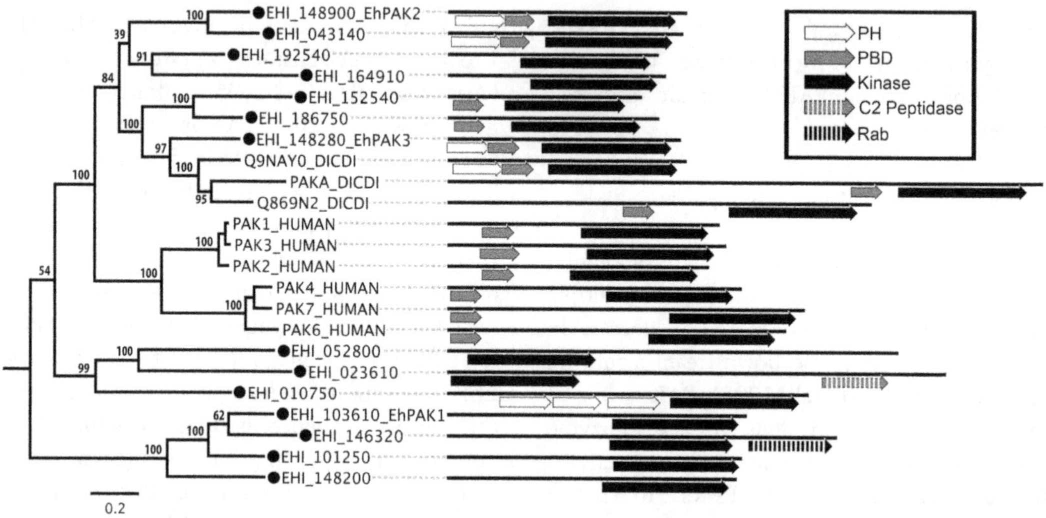

Figure 4.6 Bayesian phylogeny and domain architecture of PAKs from *E. histolytica, D. discoideum* and humans. PAK sequences (kinase domain only) were retrieved, aligned, and a Bayesian phylogeny constructed as described in the legend to Fig. 4.1. The corresponding domain architecture of the PAKs is summarized in the right panel. EhPAKs are highlighted using black dots and the key to domains is shown in the inset box.

a relatively recent event as EHI_146320 is the only member of the cluster that possesses this Rab domain. Such heterogeneity of EhPAKs may reflect the importance of phosphorylation-mediated signalling in the Rho-mediated actin regulation in *E. histolytica.*

Phospholipids

As discussed earlier, eukaryotic cells regulate actin filament assembly and organization through the action of various ABPs. However, the mechanisms for signal transduction from plasma membrane to the underlying actin cytoskeleton are mostly unclear in *E. histolytica*. Recently, it has been shown that membrane phosphoinositides (PIs) act as second messengers in cell surface receptor-mediated signal transduction (Oude Weernink *et al.*, 2007; Roth, 2004). One of the best studied PIs with known roles in surface signalling and the actin cytoskeleton is PtdIns(4,5)P2. Hydrolysis of PtdIns(4,5)P2 by phospholipase C γ (PLC γ) generates inositol 1,4,5-triphosphate [Ins(1,4,5)P3] and diacylglycerol (DAG). Ins(1,4,5)P3 stimulates intracellular Ca^{2+} storage while DAG activates PKC and subsequently activates downstream signalling. In addition, PtdIns(4,5)P2 works as a binding partner of ABPs, e.g. profilin, vinculin, N-WASP, and directly regulates actin filament remodelling. The importance of PtdIns(4,5)P2 was further demonstrated by expression of synaptojanin, a PtdIns(4,5)P2-hydrolysing phosphatase, which alters membrane ruffling and interferes with the organization of stress fibres and focal adhesions (Sakisaka *et al.*, 1997).

Phosphatidylinositol 3-kinases (PI3Ks)

Kinases that phosphorylate PtdIns or PIs on the D-3 position can be organized into three classes based on their amino acid sequences (Fruman *et al.*, 1998; Roth, 2004). Class I PI3Ks are heterodimeric molecules composed of regulatory and catalytic subunits and can be categorized into class IA and IB based on sequence similarity. Human class IA PI3Ks consist of one of the three catalytic subunits, p110α, β or δ, attached to one of five regulatory subunits, p85α, p55α, p50α, p85β or p55γ. The three catalytic subunits are encoded by separate genes, while the five regulatory subunits are encoded by three separate genes with splicing variants. Each catalytic subunit contains a N-terminal region that interacts with the regulatory subunits, Ras protein binding (RBD), calcium-dependent lipid binding (C2),

PI3K accessory (PI3Ka) and PI3K catalytic (PI3Kc) domains (InterPro entries: IPR000341, IPR002420, IPR001263 and IPR000403, respectively). Although the regulatory subunits do not possess any known enzymatic activity, they are composed of several domains with homology to those found in other signalling proteins. In particular, p85 contains a N-terminal SH3, a GAP and two SH2 domain(s). p55 and p50 are splice variants of p85 that lack the N-terminal SH3 and GAP domains. The GAP domain present in the p85 regulatory subunit does not have GAP activity and is considered to work as a binding domain for Rho proteins. Class IB PI3Ks consist of catalytic (p110) and regulatory (p110γ) subunits encoded by separate genes. Class II PI3Ks are monomeric proteins that contain RBD, PI3Ka and PI3Kc domains, as well as a distinct C-terminal C2 domain. Mammalian class II PI3Ks have an additional PX domain between the PI3Kc and C2 domains. Class III PI3Ks are heterodimeric proteins consisting of catalytic and p150 regulatory subunits, the latter containing a serine/threonine kinase domain. The class III PI3Ks are homologues of yeast Vps34, which is the only PI3K in yeast.

Class I PI3Ks are the major enzymes that catalyse PtdIns(3,4,5)P2 generation from PtdIns(4,5)P2 *in vivo* (Fruman *et al.*, 1998). Class III PI3K phosphorylates PtdIns exclusively; therefore this enzyme is only responsible for the generation of PtdIns(3)P *in vivo*. PtdIns(3)P is an important regulator of membrane trafficking by recruiting specific PtdIns(3)P-binding proteins (Lindmo and Stenmark, 2006; Roth, 2004). Compared to class I and III PI3Ks, little is known about the substrate specificities of class II PI3Ks. *In vitro* studies suggest that PtdIns and PtdIns(4)P are the substrates of class II PI3Ks (Falasca and Maffucci, 2007; Maffucci *et al.*, 2005).

From the *E. histolytica* genome database, we identified six proteins that possess N-terminal RBD, PI3Ka and C-terminal PI3K domains, and are therefore considered orthologues of the class I PI3K catalytic subunit. Only one of them does not possess a C2 domain (EHI_188970), while the other five PI3Ks have conserved C2 domains and are similar to mammalian class I PI3Ks (EHI_188970, EHI_006130, EHI_040690, EHI_095130 and EHI_013980). However, attempts to identify the regulatory subunit using BLASTP analysis yielded no significant hit in the *E. histolytica* genome, suggesting the absence of conserved orthologues for class I PI3K regulatory subunits. It also suggests that there may be divergent regulatory subunits in *E. histolytica* that regulate PI3K activity via unique mechanisms. While no class II PI3K is found in *E. histolytica*, the parasite does possess a single copy of both the catalytic (EHI_096560) and regulatory (EHI_044190) subunit of a class III PI3K.

Other phosphatidylinositol kinases

PtdIns 4-kinases catalyse the phosphorylation of PtdIns to PtdIns4P. The generation of PtdIns4P is considered to be the first step in the formation of PtdIns(4,5)P2 and PtdIns(3,4,5)P3, the best known lipid precursors. Two types of PtdIns 4-kinase have been identified, namely types II and III PtdIns4Ks, that have different sensitivity to detergents and adenosine (Endemann *et al.*, 1987; Pike, 1992). Type II PtdIns4Ks are insensitive to wortmannin and LY-294022, but are strongly inhibited by adenosine and Ca^{2+} (Carpenter and Cantley, 1990; Endemann *et al.*, 1991). Type III PtdIns4Ks are proteins localized in both the membrane and the cytosol. In contrast to type II PtdIns4Ks, type III PtdIns4Ks are sensitive to wortmannin and LY-294002 and insensitive to adenosine and Ca^{2+} (Balla, 1998). The *E. histolytica* genome encodes one orthologue of PtdIns4K (EHI_148700).

There are three types of phosphatidylinositol phosphate kinases (PIPK), known as types I, II and III PIPKs, or PIPKI, PIPKII and PIPKIII. PIPKIs and PIPKIIs generate PtdIns(4,5)P from different substrates, using PtdIns(4)P and PtdIns(5)P, respectively (Rameh *et al.*, 1997; Zhang *et al.*, 1997). *In vitro* PtdIns(3)P-kinase activity of PIPKI was demonstrated previously, suggesting that PIPKIs may have the ability to generate PtdIns(3,4,5)P3 (Zhang *et al.*, 1997), but the *in vivo* activity of PIPKIs has not been determined yet. Key amino acid residues in the activation loops, glutamine for PIPKIs and alanine for PIPKIIs, have been defined as determining their substrate specificity (Kunz *et al.*, 2000; Zhang *et al.*, 1997). In the genome of *E. histolytica*, we identified only one PIPK (EHI_153770), whose activation loop sequence is more closely related to PIPKI than to PIPKII. Thus, PIPKII appears to be absent in *E. histolytica*, as in yeast, although PIPKIIs are generally well conserved in multicellular organisms including *D. melanogaster*, *C. elegans*, and mammals.

PIPKIIIs are phosphatidylinositol 3-phosphate 5-kinases and homologues of yeast Fab1. This enzyme utilizes PtdIns(3)P as its substrate and generates PtdIns(3,5)P2. Fab1 is a large protein comprised of several conserved domains: a N-terminal FYVE domain, which binds PtdIns(3)P, a Cpn60/CTP-1 chaperonin family conserved domain (CCT), and a lipid kinase domain. In yeast, there are two Fab1p activator proteins, Vac7p and Vac14p, which participate in the hyper-osmotic activation of Fab1p. Yeast mutants lacking Vac7 or Vac14 closely recapitulate most of the Fab1 deletion defects (Bonangelino *et al.*, 1997; Bonangelino *et al.*, 2002; Dove *et al.*, 2002; Gary *et al.*, 2002). A Fab1 deletion mutant devoid of PtdIns(3,5)P2 displayed a complex, pleiotropic phenotype consisting of a dramatically enlarged vacuole, a partial defect in multivesicular body (MVB) sorting, failure in vacuolar acidification and osmoregulation, defective inheritance of vacuoles, and failure to grow at high temperatures. The function of PtdIns(3,5)P2 is still not well understood. PtdIns(3,5)P2 recruits specific effectors and contributes to retrograde transport from MVB to the *trans*-Golgi network. The *E. histolytica* genome may encode one putative Fab1 homologue (EHI_049480). The size of this protein is relatively small (714 aa) and it does not possess FYVE or CCT domains, or an apparent PIPKIII-specific region, but its putative activation loop shows significant homology to that of FAB1. However, no clear orthologues of the Fab1p activator proteins, known as PtdIns(3,5)P2 effector proteins (Vps24p, Atg18p, or myotubularin1) can be found in the genome, so the possible roles of PtdIns(3,5)P2 in *E. histolytica* are unclear.

The redundancy of PI3Ks indicates that the conversion of PtdIns(4,5)P2 to PtdIns(3,4,5)P3 may be pivotal for *E. histolytica*. Since *E. histolytica* lacks an enzyme that converts PtdIns(5)P to PtdIns(4,5)P2, PtdIns(4)P appears to be the sole PtdIns(3,4,5)P3 precursor. Although PtdIns(3,4,5)P3 has not yet been directly examined in *E. histolytica*, results from PI3K inhibition suggest the importance of PtdIns(3,4,5)P3 during endocytosis, phagocytosis and chemotaxis. Our attempt to express a green fluorescent protein (GFP)-fused Akt-PH domain, which interacts specifically with PtdIns(3,4,5)P3, failed (Nakada-Tsukui, unpublished data). This result may reflect a dominant negative effect of the depletion of PtdIns(3,4,5)P3 by the GFP-fused Akt-PH domain and suggests an essential role for PtdIns(3,4,5)P3 in *E. histolytica*.

Actin cytoskeleton in cell motility and phagocytosis

Cell motility

Cell motility of E. histolytica

During tissue invasion, *E. histolytica* trophozoites rapidly reorganize their actin cytoskeleton and glide as polarized amoebic cells towards chemotactic signals (Guillén, 1993). Amoeboid movement is a rapid crawling motion characterized by flexible oscillatory cell shape changes with dynamic and multiple membrane protrusions and retractions. This concept of amoeboid motion was mostly established by studying the single-cell stage of *D. discoideum* on two-dimensional surfaces (Yoshida and Soldati, 2006). However, the amoeboid motion of cells, such as immune cells and *E. histolytica* trophozoites, moving through three-dimensional substrates has not been characterized extensively. The initial step of cell movement, i.e. cell polarization, is regulated through the activities of the actin-based subcortical cytoskeleton. Cells polarize and protrude membrane extensions called lamellipodia or pseudopodia, where actin filaments are arranged in cross-linked webs. In fact, the membrane of cells in amoeboid motion deforms at a high frequency, resembling cells with their plasma membrane detached from the cytoskeleton such as apoptotic cells, non-adherent carcinosarcoma cells or melanoma cells, which show prolonged blebbing while spreading. *E. histolytica* trophozoites were demonstrated to polarize and move along gradients of some chemoattractants, including the complement fragment C5a, human tumour necrosis factor (TNF) and a number of extracellular matrix degradation products such as intestine components and fibronectin moieties. In particular, in the presence of a human TNF gradient, cell migration was directional and the velocity of its amoeboid motion was increased (Blazquez *et al.*, 2006).

The transcription of some cytoskeleton-related genes was significantly up-regulated upon TNF chemotaxis, implying a possible role for the actin cytoskeleton in cell motility and chemotaxis of *E. histolytica* (Blazquez *et al.*, 2008).

Roles of the Gal/GalNAc lectin in adhesion and chemotaxis

During cell migration, adhesion complexes are formed on the membrane at the leading edge and disassembled at the rear end of a moving cell. These adhesion complexes mechanically couple the substrate to the cytoskeleton thus connecting them. In other eukaryotic cells, adhesion is mediated by integrins, a family of membrane receptors that recognize the extracellular matrix and connect to the cytoskeleton. In *E. histolytica,* adhesion relies on an amoebic surface protein that recognizes the sugars galactose (Gal) and N-acetylgalactosamine (GalNAc), the Gal/GalNAc lectin (Petri *et al.*, 2002). The cytoskeleton reorganizes itself after the engagement of Gal/GalNAc lectin with human epithelial cells. The crucial roles of myosin II (for cytoskeleton reorganization) and the Gal-GalNAc lectin (for adhesion) in tissue invasion by *E. histolytica* have been demonstrated (Coudrier *et al.*, 2005). The Gal/GalNAc lectin is a heterodimeric protein composed of a transmembrane heavy chain (Hgl) and an extracellular light chain (Lgl) linked by disulfide bonds. The Lgl subunit is a cysteine-rich protein with a hydrophobic signal sequence characteristic of glycosylphosphatidylinositol (GPI)-anchored membrane proteins (Cheng *et al.*, 2001). The Hgl subunit is an integral membrane protein with a short cytoplasmic tail that has homology to the cytoplasmic domain of β2 and β7 integrins and to motifs implicated in binding to the intracellular signalling molecules Shc and Grb2 (Petri *et al.*, 2002). Expression of the Hgl cytoplasmic domain alone in transfected trophozoites (named the HGL-2 strain) has a dominant negative effect on the endogenous lectin activity, leading to a decrease in adherence (Vines *et al.*, 1998) and to changes in cell morphology and motility during interaction with host tissues (Tavares *et al.*, 2005). In fact, this low-adherence HGL-2 strain has lost the ability to move within liver tissue and induces delayed inflammation and apoptosis of host cells (Blazquez *et al.*, 2007). Similarly, trophozoites with altered motility obtained by disrupting myosin II are non-invasive and become highly sensitive to the host immune response (Tavares *et al.*, 2005).

During chemotaxis, *E. histolytica* trophozoites form a single pseudopod at the front and uroid at the rear; the latter is characterized by high densities of ligand–receptor complexes and adhesion molecules that were capped on the cell surface in a myosin II dependent manner (Arhets *et al.*, 1998). The Gal/GalNAc lectin is one of the amoebic surface molecules that concentrates in the uroid region when trophozoites are exposed to Concanavalin A (Arhets *et al.*, 1995). In chemotaxing cells, transcription of both subunits of the Gal/GalNAc lectin is up-regulated; however, chemotaxis was abrogated in the low-adherence HGL-2 strain (Blazquez *et al.*, 2008). Gal/GalNAc lectin signalling could therefore be implicated in cytoskeleton reorganization and assembly/disassembly of adhesion plaques during chemotactic migration. All these findings suggest that both cytoskeleton-based cell motility and lectin-based adhesion are essential for tissue invasion and chemotaxis of *E. histolytica,* but the signalling pathways for communication between lectin and the cytoskeleton are yet to be characterized.

PAK as a key regulator of the actin cytoskeleton

In the general model of amoeboid motility, the driving force of membrane protrusion is generated by actin polymerizing at the front and myosin II pulling at the rear of the actin network. This process is tightly regulated by a range of ABPs and through various signalling pathways. In the past decade, several ABPs and key regulators have been identified, including p21-activated kinases (EhPAKs) and the small GTPase RacG, which play crucial roles in regulating the cytoskeleton and cell motility of *E. histolytica* (Aguilar-Rojas *et al.*, 2005; Díaz-Valencia *et al.*, 2007; Guillén *et al.*, 1998; Jain *et al.*, 2008; Labruyère *et al.*, 2003). PAKs are known to regulate the organization of the actin cytoskeleton in eukaryotic cells (Chung and Firtel, 1999; Eby *et al.*, 1998). In *E. histolytica,* over-expression of the catalytic domain of EhPAK prevents the formation of a dominant pseudopod and causes a significant reduction in amoeboid migration (Labruyère *et al.*, 2003). Furthermore, in chemotaxing *E. histolytica* cells, F-actin is reinforced subcortically at the leading edge and EhPAK is translocated to the rear of the cell (Blazquez *et al.*, 2008). In chemotaxing *D. discoideum* cells, PAKa is translocated to the rear and phosphorylates the myosin heavy chain kinase A (MHCKA). This phosphorylated MHCKA no longer phosphorylates myosin II, and thus unphosphorylated myosin II assembles into filaments for cell contraction (Chung and Firtel, 1999). The role of EhPAK in phosphorylation of myosin II in chemotaxing *E. histolytica* is an interesting topic for further investigations.

Other proteins involved in chemotaxis

Microarray analysis of chemotaxing *E. histolytica* highlighted the transcriptional up-regulation of other genes encoding proteins important for signalling and organization of the actin cytoskeleton, including the ARP2/3 complex, formins, α-actinin 2 and a cofilin-like factor (Blazquez *et al.*, 2008). Cofilin is an ABP that regulates the cycles of actin assembly and disassembly by promoting rapid actin filament turnover (Lappalainen and Drubin, 1997). Interestingly, it has been shown recently that an early event in polarization of fibroblasts requires cofilin-dependent actin disassembly, which favours a *de novo* reorganization of actin filaments oriented in bundles (Mseka *et al.*, 2007). These actin bundles are oriented for cell polarization and are organized by ABPs such as α-actinin. α-actinin has been linked to the phospholipid signalling pathways through its interactions with PI3K (Shibasaki *et al.*, 1994). In fact, it has been demonstrated that PI3K regulates chemotaxis in *E. histolytica*, but not random cell motility. In chemotaxing *E. histolytica* cells, a fraction of α-actinin 2 was translocated to the rear and sides of the cell and accumulated with the Gal/GalNAc lectin in the uroid region (Blazquez *et al.*, 2008). Theoretically, the actin bundles formed in these regions provide sufficient cortical tension to inhibit the emission of lateral pseudopods, thus leading to a persistent migratory direction. In fibroblasts, α-actinin is a central member of adhesion plaques that link the substrate to the actin cytoskeleton and is regulated by PI3K (Greenwood *et al.*, 2000). Previous work has suggested that a spectrin-like protein of 70 kDa is able to bind to the cytoplasmic domain of the Gal/GalNAc lectin heavy chain, linking the lectin to the actin cytoskeleton (Marion *et al.*, 2004a). Because α-actinin 2 has spectrin-like properties and co-localizes with Gal/GalNAc lectin during chemotaxis (Blazquez *et al.*, 2008), we therefore hypothesize that the spectrin-like protein that binds the Gal/GalNAc lectin may be α-actinin 2. The working hypothesis emerging from these observations is that α-actinin 2 interacts with the tail domain of Hgl to provide signalling scaffolds for dynamic changes in cytoskeletal adhesive structures during capping of surface receptors and directed motility.

Phagocytosis

General mechanisms of phagocytosis

Phagocytosis is a universal feature of eukaryotic cells and is conserved from protozoa to mammals. In many unicellular eukaryotes, phagocytosis is an essential process for nutrient uptake and defence against predators (in cases of free-living protozoa) and host immune cells (in the case of parasitic protozoa). In multicellular eukaryotes, phagocytosis plays an indispensable role in various processes, including removal of invading pathogens by phagocytes, tissue remodelling, and development. Phagocytosis is distinguished from other endocytic processes such as endocytosis and macropinocytosis by three major criteria: 1) particles of ≥ 0.5mm in diameter are internalized by phagocytosis; 2) the process is initiated by the interaction of plasma membrane receptors with specific ligands localized on the surface of particles; 3) the ligand–receptor complex triggers local reorganization of the submembranous actin-based cytoskeleton, which provides a driving force for the engulfment of particles (Kwiatkowska and Sobota, 1999).

Molecular mechanisms underlying receptor-mediated phagocytosis in macrophages have been thoroughly characterized, particularly Fc-receptor (FcR)-mediated phagocytosis of immunoglobulin G (IgG)-coated particles. When a small particle or an erythrocyte is coated with IgG, FcR on the macrophage plasma membrane recognizes the IgG and initiates downstream responses. The plasma membrane extends and surrounds the particle, which takes approximately seven minutes. The phagosome then interacts with endosomes and lysosomes to exchange their luminal and membrane-associated components, which finally leads to degradation of phagosome contents by various hydrolases. The interaction between IgG and FcR causes clustering of FcR, which induces tyrosine phosphorylation of the immuno-receptor tyrosine-based activation motif (ITAM) located in the cytoplasmic region of FcR and this initiates assembly of a complex associated with the cytoplasmic portion of FcR (Greenberg, 1999). Members of the Src family of tyrosine kinases, Lyn and Hck, phosphorylate the ITAM motifs of FcR, which in turn recruits Syk kinase (Suzuki *et al.*, 2000). Syk kinase phosphorylates the PI3K regulatory subunit p85 which in turn activates the catalytic subunit p110, generating PtdIns(3,4,5)P3 on the membrane close to the FcR complex (Crowley *et al.*, 1997; Leverrier *et al.*, 2003). In addition, SH2-containing tyrosine phosphatase-1 (SHP-1), Grb2, SH2 domain-containing leucocyte phosphoprotein of 76kDa (LCP2), Cbl, Nck, PI 5-phosphatase, and the SH2 domain-containing polyinositol-5′-phosphatase 1 (SHIP-1) are recruited to the FcR clustering site (Coppolino *et al.*, 2001; Cox *et al.*, 2001; Gu *et al.*, 2003; Huang *et al.*, 2003; Izadi *et al.*, 1998; Kant *et al.*, 2002). Expression of FcR and Syk in a non-phagocytic cell induces efficient FcR-mediated phagocytosis, suggesting that Syk plays an important role as a downstream effector of FcR (Indik *et al.*, 1995).

Although the exact role of Syk in phagocytosis is still controversial, it is widely accepted that Syk signals regulate actin remodelling by activating at least two pathways. First, activation of class I PI3K through phosphorylation of the p85 regulatory subunit causes production of PtdIns(3,4,5)P3 from PtdIns(4,5)P2 beneath the phagocytic cup (Marshall *et al.*, 2001). The raised level of PtdIns(3,4,5)P3 then recruits RhoGEFs including Tiam1, Vav1, and ARF nucleotide binding site opener, which in turn activate Rho proteins at the phagocytic cup and lead to the actin rearrangements necessary for the engulfment of the particle (Baumeister *et al.*, 2003; Takai *et al.*, 2001; Venkateswarlu and Cullen, 2000). Second, Syk directly phosphorylates and activates Vav and Rac (Patel *et al.*, 2002). Activated Rho proteins coordinately activate the actin remodelling machinery. CDC42 interacts with and stimulates WASP, and then activates ARP2/3 complex to induce actin polymerization (May and Machesky, 2001; Takai *et al.*, 2001). GTP-bound Rac proteins activate a number of proteins important for phagocytosis, including PAK1, p67phox, WAVE, and PIPKI (Bishop and Hall, 2000; Bokoch, 2003; Etienne-Manneville and Hall, 2002; Kitayama and Uyeda, 2003). Activated PAK1 stimulates several downstream signals, including: (1) phosphorylation of LIM kinase, which in turn phosphorylates actin depolymerizing factor and cofilin, and regulates actin nucleation; (2) phosphorylation of regulatory myosin light chain or MLCK to generate contractile force; (3) phosphorylation of RhoGDI, which disrupts the RhoGDI–Rho complex and releases active Rho proteins; and (4) phosphorylation of p47phox, which regulates phagocyte oxidase (Bokoch, 2003; DerMardirossian *et al.*, 2004; Ghosh *et al.*, 2004; Sanders *et al.*, 1999).

Membrane lipids also play important roles in phagocytosis and phagosome maturation. Recent studies have revealed the involvement of various lipid mediators in phagosome biogenesis in mammals (Swanson and Hoppe, 2004; Yeung and Grinstein, 2007; Yeung *et al.*, 2006). The clustering of FcR not only activates class I PI3Ks and PIPKIs but also stimulates *de novo* synthesis of PtdIns(4,5)P2 from PtdIns(4)P. PtdIns(4,5)P2 is then hydrolysed to generate two well-known second messengers as described above. Phosphatidylcholine (PC) is hydrolysed by phospholipase D (PLD) to form phosphatidic acid (PA), a cone-shaped lipid that promotes membrane fission. Phosphatidylserine (PS) on the external leaflet of the plasma membrane is also involved in the ingestion of the particles by phagocytes. It is well known that externalized PS on apoptotic cells works as an 'eat-me' signal (Hoffmann *et al.*, 2001; Ravichandran and Lorenz, 2007). However, macrophages deficient in ATP-binding cassette transporter 1 (ABCA1), which regulates externalization of PS, failed to expose PS and ingest apoptotic cells (Hamon *et al.*, 2000; Williamson *et al.*, 2007). Although the link between PS and ingestion is still unclear, these observations indicate the complexity of the molecular mechanisms of phagocytosis.

Molecular mechanisms of phagocytosis in E. histolytica

In *E. histolytica*, phagocytosis does not simply represent a basic process to obtain nutrients, it is also an important determinant of its virulence. In fact, *E. histolytica* trophozoites phagocytose erythrocytes, immune cells, non-immune cells and commensal bacteria. Although *E. histolytica* does not posses orthologues of FcR, it is well known that Gal/GalNAc-specific lectin is important for ligand binding. However, the interaction of the lectin does not always trigger ingestion, suggesting the possible existence of a coreceptor for ingestion or an independent receptor that transduces specific signals leading to actin rearrangement. Recent studies have identified phagosome-associated transmembrane kinase (PATMK) and serine-rich *E. histolytica* protein (SREHP) as being involved in phagocytosis of apoptotic host cells (Boettner *et al.*, 2008; Teixeira and Huston, 2008). PATMK has a tyrosine kinase domain in the cytoplasmic portion but its kinase activity has not been demonstrated. In contrast, SREHP does not have any discernible catalytic domain. Thus, it remains unknown whether any signal transduces from these potential phagocytosis receptors to cytoskeletal reorganization.

Although a specific receptor associated with phagocytosis, such as FcR in mammals, has not been demonstrated in *E. histolytica*, several downstream molecules related to phagocytosis are conserved in the parasite. The EhRho proteins RacA and RacG are involved in erythrophagocytosis and internalization of human serum opsonized beads (Ghosh and Samuelson, 1997; Marion and Guillén, 2006). It has also been shown that overexpression of EhGEF2 inhibits erythrophagocytosis, suggesting its involvement in phagocytosis (Gonzalez De la Rosa *et al.*, 2007). Furthermore, EhGEF1 and EhGEF3 have been demonstrated to possess *in vitro* RhoGEF activity towards RacG and RacA, respectively (Aguilar-Rojas *et al.*, 2005; Arias-Romero *et al.*, 2007), suggesting that these regulators of Rho proteins are involved in phagocytosis. EhPAK1 has been shown to be involved in phagocytosis as a downstream effector of Rho proteins (Labruyère *et al.*, 2003). As described above, EhPAK1 does not have a conserved PBD and binds to human Rac1 in a GTP-dependent manner. Downstream effectors of the apparently redundant EhPAKs have not yet been demonstrated. RhoGDI, MLCK or myosin IB heavy chain may be candidates for direct phosphorylation by EhPAKs, although myosin IB heavy chain does not have a conserved serine or threonine residue at the usual PAK target site. In mammals, the WASP/WAVE protein, which is activated by Rho proteins, together with PtdIns(4,5)P2 are important for the activation of the ARP2/3 complex to initiate actin nucleation. Although *E. histolytica* possesses a conserved ARP2/3 complex, it apparently lacks a WASP/WAVE homologue (Clark *et al.*, 2007). Thus, there may be a unique regulation system that links Rho proteins and actin nucleation. A calmodulin-like calcium binding protein (EhCaBP1) has been shown to be involved in erythrophagocytosis by direct interaction with F-actin (Sahoo *et al.*, 2004). Recently, it has been demonstrated that EhCaBP1 also plays a role in phagosome maturation using GFP-fused EhCaBP1 and live cell imaging (Jain *et al.*, 2008). While the molecular mechanisms of EhCaBP1 in phagocytosis are not yet clear, a transformant expressing dominant negative EhCaBP1 showed a reduction in protein phosphorylation and cytotoxicity, suggesting multiple roles for EhCaBP1 in cytoskeletal regulation (Jain *et al.*, 2008). Although PI3K was suggested to be involved in phagocytosis by studies in which wortmannin was used to inhibit PI3K (Ghosh and Samuelson, 1997), the premise has not been proven properly with either biochemical or genetic approaches. Since *E. histolytica* possesses type I and III PI3Ks and other phosphatidylinositol kinases, as described above, the cause of the inhibition of phagocytosis by wortmannin needs to be interpreted carefully.

A number of RhoGEFs and RhoGAPs in *E. histolytica* were found to possess PH domains (Vargas and Gonzalez-de la Rosa, 2007). Although the binding specificity of individual PH domains present in RhoGEFs and RhoGAPs in *E. histolytica* cannot be predicted based on amino acid sequences alone, these domains usually bind to PIs. In addition, there are 11 RhoGEFs possessing a FYVE domain, which generally binds to PtdIns(3)P. In mammals, it has been demonstrated that PtdIns(3)P is involved in phagosome or endosome maturation by recruiting specific binding proteins such as Hrs, EEA1 and Rabenosin. In contrast, *E. histolytica* apparently lacks homologues of all of the above mentioned mammalian PtdIns(3)P-binding proteins. Furthermore, 11 out of 12 FYVE domain-containing proteins in

E. histolytica also possess a RhoGEF domain. The large repertoire of proteins containing both FYVE and RhoGEF domains suggests a close relationship between PtdIns(3)P and actin rearrangement in *E. histolytica*. Recently, it has been demonstrated that one such FYVE and RhoGEF domain-containing protein, named EhFP4, is recruited to a tunnel-like structure that connects newly formed phagosomes to the site of engulfment at the plasma membrane during CHO cell phagocytosis (Nakada-Tsukui, unpublished data). Although it was not possible to demonstrate RhoGEF activity *in vitro*, EhFP4 binds to four out of ten Racs, suggesting that it works as a specific RhoGEF. Overexpression of the EhFP4-FYVE domain caused a reduction in CHO cell phagocytosis. In addition, membrane binding assays have shown that the FYVE domain preferentially binds to PtdIns(4)P (Nakada-Tsukui, unpublished data). There are several explanations for this observation. Since PtdIns(4)P is the only precursor to PtdIns(4,5)P2 and PtdIns(3,4,5)P3 in *E. histolytica*, which lacks PIPKII, overexpression of the EhFP4-FYVE domain may affect generation of PtdIns(4,5)P2 and PtdIns(3,4,5)P3 and inhibit the actin rearrangement necessary for phagocytosis. Taken together, although lipid mediators and their binding proteins certainly play various important roles in phagocytosis, their precise functions and mechanisms need to be examined in detail.

Summary

The information derived from the readily accessible *Entamoeba* spp. genomes has undoubtedly accelerated the identification of conserved physiological pathways as well as the discovery of novel gene functions. However, automatic annotations of orthologous genes can be inaccurate, particularly for distinct eukaryotic lineages that possess divergent or novel members of known protein families. Therefore, we have here described a manually annotated list of structural and signalling components of the actin cytoskeleton from the three *Entamoeba* genomes and at the same time reviewed recent advances. Host tissue invasion by *E. histolytica* is driven by motility and phagocytosis (Labruyère and Guillén, 2006), which are both regulated primarily by reorganization of the cytoskeleton (Guillén, 1996). This chapter aims to provide useful information for researchers wishing to investigate the many uncharacterized actin-cytoskeletal components and their regulators in *E. histolytica*, which should shed light on the molecular mechanisms of the parasite's motility and its phagocytosis-driven tissue invasion.

References

Agarwal, M., Guerin, N.A., and Larochelle, D.A. (2004). Chimeric analysis of the small GTPase RacE in cytokinesis signaling in *Dictyostelium discoideum*. Exp. Cell. Res. *295*, 226–235.

Aguilar-Rojas, A., Almaraz-Barrera, M. de J., Krzeminski, M., Robles-Flores, M., Hernández-Rivas, R., Guillén, N., Maroun, R.C., and Vargas, M.A. (2005). *Entamoeba histolytica*: inhibition of cellular functions by overexpression of EhGEF1, a novel Rho/Rac guanine nucleotide exchange factor. Exp. Parasitol. *109*, 150–162.

Anamika, K., Bhattacharya, A., and Srinivasan, N. (2008). Analysis of the protein kinome of *Entamoeba histolytica*. Proteins *71*, 995–1006.

Archer, S.K., Claudianos, C., and Campbell, H.D. (2005). Evolution of the gelsolin family of actin-binding proteins as novel transcriptional coactivators. Bioessays *27*, 388–396.

Arhets, P., Gounon, P., Sansonetti, P., and Guillén, N. (1995). Myosin II is involved in capping and uroid formation in the human pathogen *Entamoeba histolytica*. Infect. Immun. *63*, 4358–4367.

Arhets, P., Olivo, J.C., Gounon, P., Sansonetti, P., and Guillén, N. (1998). Virulence and functions of myosin II are inhibited by overexpression of light meromyosin in *Entamoeba histolytica*. Mol. Biol. Cell *9*, 1537–1547.

Arhets, P., Rahim, Z., Raymond-Denise, A., Sansonetti, P., and Guillén, N. (1992). Identification of a myosin heavy chain gene from *Entamoeba histolytica*. Arch. Med. Res. *23*, 41–43.

Arias-Romero, L.E., Almaraz-Barrera, M. de J., Diaz-Valencia, J.D., Rojo-Dominguez, A., Hernandez-Rivas, R., and Vargas, M. (2006). EhPAK2, a novel p21-activated kinase, is required for collagen invasion and capping in *Entamoeba histolytica*. Mol. Biochem. Parasitol. *149*, 17–26.

Arias-Romero, L.E., de la Rosa, C.H., Almaraz-Barrera Mde, J., Diaz-Valencia, J.D., Sosa-Peinado, A., and Vargas, M. (2007). EhGEF3, a novel Dbl family member, regulates EhRacA activation during chemotaxis and capping in *Entamoeba histolytica*. Cell Motil. Cytoskeleton *64*, 390–404.

Aspenstrom, P., Ruusala, A., and Pacholsky, D. (2007). Taking Rho GTPases to the next level: the cellular functions of atypical Rho GTPases. Exp. Cell. Res. *313*, 3673–3679.

Attree, O., Olivos, I.M., Okabe, I., Bailey, L.C., Nelson, D.L., Lewis, R.A., McInnes, R.R., and Nussbaum, R.L. (1992). The Lowe's oculocerebrorenal syndrome gene encodes a protein highly homologous to inositol polyphosphate-5-phosphatase. Nature *358*, 239–242.

Balla, T. (1998). Phosphatidylinositol 4-kinases. Biochim. Biophys. Acta *1436*, 69–85.

Bandala-Sanchez, E., Annesley, S.J., and Fisher, P.R. (2006). A phototaxis signalling complex in *Dictyostelium discoideum*. Eur. J. Cell Biol. *85*, 1099–1106.

Bao, Y., and Shen, X. (2007). INO80 subfamily of chromatin remodeling complexes. Mutat. Res. *618*, 18–29.

Baum, B., and Kunda, P. (2005). Actin nucleation: spire – actin nucleator in a class of its own. Curr. Biol. *15*, R305–308.

Baumeister, M.A., Martinu, L., Rossman, K.L., Sondek, J., Lemmon, M.A., and Chou, M.M. (2003). Loss of phosphatidylinositol 3-phosphate binding by the C-terminal Tiam-1 pleckstrin homology domain prevents *in vivo* Rac1 activation without affecting membrane targeting. J. Biol. Chem. *278*, 11457–11464.

Beder, L.B., Gunduz, M., Ouchida, M., Gunduz, E., Sakai, A., Fukushima, K., Nagatsuka, H., Ito, S., Honjo, N., Nishizaki, K., *et al.* (2006). Identification of a candidate tumor suppressor gene RHOBTB1 located at a novel allelic loss region 10q21 in head and neck cancer. J. Cancer Res. Clin. Oncol. *132*, 19–27.

Berezuk, M.A., and Schroer, T.A. (2007). Dynactin enhances the processivity of kinesin-2. Traffic *8*, 124–129.

Bishop, A.L., and Hall, A. (2000). Rho GTPases and their effector proteins. Biochem. J. *348*, 241–255.

Blazquez, S., Guigon, G., Weber, C., Syan, S., Sismeiro, O., Coppée, J.Y., Labruyère, E., and Guillén, N. (2008). Chemotaxis of *Entamoeba histolytica* towards the pro-inflammatory cytokine TNF is based on PI3K signalling, cytoskeleton reorganization and the Galactose/N-acetylgalactosamine lectin activity. Cell. Microbiol. *10*, 1676–1686.

Blazquez, S., Rigothier, M.C., Huerre, M., and Guillén, N. (2007). Initiation of inflammation and cell death during liver abscess formation by *Entamoeba histolytica* depends on activity of the galactose/N-acetyl-D-galactosamine lectin. Int. J. Parasitol. 37, 425–433.

Blazquez, S., Zimmer, C., Guigon, G., Olivo-Marin, J.C., Guillén, N., and Labruyère, E. (2006). Human tumor necrosis factor is a chemoattractant for the parasite *Entamoeba histolytica*. Infect. Immun. 74, 1407–1411.

Blessing, C.A., Ugrinova, G.T., and Goodson, H.V. (2004). Actin and ARPs: action in the nucleus. Trends Cell Biol. *14*, 435–442.

Boettner, D.R., Huston, C.D., Linford, A.S., Buss, S.N., Houpt, E., Sherman, N.E., and Petri, W.A., Jr. (2008). *Entamoeba histolytica* phagocytosis of human erythrocytes involves PATMK, a member of the transmembrane kinase family. PLoS Pathog. *4*, e8.

Bokoch, G.M. (2003). Biology of the p21-activated kinases. Annu. Rev. Biochem. 72, 743–781.

Bonangelino, C.J., Catlett, N.L., and Weisman, L.S. (1997). Vac7p, a novel vacuolar protein, is required for normal vacuole inheritance and morphology. Mol. Cell Biol. *17*, 6847–6858.

Bonangelino, C.J., Nau, J.J., Duex, J.E., Brinkman, M., Wurmser, A.E., Gary, J.D., Emr, S.D., and Weisman, L.S. (2002). Osmotic stress-induced increase of phosphatidylinositol 3,5-bisphosphate requires Vac14p, an activator of the lipid kinase Fab1p. J. Cell Biol. *156*, 1015–1028.

Boureux, A., Vignal, E., Faure, S., and Fort, P. (2007). Evolution of the Rho family of ras-like GTPases in eukaryotes. Mol. Biol. Evol. *24*, 203–216.

Bricheux, G., and Brugerolle, G. (1997). Molecular cloning of actin genes in *Trichomonas vaginalis* and phylogeny inferred from actin sequences. FEMS Microbiol. Lett. *153*, 205–213.

Carpenter, C.L., and Cantley, L.C. (1990). Phosphoinositide kinases. Biochemistry *29*, 11147–11156.

Cerione, R.A., and Zheng, Y. (1996). The Dbl family of oncogenes. Curr. Opin. Cell Biol. *8*, 216–222.

Chardin, P. (2006). Function and regulation of Rnd proteins. Nat. Rev. Mol. Cell Biol. 7, 54–62.

Cheng, X.J., Hughes, M.A., Huston, C.D., Loftus, B., Gilchrist, C.A., Lockhart, L.A., Ghosh, S., Miller-Sims, V., Mann, B.J., Petri, W.A., *et al.* (2001). Intermediate subunit of the Gal/GalNAc lectin of *Entamoeba histolytica* is a member of a gene family containing multiple CXXC sequence motifs. Infect. Immun. *69*, 5892–5898.

Chuang, T.H., Xu, X., Kaartinen, V., Heisterkamp, N., Groffen, J., and Bokoch, G.M. (1995). Abr and Bcr are multifunctional regulators of the Rho GTP-binding protein family. Proc. Natl. Acad. Sci. U.S.A. *92*, 10282–10286.

Chung, C.Y., and Firtel, R.A. (1999). PAKa, a putative PAK family member, is required for cytokinesis and the regulation of the cytoskeleton in *Dictyostelium discoideum* cells during chemotaxis. J. Cell Biol. *147*, 559–576.

Clark, C.G., Alsmark, U.C., Tazreiter, M., Saito-Nakano, Y., Ali, V., Marion, S., Weber, C., Mukherjee, C., Bruchhaus, I., Tannich, E., *et al.* (2007). Structure and content of the *Entamoeba histolytica* genome. Adv. Parasitol. *65*, 51–190.

Coppolino, M.G., Krause, M., Hagendorff, P., Monner, D.A., Trimble, W., Grinstein, S., Wehland, J., and Sechi, A.S. (2001). Evidence for a molecular complex consisting of Fyb/SLAP, SLP-76, Nck, VASP and WASP that links the actin cytoskeleton to Fcgamma receptor signalling during phagocytosis. J. Cell Sci. *114*, 4307–4318.

Coudrier, E., Amblard, F., Zimmer, C., Roux, P., Olivo-Marin, J.C., Rigothier, M.C., and Guillén, N. (2005). Myosin II and the Gal-GalNAc lectin play a crucial role in tissue invasion by *Entamoeba histolytica*. Cell. Microbiol. 7, 19–27.

Cox, D., Dale, B.M., Kashiwada, M., Helgason, C.D., and Greenberg, S. (2001). A regulatory role for Src homology 2 domain-containing inositol 5′-phosphatase (SHIP) in phagocytosis mediated by Fc gamma receptors and complement receptor 3 (alpha(M)beta(2); CD11b/CD18). J. Exp. Med. *193*, 61–71.

Crowley, M.T., Costello, P.S., Fitzer-Attas, C.J., Turner, M., Meng, F., Lowell, C., Tybulewicz, V.L., and DeFranco, A.L. (1997). A critical role for Syk in signal transduction and phagocytosis mediated by Fcgamma receptors on macrophages. J. Exp. Med. *186*, 1027–1039.

Dan, C., Kelly, A., Bernard, O., and Minden, A. (2001). Cytoskeletal changes regulated by the PAK4 serine/threonine kinase are mediated by LIM kinase 1 and cofilin. J. Biol. Chem. *276*, 32115–32121.

Dastidar, P.G., Majumder, S., and Lohia, A. (2007). Eh Klp5 is a divergent member of the kinesin 5 family that regulates genome content and microtubular assembly in *Entamoeba histolytica*. Cell. Microbiol. *9*, 316–328.

De Melo, L.D., Sant'Anna, C., Reis, S.A., Lourenco, D., De Souza, W., Lopes, U.G., and Cunha-e-Silva, N.L. (2008). Evolutionary conservation of actin-binding proteins in *Trypanosoma cruzi* and unusual subcellular localization of the actin homologue. Parasitology *135*, 955–965.

Del Pozo, M.A., Kiosses, W.B., Alderson, N.B., Meller, N., Hahn, K.M., and Schwartz, M.A. (2002). Integrins regulate GTP-Rac localized effector interactions through dissociation of Rho-GDI. Nat. Cell Biol. *4*, 232–239.

Der Mardirossian, C., Schnelzer, A., and Bokoch, G.M. (2004). Phosphorylation of RhoGDI by Pak1 mediates dissociation of Rac GTPase. Mol. Cell *15*, 117–127.

Díaz-Valencia, J.D., Almaraz-Barrera, M. de J., Arias-Romero, L.E., Hernandez-Rivas, R., Rojo-Domínguez, A., Guillén, N., and Vargas, M.A. (2005). The ABP-120 C-end region from *Entamoeba histolytica* interacts with sulfatide, a new lipid target. Biochem. Biophys. Res. Commun. *338*, 1527–1536.

Díaz-Valencia, J.D., Almaraz-Barrera, M. de J., Jay, D., Hernández-Cuevas, N.A., García, E., González De la Rosa, C.H., Arias-Romero, L.E., Hernandez-Rivas, R., Rojo-Domínguez, A., Guillén, N., *et al.* (2007). Novel structural and functional findings of the ehFLN protein from *Entamoeba histolytica*. Cell. Motil. Cytoskeleton *64*, 880–896.

Didsbury, J., Weber, R.F., Bokoch, G.M., Evans, T., and Snyderman, R. (1989). rac, a novel ras-related family of proteins that are botulinum toxin substrates. J. Biol. Chem. *264*, 16378–16382.

Dominguez, R. (2004). Actin-binding proteins: a unifying hypothesis. Trends Biochem. Sci. *29*, 572–578.

Dovas, A., and Couchman, J.R. (2005). RhoGDI: multiple functions in the regulation of Rho family GTPase activities. Biochem. J. *390*, 1–9.

Dove, S.K., McEwen, R.K., Mayes, A., Hughes, D.C., Beggs, J.D., and Michell, R.H. (2002). Vac14 controls PtdIns(3,5)P(2) synthesis and Fab1-dependent protein trafficking to the multivesicular body. Curr. Biol. *12*, 885–893.

Drouin, G., Moniz de Sa, M., and Zuker, M. (1995). The *Giardia lamblia* actin gene and the phylogeny of eukaryotes. J. Mol. Evol. *41*, 841–849.

Drummond, A.J., Ho, S.Y., Phillips, M.J., and Rambaut, A. (2006). Relaxed phylogenetics and dating with confidence. PLoS Biol. *4*, e88.

Drummond, A.J., and Rambaut, A. (2007). BEAST: Bayesian evolutionary analysis by sampling trees. BMC Evol. Biol. *7*, 214.

Dutta, S., Sardar, A., Ray, D., and Raha, S. (2007). Molecular and functional characterization of EhPAK3, a p21 activated kinase from *Entamoeba histolytica*. Gene *402*, 57–67.

Ebert, F., Guillén, N., Leippe, M., and Tannich, E. (2000). Molecular cloning and cellular localization of an unusual bipartite *Entamoeba histolytica* polypeptide with similarity to actin binding proteins. Mol. Biochem. Parasitol. *111*, 459–464.

Eby, J.J., Holly, S.P., van Drogen, F., Grishin, A.V., Peter, M., Drubin, D.G., and Blumer, K.J. (1998). Actin cytoskeleton organization regulated by the PAK family of protein kinases. Curr. Biol. *8*, 967–970.

Edgar, R.C. (2004). MUSCLE: multiple sequence alignment with high accuracy and high throughput. Nucleic Acids Res. *32*, 1792–1797.

Edman, U., Meza, I., and Agabian, N. (1987). Genomic and cDNA actin sequences from a virulent strain of *Entamoeba histolytica*. Proc. Natl. Acad. Sci. U.S.A. *84*, 3024–3028.

Endemann, G., Dunn, S.N., and Cantley, L.C. (1987). Bovine brain contains two types of phosphatidylinositol kinase. Biochemistry *26*, 6845–6852.

Endemann, G.C., Graziani, A., and Cantley, L.C. (1991). A monoclonal antibody distinguishes two types of phosphatidylinositol 4-kinase. Biochem. J. *273*, 63–66.

Etienne-Manneville, S., and Hall, A. (2002). Rho GTPases in cell biology. Nature *420*, 629–635.

Falasca, M., and Maffucci, T. (2007). Role of class II phosphoinositide 3-kinase in cell signalling. Biochem. Soc. Trans. *35*, 211–214.

Faucherre, A., Desbois, P., Satre, V., Lunardi, J., Dorseuil, O., and Gacon, G. (2003). Lowe syndrome protein OCRL1 interacts with Rac GTPase in the trans-Golgi network. Hum. Mol. Genet. *12*, 2449–2456.

Fox, J.E., Dockter, M.E., and Phillips, D.R. (1981). An improved method for determining the actin filament content of nonmuscle cells by the DNase I inhibition assay. Anal. Biochem. *117*, 170–177.

Fruman, D.A., Meyers, R.E., and Cantley, L.C. (1998). Phosphoinositide kinases. Annu. Rev. Biochem. *67*, 481–507.

Fukumoto, Y., Kaibuchi, K., Hori, Y., Fujioka, H., Araki, S., Ueda, T., Kikuchi, A., and Takai, Y. (1990). Molecular cloning and characterization of a novel type of regulatory protein (GDI) for the rho proteins, ras p21-like small GTP-binding proteins. Oncogene *5*, 1321–1328.

Gadasi, H. (1982). Isolated *Entamoeba histolytica* actin does not inhibit DNAse-I activity. Biochem. Biophys. Res. Commun. *104*, 158–164.

Gandhi, P.N., Gibson, R.M., Tong, X., Miyoshi, J., Takai, Y., Konieczkowski, M., Sedor, J.R., and Wilson-Delfosse, A.L. (2004). An activating mutant of Rac1 that fails to interact with Rho GDP-dissociation inhibitor stimulates membrane ruffling in mammalian cells. Biochem. J. *378*, 409–419.

Gangopadhyay, S.S., Ray, S.S., Sinha, P., and Lohia, A. (1997). Unusual genome organisation in *Entamoeba histolytica* leads to two overlapping transcripts. Mol. Biochem. Parasitol. *89*, 73–83.

Gary, J.D., Sato, T.K., Stefan, C.J., Bonangelino, C.J., Weisman, L.S., and Emr, S.D. (2002). Regulation of Fab1 phosphatidylinositol 3-phosphate 5-kinase pathway by Vac7 protein and Fig4, a polyphosphoinositide phosphatase family member. Mol. Biol. Cell *13*, 1238–1251.

Gasman, S., Kalaidzidis, Y., and Zerial, M. (2003). RhoD regulates endosome dynamics through Diaphanous-related Formin and Src tyrosine kinase. Nat. Cell Biol. *5*, 195–204.

Gerald, N., Dai, J., Ting-Beall, H.P., and De Lozanne, A. (1998). A role for *Dictyostelium* racE in cortical tension and cleavage furrow progression. J. Cell Biol. *141*, 483–492.

Ghosh, M., Song, X., Mouneimne, G., Sidani, M., Lawrence, D.S., and Condeelis, J.S. (2004). Cofilin promotes actin polymerization and defines the direction of cell motility. Science *304*, 743–746.

Ghosh, S.K., and Samuelson, J. (1997). Involvement of p21racA, phosphoinositide 3-kinase, and vacuolar ATPase in phagocytosis of bacteria and erythrocytes by *Entamoeba histolytica*: suggestive evidence for coincidental evolution of amebic invasiveness. Infect. Immun. *65*, 4243–4249.

Gibson, R.M., and Wilson-Delfosse, A.L. (2001). RhoGDI-binding-defective mutant of Cdc42Hs targets to membranes and activates filopodia formation but does not cycle with the cytosol of mammalian cells. Biochem. J. *359*, 285–294.

Gloss, A., Rivero, F., Khaire, N., Müller, R., Loomis, W.F., Schleicher, M., and Noegel, A.A. (2003). Villidin, a novel WD-repeat and villin-related protein from *Dictyostelium*, is associated with membranes and the cytoskeleton. Mol. Biol. Cell *14*, 2716–2727.

Godbold, G.D., Corbett, K.D., and Mann, B.J. (2002). A Rho-like small GTPase of *Entamoeba histolytica* contains an unusual amino acid residue in a conserved GDP-stabilization region and is not a substrate for C3 exoenzyme. Exp. Parasitol. *101*, 107–110.

Goley, E.D., and Welch, M.D. (2006). The ARP2/3 complex: an actin nucleator comes of age. Nat. Rev. Mol. Cell Biol. *7*, 713–726.

Golovanov, A.P., Chuang, T.H., DerMardirossian, C., Barsukov, I., Hawkins, D., Badii, R., Bokoch, G.M., Lian, L.Y., and Roberts, G.C. (2001). Structure–activity relationships in flexible protein domains: regulation of rho GTPases by RhoGDI and D4 GDI. J. Mol. Biol. *305*, 121–135.

Gómez-Conde, E., López-Robles, M.C., Hernández-Rivas, R., Hernández-Jáuregui, P., and Vargas-Mejía, M. (2000). Structural organization of gamma-tubulin in the microtubule organizing center (MTOC) during the nuclear division of *Entamoeba histolytica* trophozoites. Arch. Med. Res. *31*, S205–206.

Gonzalez De la Rosa, C.H., Arias-Romero, L.E., Almaraz-Barrera, M. de J., Hernandez-Rivas, R., Sosa-Peinado, A., Rojo-Dominguez, A., Robles-Flores, M., and Vargas, M. (2007). EhGEF2, a Dbl-RhoGEF from *Entamoeba histolytica* has atypical biochemical properties and participates in essential cellular processes. Mol. Biochem. Parasitol. *151*, 70–80.

Goodson, H.V., and Hawse, W.F. (2002). Molecular evolution of the actin family. J. Cell Sci. *115*, 2619–2622.

Gordon, J.L., and Sibley, L.D. (2005). Comparative genome analysis reveals a conserved family of actin-like proteins in apicomplexan parasites. BMC Genomics *6*, 179.

Götz, S., García-Gómez, J.M., Terol, J., Williams, T.D., Nagaraj, S.H., Nueda, M.J., Robles, M., Talón, M., Dopazo, J., and Conesa, A. (2008). High-throughput functional annotation and data mining with the Blast2GO suite. Nucleic Acids Res. *36*, 3420–3435.

Greenberg, S. (1999). Modular components of phagocytosis. J. Leukoc. Biol. *66*, 712–717.

Greenwood, J.A., Theibert, A.B., Prestwich, G.D., and Murphy-Ullrich, J.E. (2000). Restructuring of focal adhesion plaques by PI 3-kinase. Regulation by PtdIns (3,4,5)-p(3) binding to alpha-actinin. J. Cell Biol. *150*, 627–642.

Gross, S.P. (2003). Dynactin: coordinating motors with opposite inclinations. Curr. Biol. *13*, R320–322.

Grunt, M., Zárský, V., and Cvrcková, F. (2008). Roots of angiosperm formins: the evolutionary history of plant FH2 domain-containing proteins. BMC Evol. Biol. *8*, 115.

Gu, H., Botelho, R.J., Yu, M., Grinstein, S., and Neel, B.G. (2003). Critical role for scaffolding adapter Gab2 in Fc gamma R-mediated phagocytosis. J. Cell Biol. *161*, 1151–1161.

Guillén, N. (1996). Role of signalling and cytoskeletal rearrangements in the pathogenesis of *Entamoeba histolytica*. Trends Microbiol. *4*, 191–197.

Guillén, N. (1993). Cell signalling and motility in *Entamoeba histolytica*. Parasitol. Today *9*, 364–369.

Guillén, N., Boquet, P., and Sansonetti, P. (1998). The small GTP-binding protein RacG regulates uroid formation in the protozoan parasite *Entamoeba histolytica*. J. Cell Sci. *111*, 1729–1739.

Hamon, Y., Broccardo, C., Chambenoit, O., Luciani, M.F., Toti, F., Chaslin, S., Freyssinet, J.M., Devaux, P.F., McNeish, J., Marguet, D., *et al.* (2000). ABC1 promotes engulfment of apoptotic cells and transbilayer redistribution of phosphatidylserine. Nat. Cell Biol. *2*, 399–406.

Han, J.W., Leeper, L., Rivero, F., and Chung, C.Y. (2006). Role of RacC for the regulation of WASP and phosphatidylinositol 3-kinase during chemotaxis of *Dictyostelium*. J. Biol. Chem. *281*, 35224–35234.

Hart, M.J., Maru, Y., Leonard, D., Witte, O.N., Evans, T., and Cerione, R.A. (1992). A GDP dissociation inhibitor that serves as a GTPase inhibitor for the Ras-like protein CDC42Hs. Science *258*, 812–815.

Higgs, H.N., and Peterson, K.J. (2005). Phylogenetic analysis of the formin homology 2 domain. Mol. Biol. Cell *16*, 1–13.

Hirokawa, N. (1998). Kinesin and dynein superfamily proteins and the mechanism of organelle transport. Science *279*, 519–526.

Hoffmann, P.R., deCathelineau, A.M., Ogden, C.A., Leverrier, Y., Bratton, D.L., Daleke, D.L., Ridley, A.J., Fadok, V.A., and Henson, P.M. (2001). Phosphatidylserine (PS) induces PS receptor-mediated macropinocytosis and promotes clearance of apoptotic cells. J. Cell Biol. *155*, 649–659.

Huang, Z.Y., Hunter, S., Kim, M.K., Indik, Z.K., and Schreiber, A.D. (2003). The effect of phosphatases SHP-1 and SHIP-1 on signaling by the ITIM- and

ITAM-containing Fcgamma receptors FcgammaRIIB and FcgammaRIIA. J. Leukoc. Biol. *73*, 823–829.

Huber, M., Garfinkel, L., Gitler, C., Mirelman, D., Revel, M., and Rozenblatt, S. (1987). *Entamoeba histolytica*: cloning and characterization of actin cDNA. Mol. Biochem. Parasitol. *24*, 227–235.

Indik, Z.K., Park, J.G., Pan, X.Q., and Schreiber, A.D. (1995). Induction of phagocytosis by a protein tyrosine kinase. Blood *85*, 1175–1180.

Izadi, K.D., Erdreich-Epstein, A., Liu, Y., and Durden, D.L. (1998). Characterization of Cbl-Nck and Nck–Pak1 interactions in myeloid FcgammaRII signaling. Exp. Cell. Res. *245*, 330–342.

Jaffe, A.B., and Hall, A. (2005). Rho GTPases: biochemistry and biology. Annu. Rev. Cell Dev. Biol. *21*, 247–269.

Jaffer, Z.M., and Chernoff, J. (2002). p21-activated kinases: three more join the Pak. Int. J. Biochem. Cell Biol. *34*, 713–717.

Jain, R., Santi-Rocca, J., Padhan, N., Bhattacharya, S., Guillén, N., and Bhattacharya, A. (2008). Calcium-binding protein 1 of *Entamoeba histolytica* transiently associates with phagocytic cups in a calcium-independent manner. Cell. Microbiol. *10*, 1373–1389.

Joseph, J.M., Fey, P., Ramalingam, N., Liu, X.I., Rohlfs, M., Noegel, A.A., Müller-Taubenberger, A., Glöckner, G., and Schleicher, M. (2008). The actinome of *Dictyostelium discoideum* in comparison to actins and actin-related proteins from other organisms. PLoS ONE *3*, e2654.

Jung, G., Remmert, K., Wu, X., Volosky, J.M., and Hammer, J.A. (2001). The *Dictyostelium* CARMIL protein links capping protein and the Arp2/3 complex to type I myosins through their SH3 domains. J. Cell Biol. *153*, 1479–1497.

Kabsch, W., Mannherz, H.G., Suck, D., Pai, E.F., and Holmes, K.C. (1990). Atomic structure of the actin:DNase I complex. Nature *347*, 37–44.

Kant, A.M., De, P., Peng, X., Yi, T., Rawlings, D.J., Kim, J.S., and Durden, D.L. (2002). SHP-1 regulates Fcgamma receptor-mediated phagocytosis and the activation of RAC. Blood *100*, 1852–1859.

Kapoor, P., Sahasrabuddhe, A.A., Kumar, A., Mitra, K., Siddiqi, M.I., and Gupta, C.M. (2008). An unconventional form of actin in protozoan hemoflagellate, *Leishmania*. J. Biol. Chem. *283*, 22760–22773.

Karakozova, M., Kozak, M., Wong, C.C., Bailey, A.O., Yates, J.R., Mogilner, A., Zebroski, H., and Kashina, A. (2006). Arginylation of beta-actin regulates actin cytoskeleton and cell motility. Science *313*, 192–196.

Kashina, A.S. (2006). Differential arginylation of actin isoforms: the mystery of the actin N-terminus. Trends Cell Biol. *16*, 610–615.

Kettis, A.A., Lidman, K., and Fagraeus, A. (1977). Actin in *Entamoeba histolytica* trophozoites revealed by human actin antibodies. J. Parasitol. *63*, 581–583.

Kitayama, C., and Uyeda, T.Q. (2003). ForC, a novel type of formin family protein lacking an FH1 domain, is involved in multicellular development in *Dictyostelium discoideum*. J. Cell Sci. *116*, 711–723.

Kollmar, M. (2006). Thirteen is enough: the myosins of *Dictyostelium discoideum* and their light chains. BMC Genomics 7, 183.

Korenbaum, E., and Rivero, F. (2002). Calponin homology domains at a glance. J. Cell Sci. *115*, 3543–3545.

Krugmann, S., Anderson, K.E., Ridley, S.H., Risso, N., McGregor, A., Coadwell, J., Davidson, K., Eguinoa, A., Ellson, C.D., Lipp, P., *et al.* (2002). Identification of ARAP3, a novel PI3K effector regulating both Arf and Rho GTPases, by selective capture on phosphoinositide affinity matrices. Mol. Cell *9*, 95–108.

Kwiatkowska, K., and Sobota, A. (1999). Signaling pathways in phagocytosis. Bioessays *21*, 422–431.

Labruyère, E., and Guillén, N. (2006). Host tissue invasion by *Entamoeba histolytica* is powered by motility and phagocytosis. Arch. Med. Res. *37*, 253–258.

Labruyère, E., Zimmer, C., Galy, V., Olivo-Marin, J.C., and Guillén, N. (2003). EhPAK, a member of the p21-activated kinase family, is involved in the control of *Entamoeba histolytica* migration and phagocytosis. J. Cell Sci. *116*, 61–71.

Lappalainen, P., and Drubin, D.G. (1997). Cofilin promotes rapid actin filament turnover *in vivo*. Nature *388*, 78–82.

Larochelle, D.A., Vithalani, K.K., and De Lozanne, A. (1996). A novel member of the rho family of small GTP-binding proteins is specifically required for cytokinesis. J. Cell Biol. *133*, 1321–1329.

Le Clainche, C., and Carlier, M.F. (2008). Regulation of actin assembly associated with protrusion and adhesion in cell migration. Physiol. Rev. *88*, 489–513.

Lee, S.H., Kerff, F., Chereau, D., Ferron, F., Klug, A., and Dominguez, R. (2007). Structural basis for the actin-binding function of missing-in-metastasis. Structure *15*, 145–155.

Leverrier, Y., Okkenhaug, K., Sawyer, C., Bilancio, A., Vanhaesebroeck, B., and Ridley, A.J. (2003). Class I phosphoinositide 3-kinase p110beta is required for apoptotic cell and Fcgamma receptor-mediated phagocytosis by macrophages. J. Biol. Chem. *278*, 38437–38442.

Linardopoulou, E.V., Parghi, S.S., Friedman, C., Osborn, G.E., Parkhurst, S.M., and Trask, B.J. (2007). Human subtelomeric WASH genes encode a new subclass of the WASP family. PLoS Genet. 3, e237.

Lindmo, K., and Stenmark, H. (2006). Regulation of membrane traffic by phosphoinositide 3-kinases. J. Cell Sci. *119*, 605–614.

Loftus, B., Anderson, I., Davies, R., Alsmark, U.C., Samuelson, J., Amedeo, P., Roncaglia, P., Berriman, M., Hirt, R.P., Mann, B.J., *et al.* (2005). The genome of the protist parasite *Entamoeba histolytica*. Nature *433*, 865–868.

Lohia, A., and Samuelson, J. (1993). Molecular cloning of a rho family gene of *Entamoeba histolytica*. Mol. Biochem. Parasitol. *58*, 177–180.

Lohia, A., and Samuelson, J. (1996). Heterogeneity of *Entamoeba histolytica* rac genes encoding p21rac homologues. Gene *173*, 205–208.

Machesky, L.M., and Johnston, S.A. (2007). MIM: a multifunctional scaffold protein. J. Mol. Med. *85*, 569–576.

MacQueen, A.J., Baggett, J.J., Perumov, N., Bauer, R.A., Januszewski, T., Schriefer, L., and Waddle, J.A. (2005). ACT-5 is an essential *Caenorhabditis elegans* actin required for intestinal microvilli formation. Mol. Biol. Cell *16*, 3247–3259.

Maffucci, T., Cooke, F.T., Foster, F.M., Traer, C.J., Fry, M.J., and Falasca, M. (2005). Class II phosphoinositide 3-kinase defines a novel signaling pathway in cell migration. J. Cell Biol. *169*, 789–799.

Majumder, S., and Lohia, A. (2008). *Entamoeba histolytica* encodes unique formins, a subset of which regulates DNA content and cell division. Infect. Immun. *76*, 2368–2378.

Manser, E., Leung, T., Salihuddin, H., Zhao, Z.S., and Lim, L. (1994). A brain serine/threonine protein kinase activated by Cdc42 and Rac1. Nature *367*, 40–46.

Marion, S., and Guillén, N. (2006). Genomic and proteomic approaches highlight phagocytosis of living and apoptotic human cells by the parasite *Entamoeba histolytica*. Int. J. Parasitol. *36*, 131–139.

Marion, S., Laurent, C., and Guillén, N. (2005). Signalization and cytoskeleton activity through myosin IB during the early steps of phagocytosis in *Entamoeba histolytica*: a proteomic approach. Cell. Microbiol. *7*, 1504–1518.

Marion, S., Tavares, P., Arhets, P., and Guillén, N. (2004a). Signal transduction through the Gal-GalNAc lectin of *Entamoeba histolytica* involves a spectrin-like protein. Mol. Biochem. Parasitol. *135*, 31–38.

Marion, S., Wilhelm, C., Voigt, H., Bacri, J.C., and Guillén, N. (2004b). Overexpression of myosin IB in living *Entamoeba histolytica* enhances cytoplasm viscosity and reduces phagocytosis. J. Cell Sci. *117*, 3271–3279.

Marshall, J.G., Booth, J.W., Stambolic, V., Mak, T., Balla, T., Schreiber, A.D., Meyer, T., and Grinstein, S. (2001). Restricted accumulation of phosphatidylinositol 3-kinase products in a plasmalemmal subdomain during Fc gamma receptor-mediated phagocytosis. J. Cell Biol. *153*, 1369–1380.

May, R.C., and Machesky, L.M. (2001). Phagocytosis and the actin cytoskeleton. J. Cell Sci. *114*, 1061–1077.

McGough, A.M., Staiger, C.J., Min, J.K., and Simonetti, K.D. (2003). The gelsolin family of actin regulatory proteins: modular structures, versatile functions. FEBS Lett. *552*, 75–81.

McKinney, E.C., and Meagher, R.B. (1998). Members of the *Arabidopsis* actin gene family are widely dispersed in the genome. Genetics *149*, 663–675.

Meza, I., Sabanero, M., Cazares, F., and Bryan, J. (1983). Isolation and characterization of actin from *Entamoeba histolytica*. J. Biol. Chem. *258*, 3936–3941.

Meza, I., Talamas-Rohana, P., and Vargas, M.A. (2006). The cytoskeleton of *Entamoeba histolytica*: structure, function, and regulation by signaling pathways. Arch. Med. Res. *37*, 234–243.

Michaelson, D., Silletti, J., Murphy, G., D'Eustachio, P., Rush, M., and Philips, M.R. (2001). Differential localization of Rho GTPases in live cells: regulation by hypervariable regions and RhoGDI binding. J. Cell Biol. *152*, 111–126.

Mizuguchi, G., Shen, X., Landry, J., Wu, W.H., Sen, S., and Wu, C. (2004). ATP-driven exchange of histone H2AZ variant catalyzed by SWR1 chromatin remodeling complex. Science *303*, 343–348.

Mseka, T., Bamburg, J.R., and Cramer, L.P. (2007). ADF/cofilin family proteins control formation of oriented actin-filament bundles in the cell body to trigger fibroblast polarization. J. Cell Sci. *120*, 4332–4344.

Muller, J., Oma, Y., Vallar, L., Friederich, E., Poch, O., and Winsor, B. (2005). Sequence and comparative genomic analysis of actin-related proteins. Mol. Biol. Cell *16*, 5736–5748.

Odronitz, F., and Kollmar, M. (2007). Drawing the tree of eukaryotic life based on the analysis of 2,269 manually annotated myosins from 328 species. Genome Biol. *8*, R196.

Okada, M., Huston, C.D., Oue, M., Mann, B.J., Petri, W.A., Kita, K., and Nozaki, T. (2006). Kinetics and strain variation of phagosome proteins of *Entamoeba histolytica* by proteomic analysis. Mol. Biochem. Parasitol. *145*, 171–183.

Olave, I.A., Reck-Peterson, S.L., and Crabtree, G.R. (2002). Nuclear actin and actin-related proteins in chromatin remodeling. Annu. Rev. Biochem. *71*, 755–781.

Olofsson, B. (1999). Rho guanine dissociation inhibitors: pivotal molecules in cellular signalling. Cell. Signal. *11*, 545–554.

Ortiz, D., Dominguez-Robles, M. del C., Villegas-Sepulveda, N., and Meza, I. (2000). Actin induction during PMA and cAMP-dependent signal pathway activation in *Entamoeba histolytica* trophozoites. Cell. Microbiol. *2*, 391–400.

Oude Weernink, P.A., Lopez de Jesus, M., and Schmidt, M. (2007). Phospholipase D signaling: orchestration by PIP2 and small GTPases. Naunyn Schmiedebergs Arch. Pharmacol. *374*, 399–411.

Patel, J.C., Hall, A., and Caron, E. (2002). Vav regulates activation of Rac but not Cdc42 during FcgammaR-mediated phagocytosis. Mol. Biol. Cell *13*, 1215–1226.

Pechlivanis, M., and Kuhlmann, J. (2006). Hydrophobic modifications of Ras proteins by isoprenoid groups and fatty acids: More than just membrane anchoring. Biochim. Biophys. Acta *1764*, 1914–1931.

Petri, W.A., Haque, R., and Mann, B.J. (2002). The bittersweet interface of parasite and host: lectin–carbohydrate interactions during human invasion by the parasite *Entamoeba histolytica*. Annu. Rev. Microbiol. *56*, 39–64.

Pike, L.J. (1992). Phosphatidylinositol 4-kinases and the role of polyphosphoinositides in cellular regulation. Endocr. Rev. *13*, 692–706.

Podolski, J.L., and Steck, T.L. (1988). Association of deoxyribonuclease I with the pointed ends of actin filaments in human red blood cell membrane skeletons. J. Biol. Chem. *263*, 638–645.

Popowicz, G.M., Schleicher, M., Noegel, A.A., and Holak, T.A. (2006). Filamins: promiscuous organizers of the cytoskeleton. Trends Biochem. Sci. *31*, 411–419.

Pruyne, D., Evangelista, M., Yang, C., Bi, E., Zigmond, S., Bretscher, A., and Boone, C. (2002). Role of formins in actin assembly: nucleation and barbed-end association. Science *297*, 612–615.

Rameh, L.E., Tolias, K.F., Duckworth, B.C., and Cantley, L.C. (1997). A new pathway for synthesis of phosphatidylinositol-4,5-bisphosphate. Nature *390*, 192–196.

Ravichandran, K.S., and Lorenz, U. (2007). Engulfment of apoptotic cells: signals for a good meal. Nat. Rev. Immunol. *7*, 964–974.

dos Remedios, C.G., Chhabra, D., Kekic, M., Dedova, I.V., Tsubakihara, M., Berry, D.A., and Nosworthy, N.J. (2003). Actin binding proteins: regulation of cytoskeletal microfilaments. Physiol. Rev. *83*, 433–473.

Richards, T.A., and Cavalier-Smith, T. (2005). Myosin domain evolution and the primary divergence of eukaryotes. Nature *436*, 1113–1118.

Rivero, F., Dislich, H., Glockner, G., and Noegel, A.A. (2001). The *Dictyostelium discoideum* family of Rho-related proteins. Nucleic Acids Res. *29*, 1068–1079.

Rivero, F., Illenberger, D., Somesh, B.P., Dislich, H., Adam, N., and Meyer, A.K. (2002). Defects in cytokinesis, actin reorganization and the contractile vacuole in cells deficient in RhoGDI. EMBO J. *21*, 4539–4549.

Rivero, F., and Somesh, B.P. (2002). Signal transduction pathways regulated by Rho GTPases in *Dictyostelium*. J. Muscle Res. Cell Motil. *23*, 737–749.

Roberts, C.W., and Orkin, S.H. (2004). The SWI/SNF complex: chromatin and cancer. Nat. Rev. Cancer *4*, 133–142.

Roth, M.G. (2004). Phosphoinositides in constitutive membrane traffic. Physiol. Rev. *84*, 699–730.

Roy, D., and Lohia, A. (2004). Sequence divergence of *Entamoeba histolytica* tubulin is responsible for its altered tertiary structure. Biochem. Biophys. Res. Commun. *319*, 1010–1016.

Sahoo, N., Labruyère, E., Bhattacharya, S., Sen, P., Guillén, N., and Bhattacharya, A. (2004). Calcium binding protein 1 of the protozoan parasite *Entamoeba histolytica* interacts with actin and is involved in cytoskeleton dynamics. J. Cell Sci. *117*, 3625–3634.

Saito-Nakano, Y., Loftus, B.J., Hall, N., and Nozaki, T. (2005). The diversity of Rab GTPases in *Entamoeba histolytica*. Exp. Parasitol. *110*, 244–252.

Sakisaka, T., Itoh, T., Miura, K., and Takenawa, T. (1997). Phosphatidylinositol 4,5-bisphosphate phosphatase regulates the rearrangement of actin filaments. Mol. Cell Biol. *17*, 3841–3849.

Sanders, L.C., Matsumura, F., Bokoch, G.M., and de Lanerolle, P. (1999). Inhibition of myosin light chain kinase by p21-activated kinase. Science *283*, 2083–2085.

Santiago, A., Carbajal, M.E., Benitez-King, G., and Meza, I. (1994). *Entamoeba histolytica*: PKC transduction pathway activation in the trophozoite–fibronectin interaction. Exp. Parasitol. *79*, 436–444.

Schroer, T.A. (2004). Dynactin. Annu. Rev. Cell Dev. Biol. *20*, 759–779.

Sehring, I.M., Mansfeld, J., Reiner, C., Wagner, E., Plattner, H., and Kissmehl, R. (2007). The actin multigene family of *Paramecium tetraurelia*. BMC Genomics *8*, 82.

Seigneur, M., Mounier, J., Prevost, M.C., and Guillén, N. (2005). A lysine- and glutamic acid-rich protein, KERP1, from *Entamoeba histolytica* binds to human enterocytes. Cell. Microbiol. *7*, 569–579.

Sekine, A., Fujiwara, M., and Narumiya, S. (1989). Asparagine residue in the *rho* gene product is the modification site for botulinum ADP-ribosyltransferase. J. Biol. Chem. *264*, 8602–8605.

Seoh, M.L., Ng, C.H., Yong, J., Lim, L., and Leung, T. (2003). ArhGAP15, a novel human RacGAP protein with GTPase binding property. FEBS Lett. *539*, 131–137.

Shibasaki, F., Fukami, K., Fukui, Y., and Takenawa, T. (1994). Phosphatidylinositol 3-kinase binds to alpha-actinin through the p85 subunit. Biochem. J. *302*, 551–557.

Shortle, D., Haber, J.E., and Botstein, D. (1982). Lethal disruption of the yeast actin gene by integrative DNA transformation. Science *217*, 371–373.

Silacci, P., Mazzolai, L., Gauci, C., Stergiopulos, N., Yin, H.L., and Hayoz, D. (2004). Gelsolin superfamily proteins: key regulators of cellular functions. Cell. Mol. Life Sci. *61*, 2614–2623.

Stocker, S., Hiery, M., and Marriott, G. (1999). Phototactic migration of *Dictyostelium* cells is linked to a new type of gelsolin-related protein. Mol. Biol. Cell *10*, 161–178.

Suzuki, T., Kono, H., Hirose, N., Okada, M., Yamamoto, T., Yamamoto, K., and Honda, Z. (2000). Differential involvement of Src family kinases in Fc gamma receptor-mediated phagocytosis. J. Immunol. *165*, 473–482.

Swanson, J.A., and Hoppe, A.D. (2004). The coordination of signaling during Fc receptor-mediated phagocytosis. J. Leukoc. Biol. *76*, 1093–1103.

Takai, Y., Sasaki, T., and Matozaki, T. (2001). Small GTP-binding proteins. Physiol. Rev. *81*, 153–208.

Tavares, P., Rigothier, M.C., Khun, H., Roux, P., Huerre, M., and Guillén, N. (2005). Roles of cell adhesion and cytoskeleton activity in *Entamoeba histolytica* pathogenesis: a delicate balance. Infect. Immun. *73*, 1771–1778.

Tcherkezian, J., and Lamarche-Vane, N. (2007). Current knowledge of the large RhoGAP family of proteins. Biol. Cell *99*, 67–86.

Teixeira, J.E., and Huston, C.D. (2008). Participation of the serine-rich *Entamoeba histolytica* protein in amebic phagocytosis of apoptotic host cells. Infect. Immun. *76*, 959–966.

Thompson, R.F., and Langford, G.M. (2002). Myosin superfamily evolutionary history. Anat. Rec. *268*, 276–289.

Vandekerckhove, J., and Weber, K. (1978). At least six different actins are expressed in a higher mammal: an analysis based on the amino acid sequence of the amino-terminal tryptic peptide. J. Mol. Biol. *126*, 783–802.

Vargas, M., Gonzalez-de la Rosa, C.H. (2007). Structural and functional organization of the RhoGEF proteins from *Entamoeba hisolytica*. In Advances in the Immunobiology of Parasitic Diseases, L.I. Terrazasm, ed. (Kerala, Research Signpost), pp. 339–359.

Vargas, M., Sansonetti, P., and Guillén, N. (1996). Identification and cellular localization of the actin-binding protein ABP-120 from *Entamoeba histolytica*. Mol. Microbiol. *22*, 849–857.

Vargas, M., Voigt, H., Sansonetti, P., and Guillén, N. (1997a). Molecular characterization of myosin IB from the lower eukaryote *Entamoeba histolytica*, a human parasite. Mol. Biochem. Parasitol. *86*, 61–73.

Vargas, M.A., Voigt, H., Sansonetti, P., and Guillén, N. (1997b). The tail domain of *Entamoeba histolytica* myosin IB bind F-actin. Arch. Med. Res. *28 Spec No*, 137–138.

Vartiainen, M.K., and Machesky, L.M. (2004). The WASP-Arp2/3 pathway: genetic insights. Curr. Opin. Cell Biol. *16*, 174–181.

Vayssié, L., Vargas, M.A., Weber, C., and Guillén, N. (2004). Double-stranded RNA mediates homology-dependent gene silencing of gamma-tubulin in the human parasite *Entamoeba histolytica*. Mol. Biochem. Parasitol. *138*, 21–28.

Venkateswarlu, K., and Cullen, P.J. (2000). Signalling via ADP-ribosylation factor 6 lies downstream of phosphatidylinositide 3-kinase. Biochem. J. *345*, 719–724.

Vines, R.R., Ramakrishnan, G., Rogers, J.B., Lockhart, L.A., Mann, B.J., and Petri, W.A., Jr. (1998). Regulation of adherence and virulence by the *Entamoeba histolytica* lectin cytoplasmic domain, which contains a beta2 integrin motif. Mol. Biol. Cell *9*, 2069–2079.

Virel, A., Addario, B., and Backman, L. (2007). Characterization of *Entamoeba histolytica* α-actinin2. Mol. Biochem. Parasitol. *154*, 82–89.

Virel, A., and Backman, L. (2006). Characterization of *Entamoeba histolytica* α-actinin. Mol. Biochem. Parasitol. *145*, 11–17.

Virel, A., and Backman, L. (2007). A comparative and phylogenetic analysis of the α-actinin rod domain. Mol. Biol. Evol. *24*, 2254–2265.

Vlahou, G., and Rivero, F. (2006). Rho GTPase signaling in *Dictyostelium discoideum*: insights from the genome. Eur. J. Cell Biol. *85*, 947–959.

Wallar, B.J., and Alberts, A.S. (2003). The formins: active scaffolds that remodel the cytoskeleton. Trends Cell Biol. *13*, 435–446.

Williamson, P., Halleck, M.S., Malowitz, J., Ng, S., Fan, X., Krahling, S., Remaley, A.T., and Schlegel, R.A. (2007). Transbilayer phospholipid movements in ABCA1-deficient cells. PLoS ONE *2*, e729.

Yeung, T., and Grinstein, S. (2007). Lipid signaling and the modulation of surface charge during phagocytosis. Immunol. Rev. *219*, 17–36.

Yeung, T., Ozdamar, B., Paroutis, P., and Grinstein, S. (2006). Lipid metabolism and dynamics during phagocytosis. Curr. Opin. Cell Biol. *18*, 429–437.

Yoshida, K., and Soldati, T. (2006). Dissection of amoeboid movement into two mechanically distinct modes. J. Cell Sci. *119*, 3833–3844.

Yumura, S., and Uyeda, T.Q. (2003). Myosins and cell dynamics in cellular slime molds. Int. Rev. Cytol. *224*, 173–225.

Zaki, M., King, J., Futterer, K., and Insall, R.H. (2007). Replacement of the essential *Dictyostelium* Arp2 gene by its *Entamoeba* homologue using parasexual genetics. BMC Genet. *8*, 28.

Zhang, X., Jefferson, A.B., Auethavekiat, V., and Majerus, P.W. (1995). The protein deficient in Lowe syndrome is a phosphatidylinositol-4,5-bisphosphate 5-phosphatase. Proc. Natl. Acad. Sci. U.S.A. *92*, 4853–4856.

Zhang, X., Loijens, J.C., Boronenkov, I.V., Parker, G.J., Norris, F.A., Chen, J., Thum, O., Prestwich, G.D., Majerus, P.W., and Anderson, R.A. (1997). Phosphatidylinositol-4-phosphate 5-kinase isozymes catalyze the synthesis of 3-phosphate-containing phosphatidylinositol signaling molecules. J. Biol. Chem. *272*, 17756–17761.

New Insights into the Composition and Function of the Cytoskeleton in *Giardia lamblia* and *Trichomonas vaginalis*

5

Heidi G. Elmendorf, Richard D. Hayes, Shweta Srivastava and Patricia J. Johnson

Abstract

In this chapter, we discuss insights the recent publication of the *Giardia lamblia* and *Trichomonas vaginalis* genome projects have brought to our understanding of the unique cytoskeletal protein profiles of these two parasites and integrate this new analysis of protein composition with recent structural and functional studies. Although both parasites are flagellated, the differences in both the structural and protein composition of their cytoskeletons are more numerous than their similarities. Consistent with general patterns across its genome, the cytoskeletal gene families in *Trichomonas* are greatly amplified in number, while the corresponding families in *Giardia* are – with few exceptions – represented by only one or a very few copies. While the microfilament proteome is largely complete in *Trichomonas,* it is virtually absent in *Giardia.* Thus, when similarities between the two cytoskeletons emerge, they are all the more notable. Perhaps most striking is the fact that while the dynein and kinesin motor families are well-represented in both parasites, there is an apparent absence of myosins in both, with no clear evidence of a functional substitute or altered basis for microfilament dynamics.

Introduction

The cytoskeleton is an essential feature of a living cell and is now known to play important roles in both prokaryotes and eukaryotes. It is a dynamic three-dimensional structure that contributes to cell shape, motility, intracellular transport, and cell division (Cleveland, 1999; Goldman and Chao, 1999; Pollard, 2001; Sheterline *et al.,* 1998). The eukaryotic cytoskeleton can be divided broadly into four basic components: (1) microtubules comprised of polymerized tubulin and associated proteins; (2) microfilaments comprised of polymerized actin and associated proteins; (3) intermediate filaments and associated proteins; and (4) motor proteins associated with either microfilaments or microtubules.

Microtubules are hollow cylindrical tubes, between 20 and 25 nm in diameter, comprised of alpha and beta subunits of the protein tubulin. Microtubules contribute to cell shape, form spindle fibres for separating chromosomes during cell division, and provide scaffolding for the localization and dynamics of organelles and diverse cellular cargo. When arrayed inside flagella and cilia, they are used for locomotion. In most cells, alpha and beta tubulin exist as multigene families with different isotypes often serving distinct functions. In addition to the alpha and beta tubulins, there are several newly identified forms of tubulin (gamma, delta and epsilon). Gamma tubulin is notable as a major component of microtubule organizing centres. A large number of different microtubule-associated proteins (MAPs) have been identified – often specific to cell types – and they carry out a wide range of functions, including regulating the stability of microtubules, guiding microtubules towards specific cellular locations, cross-linking microtubules, and mediating the interactions of microtubules with other proteins in the cell.

Microfilaments are fine, helically-wound thread-like protein fibres, between 3 and 6 nm in diameter, comprised predominantly of actin monomers (G-actin) assembled head-to-tail into microfilaments (F-actin). Microfilaments contribute to a diverse array of cellular processes, including cytokinesis, regulation of cell shape, crawling and gliding motility. There are numerous classes of actin-binding proteins (ABP) and microfilament-associated proteins (MFAP), typically distinguished according to whether their primary association is with monomeric or filamentous actin, respectively. Many of these classes of actin-associated protein are uncharacterized in divergent eukaryotic lineages and there is again extensive diversity observed in different cell types.

Intermediate filaments (IFs) are found predominantly in metazoan cells, where they contribute to the cell structure and tensile strength (reviewed in Goldman and Chao, 1999). There are approximately 70 different types of intermediate filament proteins, which share little sequence similarity. However, the different kinds of intermediate filament share common structural features: they all have characteristic coiled coil domains made from alpha helical rod domains surrounded by terminal globular domains, and they are all polymers that generally measure between 9–11 nm in diameter when fully assembled. Among the better known intermediate filaments are the keratins (a structural component of hair and nails), desmins (a structural component of muscle cells), vimentin (a dynamic filament that contributes to organelle positioning), and lamins (forming the nuclear lamina on the interior of the nuclear envelope).

Three superfamilies of cytoskeletal motor proteins have been recognized: myosin, kinesin and dynein (Vale and Milligan, 2000). These motor proteins are typically multi-subunit proteins with both heavy chains and light chains. Members of each motor superfamily possess either actin-binding or tubulin-binding domains and can be recognized by the presence of unique and conserved amino acid sequences that form the ATP-binding and force-producing domains of the protein. These conserved regions are combined with a diverse array of polypeptide sequences that impart functional and binding specificity to the various motors.

The myosins are an extensive superfamily of 17 different classes, although many cells possess only a subset of the full family (Sellers, 2000). Each class of myosin is responsible for a distinct dynamic activity in concert with microfilaments. Myosin activity is regulated largely through phosphorylation, controlled by both specific myosin light- and heavy-chain kinases and the more promiscuous Rac and Rho kinases.

Dyneins and kinesins are also members of large superfamilies (Hirokawa, 1998). Dyneins work as minus-oriented motors and fall into two distinct categories, cytoplasmic and axonemal. The cytoplasmic dyneins position and move various cellular cargo, such as vesicles, in a retrograde direction towards the negative pole of microtubules at the microtubule organizing centre. The axonemal dyneins drive the sliding of microtubules and therefore achieve the flexing action of cilia and flagella. In contrast, most kinesins work as plus-oriented motors, moving cargo, such as vesicles, in an anterograde direction towards the positive pole of microtubules at the cell periphery.

Although originally thought to be a defining feature of eukaryotic cells, the major classes of cytoskeletal proteins have now been identified in all three domains of life (Shih and Rothfield, 2006). The bacterial cytoplasmic elements include orthologues of the three major types of eukaryotic cytoskeletal structural proteins (actin, tubulin, and intermediate filament proteins) and a fourth group, the MinD-ParA group, that appears to be unique to bacteria. FtsZ is the tubulin orthologue in bacteria (Lowe and Amos, 1998) and MreB is the actin orthologue in bacteria (Amos *et al.*, 2004). Orthologues of FtsZ and MreB have been reported in Archaea as well (Roeben *et al.*, 2006; Vaughan *et al.*, 2004). These prokaryotic homologues of tubulin and actin show 40–50% amino acid sequence identity across prokaryotic species, and yet they are sufficiently divergent from their eukaryotic counterparts that early sequence comparison algorithms failed to identify them. Crescentrin is the known homologue of IFs in bacteria and shares 24% identity and 40% similarity with nuclear Lamin A. Crescentrin was first reported in the gram negative oligotrophic bacterium *Caulobacter crescentus* as a 8–10 nm thick cytoskeletal element and, similar to eukaryotic IFs, it organizes into a coiled coil structure forming filaments along the inner curvature of the cell to maintain the cell shape (Ausmees *et al.*, 2003).

Similar cytoskeletal proteins show a remarkable functional similarity within a domain of life, but may show significant functional differences between prokaryotes and eukaryotes. Hence, while chromosome segregation is directed by microtubules in eukaryotic cells, it is the actin orthologue, MreB, that directs this same function in bacteria. Likewise, cell growth and division is typically a microfilament-mediated process in eukaryotic cells but is controlled in bacteria by the tubulin orthologue, FtsZ. Thus, apparently early in the emergence of eukaryotes, the evolution of FtsZ and MreB – and their suite of associated proteins – was sufficiently rapid to effect a dramatic functional shift. Yet, following this period, the new sequences, associations and functions appear to have been largely optimized, such that subsequent evolution among the great diversity of eukaryotes has retained sequence and functional similarity (Erickson, 2007). The unique nature of the *Giardia lamblia* and *Trichomonas vaginalis* cytoskeletons suggests that a careful study of these flagellated parasitic protists will help to diversify our understanding of the structure–function relationships of the eukaryotic cytoskeleton.

An introduction to *Giardia*

Giardia lamblia is one of the most prevalent intestinal parasitic protists of humans and animals and causes widespread morbidity from diarrheal disease. In the US, annual cases of giardiasis from 1998 to 2002 are estimated at between 2.0 and 2.4 million (Hlavsa *et al.*, 2005; Furness *et al.*, 2000), and the prevalence of giardiasis is estimated at 20–30% in developing countries (Bouzid *et al.*, 2008). Worldwide, the WHO estimates that 3 billion people live in areas without access to clean drinking water or proper sewage treatment, and up to 1 billion of these individuals suffer from giardiasis (WHO/UNICEF, 2000).

The *Giardia* life cycle alternates between an infectious cyst stage and motile trophozoite stage (Adam, 2001; Lujan *et al.*, 1997; Gillin *et al.*, 1996). When cysts are ingested, the low pH of the stomach acid and the presence of enzymes results in excystation, with each cyst producing two trophozoites. Infections have been established in clinical trials from as few as 10 cysts (Rendtorff, 1954). Within the small intestine, the trophozoites then replicate, resisting expulsion by attaching to the intestinal walls. Infections are most frequently asymptomatic but can cause a severe diarrhoea that will typically resolve within a few weeks or, less commonly, establish a chronic infection. Changing bile, lipid and pH conditions as trophozoites move down the small intestine trigger encystation in a subset of parasites. Excreted trophozoites are unstable and not infectious, but the cysts are environmentally durable and can re-establish an infection immediately or after an extended period in the environment.

Phylogenetic inference based on analysis of ribosomal RNA, ribosomal proteins and numerous other proteins, supports the deep divergence of *Giardia*, along with *Trichomonas*, in the eukaryotic lineage (Morrison *et al.*, 2007). Ribosomal proteins in particular are believed to be reliable molecular clocks because their functional demands constrain their sequence variation. While more precise resolution of these evolutionary relationships will require genome-scale data from other related taxa, it is clear that the evolutionary distance in combination with the unique selective pressures from a parasitic lifestyle have resulted in numerous unique molecular and cellular features in this protozoan parasite. These marked differences are fascinating for the new perspectives they bring to our understanding of eukaryotic cell organization, and they also serve to direct our attention in identifying potential drug and vaccine targets against *Giardia* infections.

Giardia lamblia genome overview

The *Giardia lamblia* clone WB-C6 (ATCC 50803) whole genome shotgun sequencing project has been completed recently (Morrison *et al.*, 2007) and made publicly available at http://giardiadb.org through the EuPathDB portal (Aurrecoechea *et al.*, 2009). *Giardia*DB includes the most recent genome assembly and annotation data, as well as expression data from expressed sequence tags (ESTs), serial analysis of gene expression (SAGE), and microarrays, in addition to proteomic data and phylogenetic data. These data are searchable using both general query and data-mining tools, and pre-assembled

BLAST, Pfam, SAGE, and phylogenetic data can be retrieved (see Chapter 2 by Andersson *et al.*).

The *Giardia* genome is 11.7 million basepairs (Mb) in size and distributed over five chromosomes, coding for a predicted 6470 proteins, with transcriptional evidence for 4787 of these (Morrison *et al.*, 2007). The genome is extremely compact by many measures: 77% of the assembled sequence is defined by open reading frames (ORFs), the mean intergenic distance is <400 basepairs, there are only 4 known introns, and 1800 ORFs are overlapping (Morrison *et al.*, 2007).

An introduction to the *Giardia* cytoskeleton

Aside from having two nuclei, perhaps the most striking feature of *Giardia* is its complex and unique cytoskeleton. *Giardia* is a highly polarized protist, and the cytoskeleton is dominated by several distinctive structures, including eight flagella, the ventral disc and the median body. While all members of the family Hexamitidae are flagellated, the ventral disc and median body are unique features of the genus *Giardia* (see Chapter 2 by Andersson *et al.*).

There is a strong link between the cytoskeleton and pathogenicity of this parasite. A pathogen's ability to cause disease requires first that it can establish an infection in a host. As is the case for all intestinal pathogens, a primary challenge to establishment of infection is peristalsis, the contraction of smooth muscles that forces intestinal content forward thereby aiding digestion but also acting to expel potentially harmful microbes. *Giardia* must be able to swim within the intestine to find an optimal microenvironment and then attach tightly to the lining of the intestine to avoid being swept away. Both the motility of the parasite and its attachment ability are achieved largely by the actions of its cytoskeleton. Additionally, the cytoskeleton is presumed to play conserved roles in many other cellular processes, including determination of cell shape, internal cellular organization and trafficking, cell division, and completion of the encystation/excystation cycle.

Our initial understanding of the *Giardia* cytoskeleton was derived largely from ultrastructural, immunolocalization, biochemical, and very early molecular investigations (reviewed in Elmendorf *et al.*, 2003; Feely, 1984). A detailed characterization of the main cytoskeletal structures was performed in a series of microscopic studies from the 1950s through the 1990s by various pioneer scientists in the field, including Filice, Friend, Erlandsen, and Holberton. Classical biochemical studies by Holberton and colleagues in the 1980s and 1990s, and molecular studies by Peattie, Turner and others, identified the major cytoskeletal components. Likewise, a series of studies in the 1990s by several groups served to support earlier inferences that the cytoskeleton is important in attachment.

Now, however, with the completion of the *Giardia* genome project (Morrison *et al.*, 2007) and the development of new molecular and cellular research tools, our understanding of the unique cytoskeleton of this parasite has been strikingly transformed. We have the ability to answer various questions about the *Giardia* cytoskeleton, resolve areas of dispute, and formulate hypotheses for new areas of research – and in doing so not only improve our understanding of the role of the cytoskeleton in pathogenesis but also establish a unique insight into the diversity of the eukaryotic cytoskeleton. In the next section of the chapter, we focus on our understanding of the *Giardia* cytoskeleton based on the findings of the *Giardia* Genome Project and other recent investigations that have utilized newly available research tools to illustrate the salient features of its morphology, protein composition and regulation.

Structures of the *Giardia* cytoskeleton

The basic morphology of the *Giardia* cytoskeleton has been delineated through a long series of light and electron microscopy investigations (Erlandsen and Feely, 1984; Holberton, 1973a, 1973b; Friend, 1966; Filice, 1952). More recent use of improved imaging techniques, including field emission scanning electron microscopy, has provided new details about the cytoskeletal structures (Benchimol *et al.*, 2004; Sant'Anna *et al.*, 2005). *Giardia* trophozoites are described as having a flattened pear-shaped structure, with the broader end at the anterior of the cell and the rounded side on the dorsal surface of the parasite. Cell size ranges from 12–15 μm in length and 5–9 μm in width, depending on the genotype of the parasite (H.G. Elmendorf, unpublished data). The *Giardia* cytoskeleton contains a number of prominent microtubule-based structures, including the flagella, ventral disc and median body. Microfilaments and numerous unique cytoskeletal proteins, although less abundant, also play important role in the cytoskeletal organization.

Flagella

The *Giardia* trophozoite has four pairs of flagella (anterior, posterior-lateral, ventral, and caudal) that originate from basal bodies located between the two nuclei (Fig. 5.1). The basal bodies are typical, with a 9-by-3 array of microtubules, and the axonemes of the flagella are also typical, with a 9+2 arrangement of microtubules. The anterior flagella originate in an anterior direction from the basal bodies, cross, and loop around the anterior end of the nuclei before exiting in a posterior direction from the sides of the cell body. The other three pairs of flagella extend in a posterior direction through the cell body, exiting either from the ventral surface (ventral flagella), the sides of the cell body (posterior-lateral flagella) or the posterior tip (caudal flagella).

Examination of the ultrastructure of *Giardia* and other flagellates and ciliates reveals that while the core axoneme structure is constant, it is associated with many unique associated structures that presumably play unique roles in the regulation and motion of the axonemes. Unlike the model flagellated eukaryote *Chlamydomonas* (Silflow and Lefebvre, 2001) and other parasitic flagellates, such as trypanosomes (Landfear and Ignatushchenko, 2001), in which flagella are typically anchored in the cell by a basal body lying just under the plasma membrane, all the flagella of *Giardia* are anchored deep within the cell and thus have an extensive intracellular portion of the axoneme, which is associated with poorly-characterized accessory structures. Two structures of unknown protein composition termed the 'striated bands' (or 'marginal plates') and 'dense fibrous rods' are tightly associated with the anterior and posterior surfaces, respectively, of the intracellular portions of the axonemes of the anterior flagella. Dense fibrous rods are also seen in association with posterior-lateral flagella (Benchimol *et al.*, 2004; Filice, 1952; Sant'Anna *et al.*, 2005) and short arrays of microtubules, known as the 'funis', are associated with the caudal flagella on the dorsal and ventral sides (Benchimol *et al.*, 2004; Sant'Anna *et al.*, 2005).

During the cell cycle, the flagella undergo a profound morphological and functional rearrangement. A recent report by Nohynkova and coworkers (2006) showed that the flagella develop and mature in a process involving extensive reorganization of the flagellar apparatus, with the flagella migrating to assume different identities and functions, similar to the developmental process of flagella described in *Neophroselmis olivacea* (Melkonian *et al.*, 1987). The flagellar maturation process requires three cell cycles for each new flagellum to mature fully: the ventral and posterior-lateral flagella are formed *de novo* in cell division, migrate to become the anterior flagella in the next cell cycle, and ultimately end up

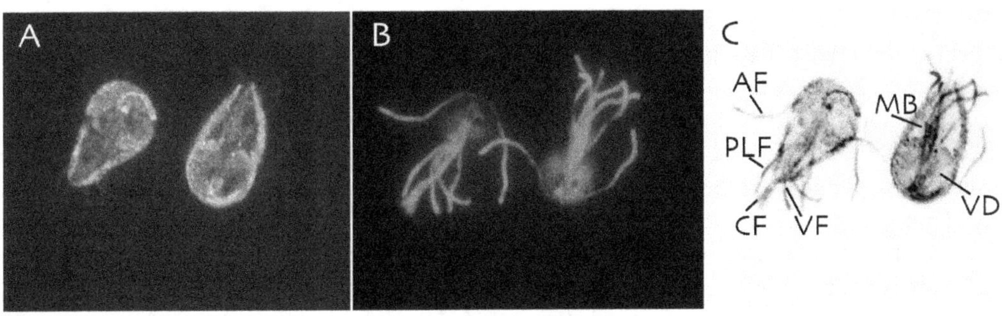

Figure 5.1 Cytoskeleton of *Giardia lamblia*. (A) Actin localization by immunofluorescence assay using a polyclonal antibody to *Giardia* actin (H.G. Elmendorf *et al.*, unpublished data). (B) Tubulin localization by immunofluorescence assay using a monoclonal antibody to *Giardia* tubulin (H.G. Elmendorf *et al.*, unpublished data). (C) Schematic representation of cytoskeletal structures: VD-ventral disc, MB-median body, AF-anterior, CF-caudal, PLF-posterior–lateral, and VF-ventral flagella.

as first the right and then the left caudal flagellum after three cell cycles. This represents a significant change in our understanding of flagellar origins and identity as it was generally accepted that each flagellum originated independently. It also raises exciting new questions for the field to answer about how the diverse associated components of each class of flagella (e.g. the striated fibres and funis) are exchanged and how this flagellar morphogenesis is regulated.

Less is understood about flagellar fate during the developmental cycle. At the time of encystment, flagella simply break into large fragments and are stored in the cytoplasm of the cyst. The fragments are quite obvious in electron micrographs of cysts (Sheffield and Bjorvat, 1977). These structures are apparently re-assembled during excystation although the exact process again awaits further research.

Ventral disc

The most striking feature of the *Giardia* cytoskeleton is a large disc located on the anterior portion of the ventral surface of the parasite, directly above the ventral plasma membrane (Fig. 5.1). The ventral disc is a unique identifying feature of the genus *Giardia* as it is not present in any of the other diplomonads. The ventral disc is comprised of a spiral array of microtubules arrayed against the plasma membrane (Feely, 1984; Mueller *et al.*, 1974; Holberton, 1973a, 1973b; Friend, 1966). Associated with the microtubules and extending dorsally into the cell are microribbons, composed of beta-giardin (Peattie *et al.*, 1989) and delta-giardin (Jenkins *et al.*, 2009) and attached by short interlocking fibres (Holberton, 1981). The disc displays increased concavity while cells are attached to an intestinal epithelial monolayer (Sousa *et al.*, 2001). Microtubules or ribbon structures are not present in the centre of the disc, a region that has therefore been termed the 'bare zone'. At the periphery of the disc a network of fibres, known as the lateral crest, replaces the microtubule array (Holberton, 1973a, 1973b). The cell body continues beyond the edge of the disc, producing a flexible overhang, termed the ventrolateral flange, which together with the lateral crest has been observed in tight apposition to epithelial cells (Campanati *et al.*, 2002; Sousa *et al.*, 2001; Mueller *et al.*, 1974; Friend, 1966). This ventrolateral flange has also been shown to be capable of dynamic protrusions, visible by light microscopy and clearly captured by scanning electron microscopy (Erlandsen *et al.*, 2004).

Recent research by Nohynkova and colleagues (Tumova *et al.*, 2007) has substantially reshaped our understanding of how *Giardia* forms new ventral discs during cell division, and has resolved several previously conflicting lines of evidence (Benchimol, 2004a; Benchimol, 2004b; Ghosh *et al.*, 2001; Solari *et al.*, 2003). In brief, the dividing cell assembles two new discs *de novo* for the daughter cells, while disassembling the parental ventral disc (Tumova *et al.*, 2007). The parental disc is slowly disassembled by first disconnecting the disc microtubules from the basal bodies, leaving the flagella free to rearrange as previously discussed. Next, a gradual shortening of the dorsal ribbons seems to loosen the overall structure of the disc, causing it to become less tightly coiled and ultimately much wider as it spreads over the broadening ventral surface of the cell. Finally, the microtubule array itself is eliminated and the dividing cell detaches from the substrate. During this time of disassembly, the two new discs are coordinately assembled on the dorsal side of the parasite in an inverted orientation, with their ventral surfaces against the plasma membrane. Assembly happens in the reverse order to disassembly, with the initial development of the microtubule arrays at the site of the new flagella organizing centres, followed by slow growth of the dorsal ribbons.

During the developmental cycle, the ventral disc is disassembled into four pieces during encystation, stored in the cyst, and then apparently reassembled during excystation (Palm *et al.*, 2005).

Median body

The median body is a prominent yet poorly defined structure with no known function (Fig. 5.1). It is an irregular array of microtubules located posterior to the nuclei and transverse above the caudal axonemes (Filice, 1952; Friend, 1966; Piva and Benchimol, 2004). Its exact size and shape is often used as a distinguishing feature among different *Giardia* species (Filice, 1952). Its location appears to be determined by small, undefined fibres that link it to adjacent cytoskeletal structures, namely the caudal axonemes and their affiliated funis (Crossley *et al.*, 1986; Piva and Benchimol, 2004). Although early reports suggested that the median body was not present in mitotic cells, more recent studies clearly demonstrate that the median body is present throughout the cell cycle (Nohynkova *et al.*, 2006; Piva and Benchimol, 2004; Sagolla *et al.*, 2006). Its function remains unknown, although some workers have proposed it to be a store of pre-polymerized tubulin for disc assembly (Feely *et al.*, 1990), a novel microtubule-organizing centre (MTOC) (Meng *et al.*, 1996), a microtubule nucleation site (Holberton, 1981), or to have a role in tail flexion (Piva and Benchimol, 2004). Interesting new data have localized both EB-1 and kinesin-13, microtubule-associated proteins involved in regulating microtubule dynamics, to the median body, confirming it as a site of microtubule turnover (Dawson *et al.*, 2007).

Proteins of the *Giardia* cytoskeleton

The *Giardia* genome database brings new resources to our analysis of the *Giardia* cytoskeleton. A detailed annotation effort has confirmed the presence and absence of various 'ubiquitous' eukaryotic cytoskeleton protein constituents – in turn informing our understanding of the function of the cytoskeleton. It has also led to the identification of unique *Giardia* cytoskeleton protein families. These cytoskeleton proteins can be roughly divided into five major categories: tubulins and associated proteins, actin and associated proteins, motor proteins, giardins, and novel coiled-coil proteins. The current state of our knowledge of each of these families is discussed below.

Tubulins and associated proteins

As discussed in the previous section, the *Giardia* cytoskeleton is particularly rich in microtubules that comprise the core structure of the flagella, ventral disc and median body (Fig. 5.1). The remarkable complexity of the microtubule system in this protozoan parasite makes it a model organism for exploring the range of diversity in composition and function of microtubule proteins and their evolution.

Tubulin was first identified as a dominant protein constituent of *Giardia* by protein analysis of purified cytoskeletons (Crossley and Holberton, 1983a; Holberton and Ward, 1981). Torian and colleagues (1984) then raised three monoclonal antibodies against *Giardia* lysate that reacted with 53 and 55 kilodalton (kDa) proteins from *G. lamblia* and cross-reacted with purified bovine tubulin. The presence of two alpha tubulin and three beta tubulin genes was finally confirmed by molecular cloning (Kirk-Mason *et al.*, 1988, 1989). From the *Giardia* genome data we now recognize that genes encoding all five tubulin forms are present, with one gene each for gamma tubulin, delta tubulin and epsilon tubulin (Table 5.1). Like gamma tubulin, delta and epsilon tubulin are associated with basal bodies, and are needed for the formation of the triplet microtubules in basal bodies and basal body duplication, respectively (Dutcher, 2003) – functions that make sense given the presence of eight basal bodies in interphase *Giardia* cells. While the full diversity of tubulin forms is present, the size of the gene families is quite small compared to many other organisms, which may contain a dozen or more tubulin gene copies. The sequences of the two alpha tubulin and three beta tubulin gene copies are strikingly similar (100% identity at the amino acid level between the alpha tubulins and 99% among the beta tubulins), despite previous reports to the contrary (Campanati *et al.*, 2003).

The presence of a wide array of genes that encode proteins involved in microtubule modification, organization and assembly indicate a high level of conservation of microtubule organization and function. Typical post-translation tubulin modifications that affect the stability of microtubules include acetylation of a conserved lysine residue at position 40 (Yang and Gregoire, 2007) and numerous carboxy-terminal modifications (polyglycylation, polyglutamylation, and a tyrosination/detyrosination cycle of the C-terminal tyrosine residue). Acetylation, polyglycylation, polyglutamylation had previously been detected biochemically, but biochemical evidence for the tyrosination/detyrosination cycle was missing (Weber *et al.*, 1997). The genome project has provided new insights into tubulin post-translational modifications in *Giardia* through the identification of genes that are likely to be responsible for the modifications. A strong candidate for the acetylation activity was identified,

Table 5.1 Tubulin and tubulin-associated genes in the *G. lamblia* genome with high similarity to known eukaryotic homologues

Tubulin cytoskeleton	
Alpha-tubulin	GL50803_112079 GL50803_103676
Beta-tubulin	GL50803_101291 GL50803_136020 GL50803_136021
Gamma-tubulin	GL50803_114218
Delta-tubulin	GL50803_5462
Epsilon-tubulin	GL50803_6336
Microtubule-associated proteins	
Tubulin-folding proteins	
Tubulin specific chaperone A	GL50803_1649
Tubulin specific chaperone B	GL50803_5374
Tubulin specific chaperone C	GL50803_15906
Tubulin specific chaperone D	GL50803_10145
Tubulin specific chaperone E	GL50803_16535
Tubulin-modifying proteins	
Tubulin acetylase (putative)	GL50803_14277
Tubulin carboxypeptidase (putative)	GL50803_10976 GL50803_15482 GL50803_17014 GL50803_17025
Tubulin tyrosine ligase	GL50803_8456 GL50803_10382 GL50803_14498 GL50803_9272 GL50803_8592 GL50803_10801 GL50803_95661
Microtubule plus-end capping/associated	
CLIP-170	None
Dynactin	None
EB-1	GL50803_14048
Microtubule severing	
Katanin	GL50803_15368
Microtubule organizing centre	
Centrin	GL50803_104685
Gamma tubulin ring complex	GL50803_12057

and the presence of four putative carboxypeptidase genes (currently annotated as Nuclear ATP/GTP-binding proteins) and seven putative tubulin tyrosine ligase (TTL) genes strongly suggest that tubulins in *Giardia* do indeed have a tyrosination/detyrosination cycle. Additionally, the recent discoveries in other cells that TTL-like proteins are responsible for polyglycylation and polyglutamylation activity (Ikegami *et al.*, 2008; Janke *et al.*, 2005) suggest that the six members of the TTL family in *Giardia* might be performing all three of these modifications.

Dimerization of alpha and beta tubulin subunits and assembly of microtubules relies on the concerted actions of a group of tubulin-specific chaperones, known as TBCA, TBCB, TBCC, TBCD, and TBCE. All five members of the family appear to be present in *Giardia*. The original genome annotation missed TBCA and erroneously annotated two copies of TBCD (Morrison *et al.*, 2007); the correct annotations are now provided in the database and in Table 5.1. Additionally, orthologues have been identified for other proteins involved in regulation of microtubule dynamics: centrin, a regulator of centriole duplication, katanin, a microtubule-severing protein, and EB-1, a plus-end tracking protein that modulates both assembly and disassembly. EB-1 has also been localized to the distal flagellar tips of *Giardia*, a location that is consistent with other cells (Dawson *et al.*, 2007). However, other broadly conserved microtubule plus-end tracking proteins (+TIPs), such as CLIP-170, Lis1, and multiple members of the dynactin complex (including p150^{Glued}), are apparently missing from *Giardia* (Morrison *et al.*, 2007).

This absence of a select subset of microtubule binding proteins has led to a closer examination. Interestingly, the missing proteins all interact not only with microtubules but also with cytoplasmic dynein and in doing so help to regulate cargo transport. As will be discussed in greater detail later, there is disagreement over the presence of cytoplasmic dyneins in *Giardia*, despite the clear presence of the full diversity of inner and outer arm axoneme dynein heavy chains (Morrison *et al.*, 2007; Wickstead and Gull, 2007). Thus, the absence of this class of microtubule-associated proteins may be correlated with a lack of cytoplasmic dynein. In either case, what remains unclear is whether novel machinery substitutes functionally in *Giardia* for the missing +TIPs.

Flagellum-specific microtubular machinery (Hao and Scholey, 2009) is apparently present and largely complete in *Giardia*. The genome project has identified orthologues of all but one member of the intraflagellar transport complexes A and B (IFTA and IFTB), as well as numerous other structurally-relevant axoneme-associated proteins (Morrison *et al.*, 2007; Wickstead and Gull, 2007). Typically, intraflagellar transport is driven by kinesin-2 (and OMC-3) in an anterograde direction and by dynein 1b (DHC1b). There is clear evidence for the functional presence of kinesin-2 (though not OMC-3) (Hoeng *et al.*, 2008), but there is again disagreement about the presence of dynein 1b in *Giardia* (Wickstead and Gull, 2006).

Microtubule organizing centre (MTOC)

The MTOC is the structure in eukaryotic cells responsible for nucleating microtubules, anchoring cilia and flagella, and organizing mitotic and meiotic spindles. The *Giardia* trophozoite has numerous MTOCs: traditional members are the eight basal bodies of interphase cells, while the ventral disc and median body appear to be unique microtubule organizing centres specific to

---

Notes to Table 5.1

To explore the cytoskeletal gene content of *G. lamblia*, BLAST (Altschul *et al.*, 1997) searches were performed for all identified open reading frames (ORFs), and any ORFs with identified similarity were explored further (Morrison, *et al.*, 2007). Pfam searches (Finn *et al.*, 2008) for protein domains and motifs were performed in each case, and reciprocal BLASTP searches against the full GenBank database (Benson *et al.*, 2005) were performed to verify protein family homology (Morrison, *et al.*, 2007). In cases where no known homologue to typical cytoskeletal proteins was identified, BLAST searches were performed using other homologues from a wide range of other eukaryotes. Genes indicated here are culled largely from the *G. lamblia* genome paper (Morrison *et al.*, 2007) with corrections as noted in the text.

Giardia. Important eukaryotic MTOC proteins are gamma-tubulin and centrin, and orthologues to these proteins are present in *Giardia*. Gamma-tubulin has been localized to the basal bodies (Nohynkova *et al.*, 2000); centrin has also been localized to the basal bodies in several studies (Belhadri, 1995; Correa *et al.*, 2004; Meng *et al.*, 1996; Sagolla *et al.*, 2006), but while a few of these studies have additionally seen it present in the ventral disc and median bodies, this secondary localization awaits confirmation.

Actin and associated proteins

Giardia shows a strikingly simplified microfilament cytoskeleton protein composition compared to other eukaryotes (Table 5.2) – even other highly divergent lineages such as *Trichomonas*. The gene for *Giardia* actin was first identified in 1995 and it shows an average of 58% identity and 68% similarity to other eukaryotic actins over its length (Drouin *et al.*, 1995). These numbers are sufficient to suggest that *Giardia* actin shares an overall structural conservation with other eukaryotic actins, a proposal confirmed by early structural homology modelling studies (S. Costanzi and H.G. Elmendorf, unpublished data). Actin is found around the periphery of the cell, suggesting a role in cell shape and structural integrity (Fig. 5.1).

Given the conservation of actin, there is a surprising absence of any of the classic eukaryotic actin-binding and microfilament-associated proteins in *Giardia*. Four actin-related proteins have been identified in the genome (Morrison *et al.*, 2007). Actin-related proteins (Arps) are a broad family comprised of 11 classes, with a numbering scheme that designates Arp1 as the sequence most similar to actin and Arp11 as the sequence most distant (Chen and Shen, 2007). The Arps include both cytoplasmic proteins that bind to actin and regulate its function (Arps 1, 2, 3, 10 and 11) and nuclear proteins that work as chromatin remodelling factors (Arps 4–9). Phylogenetic analysis of the three *Giardia* Arps indicate that they are not members of the Arp 1, 2 or 3 classes, although their significant sequence divergence and the consequent length of their branches prevents definitive placement within the chromatin remodelling classes of the Arp phylogeny (H. Goodson, personal communication and Fig. 5.2). Secondary evidence to support the claim that *Giardia* is missing the canonical Arp 1, 2 and 3 proteins comes from the fact that each of these is part of a large multi-protein complex (dynactin in the case of Arp 1 and the full Arp 2/3 complex) and no other members of the complexes can be identified either (Morrison *et al.*, 2007). The divergence of *Giardia* sequences is a recurrent problem in efforts to define identity and function of *Giardia* proteins based on sequence similarity alone. Therefore final classification of the three *Giardia* Arps as chromatin remodelling proteins awaits a functional analysis.

The absence of classical actin-binding and microfilament-associated proteins is complete and includes all known proteins responsible for actin modification, organization, and assembly (Morrison *et al.*, 2007). Hence, *Giardia* is missing all of the known proteins that are involved in regulating actin sequestration (e.g. profilin and thymosin), actin polymerization (e.g. formin), actin capping (e.g. F-actin CapA and CapB proteins), actin bundling and crosslinking (e.g. actinin and espin), and actin severing (e.g. cofilin and gelsolin). As will be discussed below, these searches also failed to reveal any member of the myosin family in *Giardia*. Based on the finding that many of these proteins (with the exception of myosin) are present in the *Trichomonas* genome, it appears that their absence in *Giardia* may be the result of secondary loss.

There is also no evidence in the genome for the large number of microfilament-associated proteins typically found in metazoa (e.g. alpha-catenin, spectrin, tropomyosin, troponin, villin, and vinculin). These results contradict those obtained in previous studies using heterologous antibodies that suggested the presence of alpha-actinin, myosin, tropomyosin, and vinculin (Feely *et al.*, 1982; Narcisi *et al.*, 1994). However, it would be quite unexpected to find these proteins outside the crown eukaryotes, and there is precedent for misleading cross-reactivity between these antibodies and proteins in yeast (H. Goodson, personal communication). It should therefore be concluded that these proteins do not exist in *Giardia*.

We do not yet have a full understanding of the consequences of the apparent absence of all

Table 5.2 Actin and actin-associated genes in the *G. lamblia* genome with high similarity to known eukaryotic homologues

Actin cytoskeleton	
Actin	GL50803_40817
Actin-related proteins	GL50803_8726 GL50803_11039 GL50803_15113 GL50803_16172
Actin-associated proteins	
CapZ (barbed-end capping protein)	
CapZ-alpha	None
CapZ-beta	None
Actin filament dynamics	
Cofilin	None
Formin	None
Gelsolin	None
Profilin	None
Thymosin	None
Actin filament bundling	
Alpha-actinin	None
Espin	None
Arp2/3 complex and related proteins	
ARP2/3 complex	None
Microfilament-associated proteins of higher eukaryotes	
Alpha-catenin	None
Spectrin	None
Tropomyosin	None
Troponin	None
Villin	None
Vinculin	None

Genes indicated here are culled largely from the *G. lamblia* genome paper (Morrison *et al.*, 2007) with corrections as noted in the text.

known actin-binding and microfilament-associated proteins. It is possible that there are functional homologues with sufficiently divergent sequences such that even the most sensitive similarity and motif searches have failed to reveal them. One argument against this is the detection of a protein (a homologue of yeast sec23) containing a highly-conserved gelsolin domain, suggesting that our inability to detect gelsolin can be attributed with greater confidence to its true absence and not to unusual evolutionary divergence. Alternatively, novel proteins may substitute functionally for

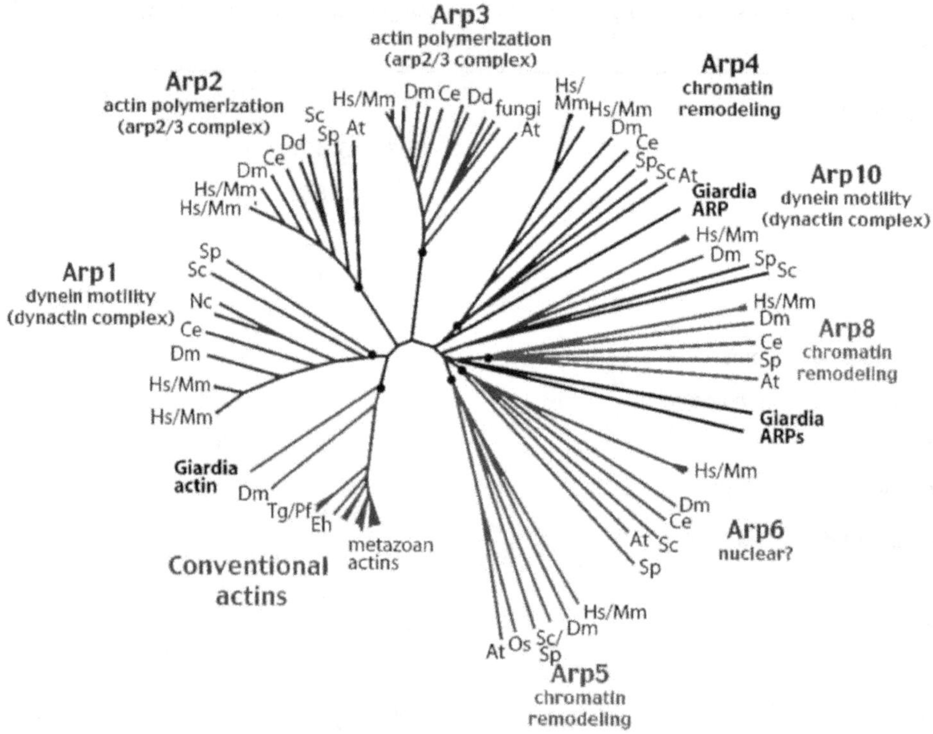

Figure 5.2 An actin-related protein (Arp) phylogeny placing *Giardia* actin and the three identified Arps onto a fuller eukaryotic Arp phylogeny (H. Goodson, unpublished data). Noteworthy is the length of the *Giardia* Arp branches, a finding that makes it difficult to define with certainty the homologues of these proteins. Black dots represent clades with bootstrap values of >90%.

those members of the microfilament family that are missing. The resolution of this question awaits further research.

Motor proteins

Although the polymerization and depolymerization cycles of tubulin and actin alone can generate cellular dynamics, most instances of movement on both the cellular and subcellular scale involve the action of motor proteins. As was the case for many of the microtubule- and microfilament-associated proteins, there is good evidence for significant conservation of the microtubule motor family, kinesins and dyneins (Table 5.3), yet no evidence for the presence of any member of the microfilament motor family, myosins (Morrison *et al.*, 2007).

Numerous detailed biochemical and molecular analyses have documented the presence of 24 kinesins and kinesin-related genes in *Giardia* (Morrison *et al.*, 2007; Dawson *et al.*, 2007; Wickstead and Gull, 2006; Iwabe and Miyata, 2002; Lawrence *et al.*, 2002). Two independent analyses have identified putative homologues for kinesins-1, 2 (two copies), 3 (two copies), 4/10, 5, and 9 by sequence similarity and phylogeny construction (Morrison *et al.*, 2007; Wickstead and Gull, 2006). Completion of the genome led to the further identification of putative kinesin-6, 7 and 8 homologues, while a detailed eukaryotic kinesin analysis reclassified several proteins annotated simply as 'kinesin-like' or 'kinesin-related' proteins as possible members of the kinesin-13, 14, 15, 16 and 17 families (Morrison *et al.*, 2007; Wickstead and Gull, 2006).

Because of the high conservation of kinesin function and its retention across broad evolutionary time periods, we can use the presence of the kinesins to make inferences about the function of microtubules in *Giardia*. Importantly, all four members of the kinesin family thought to

Table 5.3 Motor protein genes in the *G. lamblia* genome with high similarity to known eukaryotic homologues

Motor proteins			
Dyneins			
Heavy chain	GL50803_8172 GL50803_10538 GL50803_16804 GL50803_17243 GL50803_17265	GL50803_17478 GL50803_37985 GL50803_40496 GL50803_93736 GL50803_94440	GL50803_100906 GL50803_101138 GL50803_103059 GL50803_111950 GL50803_29256
Midasin	GL50803_39312		
Kinesin heavy chains			
Kinesin-1	GL50803_13825		
Kinesin-2	GL50803_16456		
Kinesin-2	GL50803_17333		
Kinesin-3	GL50803_6262		
Kinesin-3	GL50803_102101		
Kinesin-3	GL50803_112846		
Kinesin-4	GL50803_16650		
Kinesin-5	GL50803_16425		
Kinesin-6	GL50803_102455		
Kinesin-6 like	GL50803_15134		
Kinesin-7	GL50803_15962		
Kinesin-8	GL50803_4371		
Kinesin-9	GL50803_6404		
Kinesin-9	GL50803_10137		
Kinesin-13	GL50803_16945		
Kinesin-14	GL50803_8886		
Kinesin-14	GL50803_13797		
Kinesin-16	GL50803_7874		
Kinesin-16	GL50803_16161		
Kinesin-like protein	GL50803_11442		
Kinesin-like protein	GL50803_14070		
Kinesin-like protein	GL50803_16224		
Kinesin-like protein	GL50803_17264		
Kinesin-like protein	GL50803_112729		

Genes indicated here are culled largely from the *G. lamblia* genome paper (Morrison *et al.*, 2007) with corrections as noted in the text.

be associated with flagellar dynamics (kinesins-2, 9, 16 and 17) have been identified in *Giardia* (Morrison *et al.*, 2007; Wickstead and Gull, 2006). Functional roles, conserved with respect to their roles in other eukaryotes, have also been demonstrated experimentally using dominant-negative constructs for kinesin-2 – in anterograde intraflagellar transport (Hoeng *et al.*, 2008) – and kinesin-13 – as a regulator of microtubule dynamics (Dawson *et al.*, 2007).

Dynein heavy chains fall into two categories: the highly conserved cytoplasmic dynein 1 (DYNC1) and the multiple axoneme-associated dyneins – confusingly including cytoplasmic dynein 2 (DYNC2) as well as seven categories of outer and inner arm dyneins (OADs and IADs, respectively). The genome project has identified 14 dynein heavy chains and one related protein (Table 5.3), in addition to numerous dynein intermediate and light chains (Morrison *et al.*, 2007). Importantly, while all members of the true axonemal dyneins are present (Wickstead and Gull, 2007), there is no current consensus regarding the two possible cytoplasmic dynein homologues (Morrison *et al.*, 2007; Wickstead and Gull, 2007). Although one phylogenetic analysis placed members of the *Giardia* dyneins into DYNC1 and DYNC2 clusters (Wickstead and Gull, 2007), analyses performed by the *Giardia* Genome Project concluded that it simply wasn't possible to distinguish between these two categories of 'cytoplasmic' dyneins and favoured a conclusion that the true cytoplasmic dynein, DYNC1, is not present in *Giardia* (Morrison *et al.*, 2007) – a finding that would concur with the absence of the dynactin complex. One possible explanation for the divergent conclusions is the exceptionally long branch length of the two *Giardia* non-axonemal dyneins when placed into the larger dynein phylogeny (Wickstead and Gull, 2007), a characteristic that can often lead to mistaken groupings of rapidly evolving sequences. Midasin, a distant relative of dyneins, is also present in *Giardia* (Garbarino and Gibbons, 2002), a finding that signals the presence of this protein quite early in eukaryotic evolution. As is true for kinesins, dynein function appears to be well-conserved across eukaryotes (Wickstead and Gull, 2007), so the identification of the full family will prove invaluable in the coming years as the field dissects the function of these two families of motor proteins.

Perhaps most intriguing is the apparent absence of any myosin gene in the *Giardia* genome (Morrison *et al.*, 2007; Vale, 2003). Again, it is worth noting that *Trichomonas* also lacks myosins (Carlton *et al.*, 2007), which makes it more difficult to say when myosins appeared in eukaryotic lineages. This absence is perhaps particularly striking – and potentially quite important in understanding the evolution of eukaryotic cytoskeletal motors – given the fact that structural and sequence similarities have led to the hypothesis that kinesins and myosins share a common ancestor (unlike the dyneins which appear to have evolved separately from the AAA+ superfamily) (Kull *et al.*, 1998). It remains to be established whether the myosin function is also absent or is accomplished by another unique protein in *Giardia*.

Giardins

Giardins were first described by Crossley and Holberton as a group of 30 kDa proteins, associated with microtubules and were believed at that time to be unique to *Giardia* (Crossley and Holberton, 1983a; Crossley and Holberton, 1983b). Subsequent work by many investigators has shed light on what has proven to be an extensive and very diverse group of different protein families that share little beyond their common size. Indeed the group is sufficiently diverse – and clearly no longer unique to *Giardia* – that the common nomenclature of 'giardin' no longer seems appropriate.

The alpha giardins are a very large family with 21 members (Morrison *et al.*, 2007; Weiland *et al.*, 2005; Table 5.4) that have now been recognized to be members of the more general eukaryotic annexin family (Bauer *et al.*, 1999; Morgan and Fernandez, 1995; Weiland *et al.*, 2005; Weiland *et al.*, 2003). Annexins bind to phospholipids in a calcium-dependent manner (Gerke and Moss, 2002), and the alpha-giardins share the characteristic annexin repeat structure and calcium-binding site, although detailed structural studies have revealed differences in the precise nature of the annexin folds and calcium ion binding (Pathuri *et al.*, 2006; Pathuri *et al.*, 2009; Pathuri *et al.*, 2007). While alpha-1 and alpha-14 giardins have been shown to be functional analogues of annexins, as defined by calcium-dependent binding to phospholipids (Bauer *et al.*, 1999; Pathuri *et al.*, 2009; Weiland *et al.*, 2003), alpha-11 giardin is not (Pathuri *et al.*, 2007), and functional roles for most members of the family have not yet been

Table 5.4 Giardin genes in the *G. lamblia* genome

Giardins	
Alpha-1 giardin	GL50803_11654
Alpha-2 giardin	GL50803_7796
Alpha-3 giardin	GL50803_11683
Alpha-4 giardin	GL50803_7799
Alpha-5 giardin	GL50803_7797
Alpha-6 giardin	GL50803_14551
Alpha-7.1 giardin	GL50803_103373
Alpha-7.2 giardin	GL50803_114119
Alpha-7.3 giardin	GL50803_114787
Alpha-8 giardin	GL50803_11649
Alpha-9 giardin	GL50803_103437
Alpha-10 giardin	GL50803_5649
Alpha-11 giardin	GL50803_17153
Alpha-12 giardin	GL50803_10073
Alpha-13 giardin	GL50803_1076
Alpha-14 giardin	GL50803_15097
Alpha-15 giardin	GL50803_13996
Alpha-16 giardin	GL50803_10036
Alpha-17 giardin	GL50803_15101
Alpha-18 giardin	GL50803_10038
Alpha-19 giardin	GL50803_4026
Beta-giardin	GL50803_4812
Gamma-giardin	GL50803_17230
Delta-giardin	GL50803_86676

Genes indicated here are culled largely from the *G. lamblia* genome paper (Morrison *et al.*, 2007) with corrections as noted in the text.

determined. Localization data are not available for every member of the family, but the majority have been shown by epitope-tagging studies to localize to the plasma membrane or cytoskeleton elements (mainly flagella and disc) (Weiland *et al.*, 2005). Alpha-14 giardin has been shown to bind to tubulin specifically as a likely mediator of its flagellar localization (Vahrmann *et al.*, 2008). The discovery of 21 alpha-giardins in *Giardia* is interesting, given that many eukaryotes have only a few annexins, e.g. *Saccharomyces cerevisiae, Dictyostelium discoideum, Neurospora crassa* and *Caenorhabditis elegans* (Gerke and Moss, 2002). Clustering of the alpha-giardin genes on chromosomes suggests their expansion has been through gene duplication, an idea supported by the phylogenetic relationships among the members of the family, and subsequent functional divergence (Weiland *et al.*, 2005).

Beta-giardin was first identified in 1983 by Holberton and colleagues as a key component of the dorsal ribbons in the ventral disc and later characterized as a coiled-coil protein following cloning and sequencing from a cDNA library

(Crossley and Holberton, 1985; Crossley and Holberton, 1983a; Holberton *et al.*, 1988). Although it is a single copy gene (Table 5.4), it appears to be a member of a loosely-related (~20% identity and 40% similarity) small family of proteins whose members are now are known to also include delta-giardin (R.S. Khoury and H.G. Elmendorf, unpublished data; Jenkins *et al.*, 2009) and SALP-1 (Palm *et al.*, 2005), proteins that are also encoded by single-copy genes and that share both coiled-coil domains and a localization to the ventral disc (ultrastructural localization to the dorsal ribbons has not yet been confirmed). This family is not unique to *Giardia* but shares a similar level of sequence and structural conservation to SF-assemblin, a structural protein from *Chlamydomonas* that forms filaments at the flagellar base (Weber *et al.*, 1993). Given that there are many uncharacterized coiled-coil proteins in *Giardia* (see below), it remains an open questions whether any additional proteins may be members of this loose family and/or components of the ventral disc.

Gamma-giardin was originally identified in 1992 (Nohria *et al.*, 1992) and is also encoded by a single copy gene. However, little else is known about it since it has no known homologues in other cell types.

Novel coiled-coil proteins

A unique feature of *Giardia* is the presence of many novel large proteins with significant coiled-coil domains. Coiled-coil domains are structural motifs comprised of alpha-helices with a conserved pattern of hydrophobic and polar amino acids along the faces of the helices that facilitates oligomerization of the helices in an aqueous environment (Parry *et al.*, 2008). This requirement for oligomerization is believed to be the link between structure and function of coiled-coil proteins; hence, not only do the rod-like structural proteins of the cytoskeleton often possess coiled-coil domains, but so do transcription factors, cell surface receptors, and motor proteins.

What distinguishes the *Giardia* large coiled-coil proteins is the pairing of an ankyrin repeat motif with the coiled-coil domain (Elmendorf *et al.*, 2005). At least 32 members of this unique family are present in *Giardia*. Ankyrin repeats are ~33 amino acid helix-turn-helix motifs that mediate an astonishing diversity of protein–protein interactions, since the configuration of the repeats rather than the sequence of each repeat determines the binding specificity (Li *et al.*, 2006). Ankyrin repeats are one of the most common cellular motifs and this is true also for *Giardia*, where an estimated 10% of the genes contain ankyrin repeats (H.G. Elmendorf, unpublished data). Surprisingly, the pairing of ankyrin repeats with coiled-coil domains is not found in proteins of other cells. While the localization/function of most members of this family are not yet known, intriguing evidence about one member, GASP-180 (*Giardia* axoneme-associated protein), suggests that they may be an important class of largely uncharacterized cytoskeletal proteins. GASP-180 localizes to the proximal base of the anterior flagellar axonemes (Elmendorf *et al.*, 2005), and while ankyrin repeats have previously been identified as capable of binding to tubulin (Davis *et al.*, 1991) further research will be necessary to show whether the association of GASP-180 with axonemes occurs via a direct interaction with the microtubules. The combination of the localization and the structural and functional motifs of GASP-180 make it a strong candidate to participate in control of flagellar activity and motility.

In addition to this large family, there are several other even larger coiled-coil proteins in *Giardia* of presumed cytoskeletal function: a 101 kDa protein associated with the median body (Marshall and Holberton, 1993) and a 183 kDa protein (HPSR2) of undetermined cellular location (Marshall and Holberton, 1995).

Functions of the *Giardia* cytoskeleton

A recent focus by several laboratories has contributed important functional evidence to complement the recent genome studies on the role of the *Giardia* cytoskeleton in motility, attachment and detachment, and the mitotic and developmental cycles. Beyond this we know little of its function in basic cellular processes, although one recent study has demonstrated a conserved role for microfilaments in endocytosis (Hernandez *et al.*, 2007).

Motility

Cell motility is essential to the parasitic life cycle of *Giardia* trophozoites as they manoeuvre to appropriate locations along the walls of the small intestine during attachment. Four pairs of flagella control different aspects of the motility in the *Giardia* trophozoites. Studies performed using video microscopy have examined the role of the various flagella pairs during motility (Elmendorf *et al.*, 2003; Campanati *et al.*, 2002; Ghosh *et al.*, 2001). Forward motion of the parasite seems to be controlled primarily by the anterior flagella, which beat slowly and synchronously in a whip-like motion to propel the parasite forward in a push-and-glide motion, not unlike a slow breaststroke for a swimmer. The caudal flagella are largely present as intracellular axonemes with only a very short extracellular portion and do not beat; instead, they seem to act largely as a rudder to steer the parasite. The posterior-lateral flagella beat while the parasite is swimming, though their contribution to motility is not certain and they may act largely to stabilize the parasite. The ventral flagella beat continuously in a sinusoidal wave motion, but since they continue to beat while the parasite is attached it seems that their activity is unrelated to motility.

The means by which motility is regulated is less clear. Kinesin-13 has been identified as a regulator of axoneme length (Dawson *et al.*, 2007), but regulators of the actions of the flagella have not yet been identified. It is tempting to ascribe this role to the diverse axoneme-affiliated structures (dense fibrous rods, striated bands and funis) but no direct evidence links the structures and function. The localization of several kinases (one protein kinase A and two aurora kinase homologues) to the axoneme-affiliated structures provides a promising starting point for future studies in this area (Abel *et al.*, 2001; Davids *et al.*, 2008; Gibson *et al.*, 2006).

Attachment and detachment

The *Giardia* trophozoite must remain attached to the host small intestine in order to proliferate and subsequently cause disease. Because of the constant shedding of intestinal epithelial cells, there is a continual cycle of attachment/detachment that is required for maintenance of an infection. The most important cell organelle for the attachment and detachment process is the microtubule-enriched ventral disc, although the mechanism of attachment is still a subject of debate. Proposed mechanisms of attachment can be grouped into three major categories: (i) binding, (ii) clutching, and (iii) negative pressure mechanisms.

Binding models require close contact between the cell and substrate in order to generate attachment force via molecular (protein and sugar) level interactions (Gonzalez-Robles *et al.*, 1989). While there was early evidence for a charge or lectin/sugar interaction (reviewed in Elmendorf *et al.*, 2003), and more recent studies show some evidence of an 'adhesive attachment' (Erlandsen *et al.*, 2004) and a role for electrostatic forces (Roskens and Erlandsen, 2002), a recent innovative study clearly demonstrated that these forces cannot be the primary means of attachment by showing that the forces necessary to detach *Giardia* from differentially-treated glass surfaces (inert, positively charged, hydrophobic) were not significantly different (Hansen *et al.*, 2006).

Importantly, a very large number of studies have further demonstrated that the most significant determinant in parasite attachment is the dynamics of the parasite cytoskeleton (reviewed in Elmendorf *et al.*, 2003). Although there is not a complete consensus in the literature, the majority of the data indicate that it is the microfilament cytoskeleton that is most critical, as demonstrated by a specific reduction of attachment following parasite treatment with microfilament-disrupting agents (e.g. cytochalasins) over what is seen with microtubule-disrupting agents (e.g. colchicine or nocodazole). Collectively, these data direct us towards a force of attachment that can be generated by cytoskeletal dynamics.

The clutching model derives largely from ultrastructural studies showing that the edges of the ventral disc are usually deeply embedded into the epithelial cell layer. Indeed when the parasites detach, the imprint of the edges of the ventral disc is clearly left behind on the epithelial cell surface (e.g. Erlandsen and Feely, 1984). It is known that the ventral disc changes shape upon attachment by deepening its bowl-like shape, an action that is believe to be mediated by the microfilaments located at the lateral crest at the periphery of the disc. Video microscopy experiments by Campanati *et al.* (2002) provide support for ventral disc dynamics being involved in attachment/detachment by showing contraction of either the anterior or posterior end of the disc during parasite detachment. The localization of actin to the periphery of the ventral disc, among other sites, is consistent with a role for the cytoskeleton in a clutching model of attachment.

While the clutching model is consistent with much of the observed data, it cannot fully explain more recent observations concerning the effects of tonicity on parasite attachment and is probably only a contributor to – and not the primary force of – attachment. These recent observations strongly support a primary role for the negative pressure model of attachment – an interesting return to ideas first postulated by Soloviev and Holberton in the late 1960s and early 1970s. The initial suggestion derived from the observation that the ventral pair of flagella beat continuously in a sinusoidal wave motion and it was proposed that the beating of the flagella generated a flow of fluid out from under the ventral disc to create negative pressure (Holberton, 1973a, 1973b, 1974; Soloviev, 1968). However, subsequent studies have effectively argued against a role for the ventral flagella in attachment by demonstrating that the flagella beat whether the parasite is swimming, attached, or in the process of detaching (Hansen and Fletcher, 2008). With a refined understanding of the parasite ultrastructure, doubt is cast on the original biophysical calculations of the conditions necessary to achieve sufficient negative pressure. Instead, other researchers have proposed alternative functions for the flow generated by the flagella – such as enhancing nutrient availability for the parasite (Ghosh *et al.*, 2001).

Despite reconsideration of a role for the flagella, the negative pressure model has remained alive in the literature largely because of the intuitive visual analogy of the disc to a suction cup. Interestingly, even studies designed to provide evidence of 'adhesive attachment' (Erlandsen *et al.*, 2004) end up primarily supporting a negative pressure model by demonstrating that attachment was reduced by 84% if parasites were not allowed to attach to a smooth surface that would permit the generation of negative pressure. More direct evidence in support of a negative pressure model has come recently from a study by Hansen and colleagues who demonstrated that sharp changes in tonicity are sufficient to very rapidly detach *Giardia* cells, although they can remain attached in the face of smaller changes. The implications of this study are that *Giardia* can typically regulate the osmotic pressure underneath their ventral disc to maintain attachment under stable environmental conditions (Hansen and Fletcher, 2008). This model, while consistent with the available data, has not been directly confirmed and will require significant further study to illuminate the mechanism by which the parasites can regulate external osmolality and elucidate the full role of the cytoskeleton in this model.

Mitotic division

In contrast to our lingering uncertainty regarding the mechanistic function of the cytoskeleton in attachment, our enhanced understanding of the protein repertoire of *Giardia* from the genome project combined with sophisticated bench experiments has led to dramatic gains in our understanding of mitotic division in *Giardia*.

There has been a long-standing debate concerning the plane of cell division (reviewed in Elmendorf *et al.*, 2003), with multiple studies concluding that cytokinesis may be a highly irregular event (Benchimol, 2004b; Solari *et al.*, 2003). Two recent studies by Cande and colleagues (Sagolla *et al.*, 2006) and Nohynkova and colleagues (Tumova *et al.*, 2007) have used a series of technically-sophisticated approaches to resolve this debate. By GFP-labelling centromeres using a cenH3-GFP fusion protein and staining microtubules and the MTOC with anti-tubulin and anti-centrin antibodies respectively, Sagolla *et al.* (2006) were able to use three-dimensional deconvolution microscopy to visual the precise spatial and temporal relationship among cell constituents during the mitotic process. These studies led to the conclusion that the nuclei first align along the central longitudinal axis of the parasite, before the spindle fibres pull replicated sister chromatids towards the left and right sides of the parasite as each nucleus elongates laterally. As the cell begins to divide in a longitudinal plane (anterior to posterior) to produce daughter cells from the left and right halves of the parental cell, the nuclei divide to produce four nuclei – two within each lobe of the heart-shaped parasite. The parasite then completes cytokinesis by extending the plane of division, resulting in two daughter cells, each with copies of both parental nuclei. Looking instead at the replication of the ventral disc during cytokinesis in a complementary series of video, immunofluorescence, and transmission electron microscopy studies, Tumova *et al.* (2007) provide corroborating evidence for a longitudinal left-right plane of division by showing that the two daughter ventral discs form facing the dorsal surface at the left and right sides of the heart-shaped dividing parental cell in a mirror image symmetry.

Beyond the plane of cell division, a second unresolved question concerned the pattern of nuclear inheritance. While it has been clear from two previous studies that each daughter cell inherits a copy of each parental nucleus (Ghosh *et al.*, 2001; Yu *et al.*, 2002), there was disagreement about whether the left/right identity of the nuclei were maintained in the daughter cells (Yu *et al.*, 2002) or whether cytokinesis resulted in mirror image symmetry (Ghosh *et al.*, 2001). While the studies described above lent strong support to the mirror-image symmetry model, Sagolla *et al.* (2006) provided the definitive evidence using FISH to define nuclear inheritance patterns by visualizing episomally-maintained plasmids in one of the two nuclei. The authors showed that the daughter cells are mirror images of each other, with one daughter cell maintaining the same left/right orientation of the episome-labelled nucleus as the parent cell, while the other daughter cell switches the nuclear orientation.

A third open question concerned the role of spindle fibres in karyokinesis, with one study arguing that a spindle-fibre-independent mechanism was responsible for chromosome segregation (Solari *et al.*, 2003). Again, Sagolla *et al.* (2006) provide strong evidence for spindle-fibre-mediated chromosome segregation. The authors used high-resolution transmission electron microscopy to verify that the nuclear envelopes remain intact during cell division, with clear spindle fibres originating from the MTOC in the cytoplasm and penetrating the nuclear envelopes for a semi-open model of mitosis. A functional role for the spindle fibres was demonstrated by showing that treatment with the microtubule-stabilizing compound taxol results in elongated spindle fibres and a block to correct chromosome segregation.

One unexpected finding from the cell division studies concerns the highly ordered developmental progression of flagellar identity through successive rounds of cell division (Nohynkova *et al.*, 2006; Tumova *et al.*, 2007). Although flagellar development has been observed in other cell types (e.g. Melkonian *et al.*, 1987), it is perhaps particularly surprising in *Giardia* given the complexity of the axoneme-associated structures. How these structures are reassociated during mitosis and what signals regulate this complex developmental process remain to be identified.

This last point is a common theme regarding our current knowledge of the involvement of the cytoskeleton in mitotic cell division: while we now know a great deal more about the morphological aspects of the process, we still know relatively little about the regulatory mechanisms or detailed molecular processes beyond the tantalizing hints provided by the identification of the involvement of an aurora kinase (Davids *et al.*, 2008) and kinesin-13 (Dawson *et al.*, 2007) in cytoskeletal dynamics.

Encystation/excystation

Giardia encystation has been reviewed recently, describing our current morphological, cellular, biochemical and molecular understanding (Lauwaet *et al.*, 2007). Yet important gaps remain, and many of the mechanisms that coordinate responses to external signals remain enigmatic. The *Giardia* cytoskeleton changes dramatically during encystation and excystation, becoming largely irrelevant as cysts are non-motile and non-dividing, and the cyst wall assumes responsibility for structural maintenance. The flagella and disc are fragmented during encystation and stored as large pieces in the cytoplasm of the cyst and are rapidly reassembled in the newly emerging daughter trophozoites during excystation (Palm *et al.*, 2005; Sheffield and Bjorvat, 1977). The mechanism of fragmentation and stabilization of the fragments remains unclear, although perhaps particularly perplexing is the mechanism by which the parasite forms two new discs so rapidly from pieces of the single parental disc. Since the formation of daughter discs during mitosis happens *de novo* rather than from reorganization of the parental disc units, if our current understanding of the morphogenetic progression of disc formation during excystation is correct, then the machinery and mechanism involved must be wholly different in mitosis and excystation.

Introduction to *Trichomonas vaginalis*

Trichomonas vaginalis is the aetiological agent of the most prevalent non-viral sexually transmitted infection worldwide, trichomoniasis (Schwebke and Burgess, 2004). Over 170 million new cases of trichomoniasis are estimated to occur annually (Weinstock *et al.*, 2004; WHO, 2001). *T. vaginalis* colonizes the human urogenital tract, where it remains extracellular and causes lesions in the vaginal epithelium, leading to symptoms in infected women that range from inflammatory disease, fertility and pregnancy complications, to predisposition to HIV infection and cervical cancer (Schwebke and Burgess, 2004). In men, the infection is usually asymptomatic and self-limiting (Krieger *et al.*, 1993); however, *T. vaginalis* is a common cause of nongonococcal urethritis. Parasites may, however, colonize the testicles which can lead to infertility (Lloyd *et al.*, 2003). Colonization of the prostate is also correlated with an increase risk of prostate cancer (Sutcliffe *et al.*, 2006). Despite the prevalence of trichomoniasis, treatments are limited. The 5-nitroimidazole drugs, metronidazole and tinidazole, are the only drugs approved for therapy and cross-resistance to both is a growing problem (Narcisi and Secor, 1996).

The free-swimming form of *T. vaginalis* is an ovoid trophozoite cell (Benchimol, 2004). While cell dimensions vary with culture conditions and microscopic slide preparation, an average length of 9.7 μm and width of 7 μm have been reported (Honigberg and King, 1964). Cytopathogenicity has been shown to be correlated with attachment to host epithelium, whereupon a poorly understood cell structure rearrangement takes place transforming the cell to a flattened amoeboid shape (Gonzalez-Robles *et al.*, 1995). Amoeboid *T. vaginalis* cells form close associations with the plasma membrane of host cells and possible membrane fusion between parasite and host cells has been suggested (Furtado and Benchimol, 1998). Pseudocysts, a reversible compaction to a nonmotile cell form with internalized flagella but no cyst wall, have been observed in the related species *Tritrichomonas foetus* in response to environmental stress and treatment with drugs such as nocodazole (Granger *et al.*, 2000; Pereira-Neves *et al.*, 2003). Similar cellular rearrangements have been proposed in *T. vaginalis* (Pereira-Neves *et al.*, 2003), although debate continues over whether this is an artefact of *in vitro* cell culture or is biologically relevant.

T. vaginalis is a member of the Parabasalida, a diverse group of flagellated, often parasitic, protists characterized by an axostyle, striated parabasal rod structures associated with the Golgi, and closed mitosis (Hampl *et al.*, 2007; Honigberg, 1963). Notably, these organisms, like *Giardia*, lack mitochondria. In trichomonads, cytosolic glycolysis is instead supplemented with pyruvate fermentation in a double-membrane bound organelle termed the hydrogenosome (Dyall *et al.*, 2004; Grey, 2005; Lindmark and Müller, 1973; Müller, 1993). A related organelle called the mitosome is found in *Giardia* (Tovar *et al.*, 2003); however there is no evidence for its involvement in carbohydrate metabolism. Once thought to represent one of the earliest branching eukaryotic lineages along with *Giardia*, the placement of both *T. vaginalis* and *Giardia* in the eukaryotic tree of life remains a source of considerable debate (Adl *et al.*, 2005; Baldauf, 2003; Delgado-Viscogliosi *et al.*, 2000; Embley and Martin, 2006; Viscogliosi *et al.*, 1993).

Trichomonas vaginalis genome overview

The genome of a laboratory strain of *Trichomonas vaginalis* (G3) was sequenced recently (Aurrecoechea *et al.*, 2009; Carlton *et al.*, 2007). As the chapter by Carlton *et al.* in this book focuses on preliminary analyses of the *T. vaginalis* genome, this will be discussed only briefly here. The most intriguing properties to emerge from these analyses are the unpredicted large size of the genome and the common repetition of gene families. The genome, which is contained on 6 haploid chromosomes, is approximately 160 Mb in size, much larger than the genomes of other eukaryotic parasites, which typically range from 8 to 40 Mb (see the Eukaryotic Pathogens Bioinformatics Resource Centre, www.eupath.org). For example, the *T. vaginalis* genome is more than 14 fold larger than the *Giardia* genome. The unusually large size of the *T. vaginalis* genome has been attributed mainly to an extensive array of repetitive elements that make up approximately 65% of the genome sequence (Carlton *et al.*, 2007). In addition to a wide array of other transposable elements, nearly 1000 copies of the Mariner element were found, representing the first documented case of this transposon family in a protist (Silva *et al.*, 2005). The large amount of repetitive sequence has made full genome assembly problematic, and it is currently represented as a set of 17,920 contigs.

Further complicating genomic analysis, many protein-coding gene families are present in unusually high copy numbers. Examples of this phenomenon range from the most extreme examples of 927 protein kinase and 328 small GTPase genes, to 36 genes encoding the dynein heavy chain, and 19 and 12 kinesin and actin paralogues, respectively. The high degree of amino acid similarity among these protein family members often renders unique identification of expressed proteins by proteomics difficult. In total, the current 1.0 release of the *T. vaginalis* genome annotation contains close to 60,000 genes, nearly half of these with supporting EST evidence (Aurrecoechea *et al.*, 2009).

Several clinical isolates of *T. vaginalis* harbour a double-stranded RNA (dsRNA) virus (Benchimol *et al.*, 2002; Khoshnan and Alderete, 1993). The effect of this persistent viral infection on *T. vaginalis* is unknown, although some studies show alterations in protein expression in virus-laden clinical isolates that are not observed in virus-free isolates (Liu *et al.*, 1998; Wang *et al.*, 1987).

An introduction to the *Trichomonas vaginalis* cytoskeleton

Similar to *Giardia*, trichomonads are excellent models for studying a highly divergent and unique cytoskeleton in the context of eukaryotic cell biology. Our current knowledge of the *Trichomonas* cytoskeleton is based largely on ultrastructural studies employing light and electron microscopy (Benchimol, 2004; Benchimol, 2005; Honigberg and King, 1964; Smith and Stewart, 1966). Many of these studies were done on *Tritrichomonas foetus* and *Trichomonas gallinae*, which are bovine and chicken parasites, respectively. Overall ultrastructural similarity between these species is generally assumed, with minor differences manifesting, for example, in the number of anterior flagella. Only limited biochemical and molecular analyses were available prior to sequencing of the *T. vaginalis* genome. Where such analyses are available in the literature, results are discussed in the sections devoted to individual protein families and structural features.

Together with one recurrent and four anterior flagella, the basal bodies and associated filaments form a structure collectively known as the mastigont system in the anterior portion of the cell (Brugerolle, 1991; Mattern *et al.*, 1967). Several other structures are poorly characterized beyond their ultrastructural morphology, and often have unknown functions and molecular composition. The pelta and axostyle are known to be composed of microtubules (Lopes *et al.*, 2001). The pelta is viewed as a reinforcing structure for the flagellar canal and the axostyle has a putative general structural support role in the cell (Benchimol *et al.*, 2000; Honigberg and Brugerolle, 1990). Additionally, the axostyle has been proposed to contribute to karyokinesis in *T. foetus* (Ribeiro *et al.*, 2000). The costa is a striated fibre seen closely associated with the recurrent flagellum, possibly acting as structural support for the undulating membrane (Benchimol, 2004). The sigmoidal filaments are curved structures observed to connect basal bodies to the pelta. Parabasal filaments are also of unknown composition, but their close proximity to the Golgi apparatus has led to speculation that they have a structural support role for that organelle (Benchimol, 2004).

Morphology and structures of the *Trichomonas* cytoskeleton

Flagella

T. vaginalis flagellar axonemes exhibit the 9+2 microtubule organization that is typical of motile flagella (Benchimol, 2004). Four flagella emerge through a channel at the anterior end of the cell, extending from the complex network of basal bodies and other filaments localized in that region. One recurrent flagellum also extends from the anterior, making a sharp turn to run towards the posterior of the cell and remaining attached to the cell surface inside an associated undulating membrane. Several studies have examined flagellar beating in trichomonads (Monteiro-Leal *et al.*, 1995; Monteiro-Leal *et al.*, 1996). The anterior flagella beat in a ciliary fashion, with effective and recovery strokes. In *T. foetus,* one flagellum was found to beat in a different pattern relative to the other anterior flagella, extending forward during its effective stroke. It was speculated that this may be due to differences in how the emergence of the axonemes of individual flagella are coordinated by the underlying pelta structure (Monteiro-Leal *et al.*, 1995). In contrast, the recurrent flagellum was shown to exhibit a more traditional flagellar beat waveform at twice the frequency of the anterior flagella, leading to speculation that each may be coordinated and controlled via distinct processes.

Prior to the completion of the *T. vaginalis* genome, partial purification and immunological characterization of the anterior flagella was reported for both *T. vaginalis* and *T. foetus* (Jemilohun, 1998; Jemilohun *et al.*, 2005). Several prominent SDS-PAGE gel bands were enriched in these cell fractionations and immunological investigation of these putative flagellar antigens was performed, yet their molecular identification remains an open question.

The pelta–axostyle system

The pelta and axostyle are two unusual cytoskeletal structures in *T. vaginalis* that perform a putatively structural support role (Benchimol *et al.*, 2000). Both are known to be composed of microtubules.

In contrast to studies in other flagellated organisms, such as *Hexamastix, Honigbergiella, Monocercomonas, Pyrsonympha, Retortamonas, Spironympha,* and *Trichomonoides* species (Brugerolle, 2006; Hampl *et al.*, 2007; Smith and Arnott, 1974), the axostyle of *T. vaginalis* is a stable, non-contractile structure. It is a ribbon of microtubules that runs the length of the cell, beginning with a more open morphology at the anterior end of the cell and curling around to produce a progressively thinning tube as the structure extends towards the posterior end. The structure extends beyond the cell body at the posterior end, surrounded by the plasma membrane (Benchimol, 2004). Indeed, examination of *T. vaginalis* cultures by light microscopy reveals the presence of the axostyle protruding to form a spike-like structure at the base of the parasite.

Another possible function for the axostyle is participation in mitosis and cell division, although its precise role in these processes is controversial. The axostyle is closely associated with the nucleus, possibly connected by protein bridges (Benchimol, 2005). One study found the axostyle to remain stable and undergo duplication prior to mitosis (Ribeiro *et al.*, 2002). Others have reported that parental axostyles depolymerize during mitosis, and are assembled at the polar basal bodies prior to division (Viscogliosi and Brugerolle, 1994a).

The pelta structure is poorly defined. It is found closely associated with the anterior portion of the axostyle and the base of the anterior flagella, leading to a proposed structural support role in the cell (Mattern *et al.*, 1967; Warton and Honigberg, 1979). Having a genome sequence available opens the door for proteomic analyses of these unusual structures, which should provide critical clues regarding their function. Such analyses, however, await the ability to obtain subcellular fractions enriched in the pelta and/or axostyle.

Costa

The costa is a striated fibre of unknown protein composition that is unique to trichomonads. Leading out from the anterior mastigont system along the cell interior, it follows the length of the recurrent flagella and undulating membrane, leading to speculation that the costa fulfils another structural support role in the *T. vaginalis* cytoskeleton (Benchimol, 2004). Electron microscopy has strengthened this argument, showing several 'proteinaceous links' between the costa structure and the undulating membrane. Costae are also known for their strong association with hydrogenosomes in microscopy studies, so much so that the energy organelle was at one time termed the paracostal granule (Smith and Stewart, 1966).

Several studies have attempted to isolate and characterize the costa at the molecular level (Monteiro-Leal *et al.*, 1993; Viscogliosi and

Brugerolle, 1994b). While antibodies to costa-specific proteins were raised in those experiments, no molecular cloning or further characterization was reported. In this post-genomic era for *T. vaginalis*, the costa remains a structure ready for future analysis.

Proteins of the *Trichomonas* cytoskeleton

Tubulin and associated proteins

Immunofluorescence studies of diverse trichomonad species using fluorescent Taxol indicate the presence of extensive networks of microtubules (Lecke *et al.*, 2002; Vieira *et al.*, 2008). In agreement with this observation, the genes encoding components of the microtubule cytoskeleton are well conserved in the *T. vaginalis* genome (Table 5.5). All five members of the tubulin superfamily are present. The current annotation of the genome identifies seven genes encoding alpha-tubulin, five for beta-tubulin, two for gamma-tubulin, and one each for delta- and epsilon-tubulin. However, recent functional inference prediction by phylogenetic analysis identifed at least 17 full-length genes encoding one of the five tubulin subclasses, corroborated in most cases by existing PFAM domain homologies. These data indicate *T. vaginalis* to have a set of 8 alpha-tubulins, 6 beta-tubulins, and one copy each of gamma-, delta-, and epsilon-tubulin (R.D. Hayes and P. Johnson, unpublished data).

In most eukaryotes, alpha- and beta-tubulin form heterodimers that assemble into microtubules, while the other tubulins are localized to the basal body (McKean *et al.*, 2001). Gamma-tubulin is involved in filament nucleation at the centrosome, while the roles of delta- and epsilon-tubulin are less well understood. Although it remains to be verified, the conservation of the tubulin gene family in *T. vaginalis* implies similar roles for these proteins in this organism.

Genes encoding most well conserved microtubule-capping, severing, and decorating families are present. Several homologues encoding CLIP-170 and EB family proteins are found; these proteins are known to localize to the plus-end of polymerizing microtubules where they are involved in treadmilling (Carvalho *et al.*, 2003). Another plus-end-associated protein family, CLASP, is also present. CLASP binds to kinetochores and regulates microtubule dynamics during mitosis in other eukaryotes (Maiato *et al.*, 2005), although its role in *T. vaginalis* has not been investigated. Two classes of microtubule-associated proteins that bind along the microtubule filament area also found. MAP1 homologues are known to stabilize microtubules and also interact with actin filaments (Halpain and Dehmelt, 2006). MAP-65, first characterized in plants, likewise stabilizes microtubules and aids in higher-order microtubule bundling (Smertenko *et al.*, 2000). *T. vaginalis* encodes both subunits of katanin, a heterodimeric protein that severs microtubules in an ATP-dependent fashion (McNally and Vale, 1993).

Homologues of many common flagella-associated proteins are also present in the genome. These include PF16 and PF20, known to be involved in the assembly and stability of the central pair of microtubules (Smith and Lefebvre, 1997), as well as LF4, which is a MAP kinase first isolated from *Chlamydomonas reinhardtii* where it regulates flagellar length (Berman *et al.*, 2003). Four genes with high similarity to the intraflagellar transport particle proteins IFT20, IFT57 and IFT172 are found in the genome. This diverse class of proteins is required for assembly and length regulation in motile cilia and flagella, and is an integral component of molecular cargo delivery complexes trafficked by axonemal motors (Scholey, 2003). Genes encoding RIB43 and RIB72, known to be associated with microtubule protofilament ribbon structures in *C. reinhardtii* (Linck and Norrander, 2003), are also found in three and two copies, respectively.

The *T. vaginalis* genome encodes several homologues of components of the radial spoke, a T-shaped structure that extends from the central pair that is essential for dynein arm signalling and activity in *C. reinhardtii* (Yang *et al.*, 2006). Both outer dynein arm and inner dynein arm protein classes are thought to be necessary for motile flagella (Wickstead and Gull, 2007). While *T. vaginalis* possesses multiple genes similar to those that encode members of the outer dynein arm complex, no apparent homologues encoding inner dynein arm proteins could be found. It is possible that these sequences are highly divergent in *T. vaginalis*. Alternatively, as there is evidence

Table 5.5 Tubulin and tubulin-associated genes in the *T. vaginalis* genome with high similarity to known eukaryotic homologues

Tubulin cytoskeleton			
Alpha-tubulin	TVAG_196270 TVAG_206890 TVAG_312330	TVAG_359090 TVAG_360870	TVAG_448390 TVAG_467840 TVAG_519620
Beta-tubulin	TVAG_008680 TVAG_034440	TVAG_062880 TVAG_338530	TVAG_456920 TVAG_525430
Gamma-tubulin	TVAG_109820		
Delta-tubulin	TVAG_207590		
Epsilon-tubulin	TVAG_065740		
Microtubule-associated proteins			
Microtubule plus-end capping/associated			
CLIP-170	TVAG_011370	TVAG_314670	
EB family	TVAG_146390 TVAG_228400	TVAG_266680	TVAG_381030
CLASP family	TVAG_192360 TVAG_236560	TVAG_262230 TVAG_266910	TVAG_470180
Microtubule decorating			
MAP1 homologue	TVAG_239800	TVAG_486080	
MAP-65 family	TVAG_171600	TVAG_413720	
Microtubule Severing			
Katanin p60 subunit	TVAG_028490 TVAG_109940	TVAG_174770 TVAG_245050	TVAG_498860
Katanin p80 subunit	TVAG_499000		
Flagellar-associated			
PF16 homologue	TVAG_035230	TVAG_059360	TVAG_136270
PF20 homologue	TVAG_035910		
LF4 homologue	TVAG_057050	TVAG_272410	TVAG_328960
Intraflagellar Transport Particle Protein IFT20 homologue	TVAG_094430	TVAG_182360	
Intraflagellar Transport Particle Protein IFT57 homologue	TVAG_381870		
Intraflagellar Transport Particle Protein IFT172 homologue	TVAG_061760		
RIB43 family	TVAG_282620	TVAG_283250	TVAG_352320
RIB72 family	TVAG_159140	TVAG_492260	
Radial spoke protein family	TVAG_084460 TVAG_182860	TVAG_358840 TVAG_375180	TVAG_455950
Outer dynein arm docking complex	TVAG_013830 TVAG_026430	TVAG_414230	TVAG_567400
Trypanin family	TVAG_094850	TVAG_221750	TVAG_344090

Microtubule organizing centre			
GCP-2 homologue	TVAG_206250		
GCP (Spc97/Spc98 homologues)	TVAG_081520 TVAG_084700	TVAG_117310 TVAG_131250	TVAG_348130 TVAG_404950
Centrin	TVAG_017360 TVAG_026750 TVAG_029880 TVAG_038590 TVAG_062030 TVAG_068420 TVAG_120190 TVAG_119000	TVAG_164690 TVAG_174000 TVAG_238650 TVAG_252700 TVAG_306320 TVAG_311320 TVAG_311330 TVAG_320970	TVAG_335140 TVAG_357970 TVAG_367580 TVAG_376480 TVAG_391120 TVAG_404740 TVAG_416270 TVAG_476820

To explore the cytoskeletal gene content of *T. vaginalis*, BLAST (Altschul *et al.*, 1997) searches of known components in *Giardia lamblia* (Aurrecoechea *et al.*, 2009) and *Trypanosoma cruzi* (El-Sayed *et al., 2005)* were performed. In cases where no known homologue was annotated in *G. lamblia* or *T. cruzi*, BLAST searches were performed using other homologues from *S. cerevisiae* (Dwight *et al.*, 2004), GeneDB (Hertz-Fowler *et al.*, 2004), or NCBI refseq records (Pruitt *et al.*, 2005). Typically, an e-value cut-off value of e-20 or lower was employed to indicate significant homology. Matches in the *T. vaginalis* genome sequence were compiled, cross-referenced with the draft annotation, and reciprocal BLAST against the full GenBank database (Benson *et al.*, 2005) was performed to verify protein family homology. Due to the incomplete genomic assembly, several matches to annotated pseudogenes or partial sequence in degenerate contigs were removed from further consideration. The tubulin assignments have been updated due to unpublished tubulin phylogenetic analyses.

Tubulin-tyrosine ligase (46 homologues) is not included in this table.

that canonical inner dynein arms are not required for motile flagella in at least one species, the diatom *Thalassiosira pseudonana* (Wickstead and Gull, 2007), these structures may also be absent in *T. vaginalis*.

The dynein regulatory complex (DRC), a group of proteins well studied in *C. reinhardtii*, *Trypanosoma brucei*, and other kinetoplastids (Hill, 2003; Rupp and Porter, 2003), is critical for flagellar beating coordination. The most extensively characterized component of the DRC, trypanin, has been shown to be essential for motility in *T. brucei* (Ralston and Hill, 2006). Although this complex has not been investigated in *T. vaginalis*, three trypanin genes are present in the genome. Tubulin post-translational modifications in *T. vaginalis* have been studied using immunofluorescence and microscopy and there is evidence for acetylation, glutamylation, non-tyrosination, and non-glycylation of these proteins (Delgado-Viscogliosi *et al.*, 1996; Lopes *et al.*, 2001; Noël *et al.*, 2001). Several immunolabeling experiments have shown that axostyle microtubules are composed of acetylated and glutamylated tubulins, providing evidence that this is a stable structure (Delgado-Viscogliosi *et al.*, 1996; Lopes *et al.*, 2001). Pelta tubulin isoforms were also found to be glutamylated, likewise indicating this to be a stable structure (Lopes *et al.*, 2001). While *T. vaginalis* microtubules have been reported to be non-tyrosinated, the genome contains a total of 46 genes encoding TTL, the enzyme responsible for reattaching this tyrosine residue at the C-terminal end of alpha-tubulin (MacRae, 1997).

Microtubule-organizing centre

Gamma-tubulin and a host of accessory proteins regulate microtubule filament nucleation at centrosomes and basal bodies (McKean *et al.*, 2001; Raynaud-Messina and Merdes, 2007). Based on sequence analyses, *T. vaginalis* gamma-tubulin ring complex components appear to be highly divergent and difficult to classify by sequence similarity alone. A single gene for gamma–tubulin complex component 2 (GCP-2) exists, but no GCP-3 homologues were discovered. Both components are necessary for gamma-tubulin small complex assembly (Raynaud-Messina and Merdes, 2007). Six genes encoding members of the yeast Spc97/Spc98 protein family, homologues of GCP-2 and GCP-3, were found that may compensate for the lack of a direct GCP-3 homologue in *T. vaginalis*, but this has not been tested directly.

An unusual set of 24 genes encoding centrins, known to localize to centrioles and play an important role in basal body duplication (Azimzadeh and Bornens, 2007), are present in the genome. Centrin in *T. vaginalis* has been localized to basal bodies, the undulating membrane, hook fibres and sigmodial filaments at the pelta-axostyle junction, and the basal body arms that anchor the flagella to the plasma membrane (Brugerolle *et al.*, 2000). A role for centrin in cell motility has been proposed for *G. lamblia* (Correa *et al.*, 2004), although this has not been directly observed in *T. vaginalis*.

Actin and associated proteins

Based on homology to known gene families, genes encoding proteins of the actin cytoskeleton are well represented in the genome (Carlton *et al.*, 2007; Table 5.6). However, as in the *G. lamblia* genome, there is an apparent lack of myosin genes in *T. vaginalis*. The absence of this motor protein in both *Giardia* and *Trichomonas* is striking, as it is otherwise highly conserved in eukaryotes. There are 12 homologues of actin genes in the genome, with 11 genes annotated as encoding 'actin family proteins.' The actin microfibrillar network is associated with amoeboid cell adhesion, based on phalloidin and antibody staining immunofluorescence experiments (Brugerolle *et al.*, 1996).

The actin network of a cell is regulated dynamically by a host of associated protein families known to control actin filament production (dos Remedios *et al.*, 2003). Unlike *Giardia*, homologues for many of these genes appear to be present in *T. vaginalis*; however these gene families have not been studied extensively in this organism. There are several copies of genes encoding both subunits of the heterodimeric CapZ protein, which binds to the barbed-end of filaments, inhibiting the addition of new actin monomers (Casella *et al.*, 1986). The ADF/Cofilin family regulates actin filament dynamics at the plus-end, including actin monomer disassociation leading to treadmilling and steady-state filament turnover. If filaments are capped at the barbed end, differences in cofilin kinetics can result in either filament nucleation or depolymerization (dos Remedios *et al.*, 2003). Profilin, a monomeric actin binding protein, enhances the rate of filament treadmilling in the presence of cofilin by promoting the addition of actin monomers at the barbed-end (Didry *et al.*, 1998). Coactosin, although demonstrated to have structural homology to the cofilin family (Hellman *et al.*, 2004), instead promotes actin depolymerization via competitive binding at filament ends, interfering with capping proteins (Rohrig *et al.*, 1995). The gelsolin superfamily, known to regulate actin networks through processes of binding to and severing F-actin, capping the barbed end of filaments, and nucleating assembly of new filaments, is represented by three putative homologues in the *T. vaginalis* genome based on sequence similarity to the gelsolin repeat motif (Silacci *et al.*, 2004).

The Arp2/3 complex is an important regulator of actin filament nucleation and bundling (Machesky and Gould, 1999). Typically, this complex is composed of seven subunits, all of which appear to be encoded in the *T. vaginalis* genome except for ARPC2 and ARPC5. The absence of these may be explained by high sequence divergence of the *T. vaginalis* homologues or poor genome annotation, although there is precedence for a lack of ARPC5 in otherwise stable Arp2/3 complexes that promote filament nucleation poorly (Pollard and Beltzner, 2002). The presence of multiple genes encoding members of the Scar/WASp family, an important activator of the Arp2/3 complex, may also compensate (Machesky *et al.*, 1999). There is one gene for coronin, a protein that binds actin filaments and inhibits the activity of Arp2/3 complex (Uetrecht and Bear, 2006). The *T. vaginalis* genome also encodes several homologues of formin, another promoter of actin filament nucleation that operates independent of the Arp2/3 complex (Evangelista *et al.*, 2003). Alpha-actinin, an actin filament crosslinking protein, has been

Table 5.6 Actin and actin-associated genes in the *T. vaginalis* genome with high similarity to known eukaryotic homologues

Actin cytoskeleton			
Actin	TVAG_054030 TVAG_090470 TVAG_149090 TVAG_150270	TVAG_160060 TVAG_172680 TVAG_200190 TVAG_247170	TVAG_249200 TVAG_310030 TVAG_337240 TVAG_485210
Actin family protein	TVAG_027630 TVAG_043970 TVAG_094140 TVAG_161530	TVAG_167330 TVAG_189860 TVAG_225210 TVAG_276270	TVAG_354260 TVAG_434970 TVAG_495500
Actin-associated proteins			
CapZ (barbed-end capping protein)			
CapZ-alpha	TVAG_212270	TVAG_470230	TVAG_470240
CapZ-beta	TVAG_094060 TVAG_261590	TVAG_271840 TVAG_396060	TVAG_398750
Actin filament dynamics			
ADF/Cofilin	TVAG_192620 TVAG_295380	TVAG_295390	TVAG_402260
Profilin	TVAG_100620 TVAG_137920	TVAG_159160 TVAG_319300	TVAG_459130 TVAG_492250
Coactosin	TVAG_150940 TVAG_203740	TVAG_228160	TVAG_402260
Gelsolin	TVAG_237690	TVAG_244120	TVAG_496220
Formin	TVAG_029100 TVAG_348260	TVAG_397270 TVAG_436740	TVAG_457060
Actin filament bundling			
Alpha-actinin	TVAG_156680 TVAG_190450	TVAG_239310 TVAG_247460	TVAG_260390
Fimbrin	TVAG_116370	TVAG_351310	
Arp2/3 complex and related proteins			
ARP2	TVAG_354020	TVAG_354040	
ARP3	TVAG_371880		
ARPC1	TVAG_035180		
ARPC3	TVAG_070930	TVAG_083290	TVAG_205590
ARPC4	TVAG_234160	TVAG_470160	
Scar/WASp	TVAG_078700 TVAG_098540	TVAG_402650	TVAG_454160
Coronin	TVAG_021420		

demonstrated to localize to the cell periphery upon host cell attachment, and is hypothesized to be an integral component of the cytoskeletal rearrangement from free-swimming ovoid cell to the amoeboid form (Fiori *et al.*, 1999); there are five genes encoding this protein in *T. vaginalis.*

The genome also encodes homologues of fimbrin, another crosslinker that forms tight filament bundles and has been localized to filopodia, membrane ruffles, and nonintestinal microvilli in metazoans (Bretscher and Weber, 1980).

Motor proteins

The dynein family of motors and associated proteins is very well represented in the *T. vaginalis* genome (Table 5.7). A rather large set of 36 dynein heavy chains are in evidence, along with multiple copies of intermediate chains, and axonemal, cytoplasmic, and Roadblock domain light chains. Roadblock domain dyneins are known to be localized cytoplasmically in other eukaryotes (Susalka *et al.*, 2002) and, together with *T. vaginalis* cytoplasmic dynein light chain homologues, comprise an unusually large set of 15 dynein light chains that may aid the cell in compensation for the lack of myosin-directed trafficking. Although the presence of cytoplasmic dynein in *T. vaginalis* remains to be confirmed, if present this feature may set it apart from *Giardia.* As discussed above there is as yet no consensus on whether cytoplasmic dyneins are present in *Giardia*.

T. vaginalis has a total of 19 genes encoding the kinesin heavy chain (Table 5.7). Application of Hidden Markov Model search techniques used in a large scale phylogenetic survey of kinesin subfamilies (Wickstead and Gull, 2006) has enabled subclassification of *T. vaginalis* sequences beyond that available in the preliminary annotation (R.D. Hayes and P. Johnson, unpublished data).

Subfamilies that were expected include those kinesins known to be associated with cilia and flagella, kinesin-2 and kinesin 9 (Wickstead and Gull, 2006). Kinesin-2 is an intraflagellar transport motor (Marszalek and Goldstein, 2000). Kinesin-9 functionality is less well characterized, although the *C. reinhardtii* homologue localizes to the axoneme central pair (Yokoyama *et al.*, 2004). Several homologues of kinesin-6 and kinesin-7, known to be involved in mitotic processes, including cytokinesis and spindle polarity in metazoa, were found as well (Miki *et al.*, 2005).

One gene encoding a homologue of the kinesin 4/10 subfamily and two encoding kinesin-8 homologues were found; both classes are implicated in organellar transport and chromosome/nuclear migration (Pereira *et al.*, 1997; Sekine *et al.*, 1994; Zhang *et al.*, 1990). Most intriguing was the identification of a kinesin-14 gene, a subfamily characterized by a C-terminal motor domain that results in movement along microtubules with opposite polarity relative to other kinesins (McDonald *et al.*, 1990). Together, these subfamilies may help compensate for the lack of myosin in *T. vaginalis* in directing cargo around the cytoplasm.

The metazoan kinesin holoenzyme is built from two homodimers, one composed of two copies of the heavy chain and one made from two copies of the light chain kinesin proteins (DeBoer *et al.*, 2008). There is evidence that cargo specificity may be related to the combinations of heavy and light chains that are assembled (Vale, 2003). *T. vaginalis* has only one gene annotated as a putative light chain kinesin, sharing the tetratricopeptide repeat (TPR) motif found in other light chains and many other protein families that mediate protein–protein interactions (Blatch and Lassle, 1999). A set of 79 genes exists in the *T. vaginalis* genome encoding proteins that contain one or more Pfam TPR motifs; many of these have ambiguous annotation and it is possible that these may represent more divergent kinesin light chain genes. Additionally, two genes are annotated as 'kinesin-associated,' but their function in *T. vaginalis* is currently unknown. Further investigation is warranted.

As mentioned above, and similar to observations in *Giardia* (Morrison *et al.*, 2007), *T. vaginalis* apparently lacks myosin (Carlton *et al.*, 2007) indicating that cytoplasmic transport via the association of myosin with an actin network does not exist in these organisms. Although it cannot be excluded rigorously that these genes were inadvertently lost during the sequencing and assembling of these genomes, it is unlikely. In the case of *T. vaginalis*, the presence of large kinesin and cytoplasmic dynein protein families may compensate for the absence of myosin, as examples exist of cytosolic organelle transport primarily via these microtubular motors, particularly in *D. discoideum* and filamentous fungi (Vale, 2003).

Unique coiled-coil proteins

The *T. vaginalis* genome project uncovered a large collection of 62 genes annotated as encoding

Table 5.7 Motor protein genes in the *T. vaginalis* genome with high similarity to known eukaryotic homologues

Motor proteins			
Dyneins			
Light chain (axonemal)	TVAG_166690	TVAG_357850	TVAG_453920
	TVAG_240770	TVAG_432720	TVAG_481690
	TVAG_253750	TVAG_448610	
Light chain (cytoplasmic)	TVAG_020400	TVAG_164580	TVAG_390830
	TVAG_134920	TVAG_312170	TVAG_420370
Light chain (roadblock domain)	TVAG_050160	TVAG_207270	TVAG_442810
	TVAG_116140	TVAG_292720	TVAG_458050
	TVAG_160070	TVAG_354740	TVAG_486650
	TVAG_202840		
Intermediate chain	TVAG_077930	TVAG_237760	TVAG_249070
	TVAG_151730		
Heavy chain	TVAG_006480	TVAG_162980	TVAG_365640
	TVAG_011080	TVAG_174850	TVAG_376510
	TVAG_011310	TVAG_184370	TVAG_410920
	TVAG_013770	TVAG_184810	TVAG_430000
	TVAG_013270	TVAG_188120	TVAG_432890
	TVAG_032910	TVAG_207440	TVAG_440130
	TVAG_034900	TVAG_211860	TVAG_456970
	TVAG_068390	TVAG_221220	TVAG_460730
	TVAG_105170	TVAG_236550	TVAG_463220
	TVAG_116100	TVAG_262920	TVAG_471380
	TVAG_118950	TVAG_295280	TVAG_490910
	TVAG_154270	TVAG_365430	TVAG_497260
Beta chain	TVAG_004750	TVAG_263900	
Kinesin-associated			
Kinesin light chain	TVAG_348280		
Kinesin-associated protein	TVAG_230110	TVAG_311270	
Kinesin heavy chains			
Kinesin-2	TVAG_120610	TVAG_273680	TVAG_398700
	TVAG_267940	TVAG_299730	
Kinesin-4/10	TVAG_044330		
Kinesin-6	TVAG_030740	TVAG_258510	TVAG_380380
	TVAG_117650		
Kinesin-7	TVAG_143640	TVAG_431110	
Kinesin-8	TVAG_395330	TVAG_447880	
Kinesin-9	TVAG_164680	TVAG_291770	TVAG_355020
	TVAG_257720		
Kinesin-14	TVAG_200810		

coiled-coil proteins and at least 1280 encoding proteins containing ankyrin repeats, similar to those found in *Giardia*. Coiled-coil domains are found in a diverse set of proteins, some with structural roles (Burkhard *et al.*, 2001). The characteristic heptad amino acid repeat motif forms sets of alpha-helices and promotes self-oligomerization. The *T. vaginalis* sequences may represent a new, divergent set of intermediate filament proteins with structural roles similar to laminin or vimentin in other eukaryotes. Intriguingly, structural regions with coiled-coil domain similarity exist within all molecular motor families, so these sequences might also represent novel motor proteins that would help compensate for the lack of myosin in *T. vaginalis*. The ankyrin repeat is one of the most common protein motifs, forming a helix-turn-helix structure that facilitates protein–protein interactions (Li *et al.*, 2006). As described above, a novel class of cytoskeletal proteins in *Giardia* with both ankyrin and coiled-coil domains (Elmendorf *et al.*, 2005) underscores the importance of further scrutiny of this unusually large group of ankyrin repeat proteins in *Trichomonas*.

Functions of the *Trichomonas* cytoskeleton

Mitotic division

Similar to *Giardia*, *Entamoeba histolytica*, *Trypanosoma* spp. and certain fungi (Chavez-Munguia *et al.*, 2006; De Souza and Osmani, 2007; Ersfeld *et al.*, 1999; Sagolla *et al.*, 2006), *Trichomonas* and other trichomonads undergo a variant of mitosis where the nuclear envelope remains intact. Additionally, the *Trichomonas* mitotic spindle remains entirely extranuclear, making this a true closed mitosis (also termed pleuromitosis). The *Trichomonas* cytoskeleton has been shown to be duplicated prior to the onset of mitosis. As previously discussed, there is conflicting evidence for whether the axostyle remains stable during mitosis, although a recent study reports that a stable axostyle plays a role in aiding karyokinesis (Ribeiro *et al.*, 2002). In that report, daughter cell axostyles were shown with posterior tips aligned during telophase prior to cell division.

Spindle microtubules are described from their localization as pole-to-pole, pole-to-nucleus, or pole-to-cytosol (Ribeiro *et al.*, 2002). The pole-to-pole spindle forms once the basal bodies migrate to opposite poles and potentially this spindle can interact with the surface of the nuclear membrane. Pole-to-nucleus spindle microtubules exhibit a (7 + 1) configuration, and extend from the poles to the nucleus, but are not observed piercing the intact nuclear envelope. Instead, in *Trichomonas* these microtubules appear to connect to kinetochores at the nuclear membrane in attachment zones of unknown composition and are involved in the separation of daughter nuclei in the transition from anaphase to telophase. Pole-to-cytosol spindle microtubules are analogous to aster microtubules found in open mitosis. No intranuclear spindle microtubules have been found in electron microscopy studies of *Trichomonas*. Following cytoskeleton duplication, the pole-to-pole spindle polymerizes during prophase with a novel curved shape. Basal body separation continues until the duplicated nuclei arrive at their respective poles. Two axostyles form the cell into a heart shape at intermediate steps, pointing their posterior ends together. Karyokinesis follows at the onset of anaphase, accompanied by continued elongation and straightening of the pole-to-pole spindle. By late telophase, the pole-to-pole spindle has completely straightened. Daughter cells separate at their posterior ends during cytokinesis, a process thought to be mediated by flagellar motility (Ribeiro *et al.*, 2002).

Conclusions

Giardia and *Trichomonas* are useful models to study many aspects of the cytoskeleton; in addition to containing cytoskeletal components that are conserved in most eukaryotic cells, they contain many unusual structures that are poorly characterized. Classic structures such as the flagellar axonemes have been studied, primarily by microscopy, and appear to contain proteins thought to be conserved in motile flagella. At the same time, biochemical analyses aimed at defining the protein compositions and functions of the unique cytoskeletal structures are still largely unfinished.

Despite differences in their morphology, the general molecular makeup of the cytoskeleton is similar in *Giardia* and *Trichomonas*. Based on genome sequence analysis the microtubule proteome is well-conserved in both cells and includes all five tubulin isoforms, as well as proteins involved in microtubule modification, organization, and assembly. However, the classical microfilament proteome is underrepresented in *Giardia* compared to *Trichomonas*. The *Giardia* genome contains a single actin gene and does not encode other classical microfilament proteins while in *Trichomonas* there are 12 actin genes and a wide range of actin-related, actin-binding, and microfilament-associated proteins that regulate microfilament dynamics. Interestingly, myosin genes appear to be lacking in the genomes of both parasites, raising the possibility that either novel, divergent proteins substitute functionally for myosin, or altered cytoskeletal dynamics accommodate its absence. In addition to the classical cytoskeletal systems, *Giardia* and *Trichomonas* also possess numerous members of an otherwise unique cytoskeleton protein family that pairs coiled-coil domains with ankyrin repeats. Their function remains an open question and points to many interesting years ahead for research into the *Giardia* and *Trichomonas* cytoskeletons.

With the genomic sequences of *Giardia* and *Trichomonas* in hand, several large-scale, high-throughput techniques can now be employed to further elucidate the foundations of its cytoskeletal cell biology. Proteomic investigation of cell fractionations via mass spectrometry is a clear example of a way to further characterize the protein component of filaments and structures of as yet unknown composition. Proteomics coupled with genomic techniques such as microarrays and SAGE would also allow for verification of the genes present in the current annotation.

The ambiguity in the placement of *Giardia*, *Trichomonas* and other excavates in the tree of eukaryotic life (Baldauf, 2003) makes these organisms clear candidates for additional phylogenetic analyses. Work touched on briefly here involving generation of consensus trees for the dynein and kinesin motor families (Miki *et al.*, 2005; Wickstead and Gull, 2006; Wickstead and Gull, 2007) now can be extended to the homologues in *Giardia* and *Trichomonas*. As discussed here, the presence of many other families of both conserved and novel cytoskeletal genes provide numerous candidates for future informative phylogenetic and biochemical analyses.

The cytoskeleton is a striking and important feature of flagellated protozoan cells such as *Giardia* and *Trichomonas*. Because of their unique cytoskeletal systems, *Giardia* and *Trichomonas* have the potential to be useful as model organisms for the study of the cytoskeleton; the extreme divergence of their cytoskeleton protein composition, inferred by experimental and genome analysis, suggests that we still have much to learn about cytoskeletal evolution and organization in eukaryotic cells. Additionally, insights into cytoskeletal function hold the promise of giving us new insights into the pathogenesis of these two important human parasites, and possibly also tools in the fight against them.

Acknowledgements

We thank Holly Goodson for the Arp phylogeny shown in Fig. 5.2, and Scott Dawson for fruitful discussions concerning the dynein and kinesin family in *Giardia*.

References

Abel, E.S., Davids, B.J., Robles, L.D., Loflin, C.E., Gillin, F.D., and Chakrabarti, R. (2001). Possible roles of protein kinase A in cell motility and excystation of the early diverging eukaryote *Giardia lamblia*. J. Biol. Chem. *276*, 10320–10329.

Adam, R.D. (2001). Biology of *Giardia lamblia*. Clin. Microbiol. Rev. *14*, 447–475.

Adl, S.M., Simpson, A.G., Farmer, M.A., Andersen, R.A., Anderson, O.R., Barta, J.R., Bowser, S.S., Brugerolle, G., Fensome, R.A., Fredericq, S., James, T.Y., *et al.* (2005). The new higher level classification of eukaryotes with emphasis on the taxonomy of protists. J. Eukaryot. Microbiol. *52*, 399–451.

Altschul, S.F., Madden, T.L., Schaffer, A.A., Zhang, J., Zhang, Z., Miller, W., and Lipman, D.J. (1997). Gapped BLAST and PSI BLAST: a new generation of protein database search programs. Nucleic Acids Res. *25*, 3389–3402.

Amos, L.A., van den Ent, F., and Löwe, J. (2004). Structural/functional homology between the bacterial and eukaryotic cytoskeletons. Curr. Opin. Cell Biol. *16*, 24–31.

Aurrecoechea, C., Brestelli, J., Brunk, B.P., Carlton, J.M., Dommer, J., Fischer, S., Gajria, B., Gao, X., Gingle, A., Grant, G., Harb, O.S., *et al.* (2009). *Giardia*DB and TrichDB: integrated genomic resources for the eukaryotic protist pathogens *Giardia lamblia* and *Trichomonas vaginalis*. Nucleic Acids Res. *37*, D526–530.

Ausmees, N., Kuhn, J.R., and Jacobs-Wagner, C. (2003). The bacterial cytoskeleton: an intermediate filament-like function in cell shape. Cell *115*, 705–713.

Azimzadeh, J., and Bornens, M. (2007). Structure and duplication of the centrosome. J. Cell Sci. *120*, 2139–2142.

Baldauf, S.L. (2003). The deep roots of eukaryotes. Science *300*, 1703–1706.

Bauer, B., Engelbrecht, S., Bakker-Grunwald, T., and Scholze, H. (1999). Functional identification of alpha 1-giardin as an annexin of *Giardia lamblia*. FEMS Microbiol. Lett. *173*, 147–153.

Belhadri, A. (1995). Presence of centrin in the human parasite *Giardia*: a further indication of its ubiquity in eukaryotes. Biochem. Biophys. Res. Commun. *214*, 597–601.

Benchimol, M. (2004). Trichomonads under microscopy. Microsc. Microanal. *10*, 528–550.

Benchimol, M. (2004a). Participation of the adhesive disc during karyokinesis in *Giardia lamblia*. Biol. Cell *96*, 291–301.

Benchimol, M. (2004b). Mitosis in *Giardia lamblia*: multiple modes of cytokinesis. Protist *155*, 33–44.

Benchimol, M. (2005). New ultrastructural observations on the skeletal matrix of *Tritrichomonas foetus*. Parasitol. Res. *97*, 408–416.

Benchimol, M., Diniz, J.A., and Ribeiro, K. (2000). The fine structure of the axostyle and its associations with organelles in trichomonads. Tissue Cell *32*, 178–187.

Benchimol, M., Monteiro, S., Chang, T.H., and Alderete, J.F. (2002). Virus in *Trichomonas* – an ultrastructural study. Parasitol. Int. *51*, 293–298.

Benchimol, M., Piva, B., Campanati, L., and de Souza, W. (2004). Visualization of the funis of *Giardia lamblia* by high-resolution field emission scanning electron microscopy – new insights. J. Struct. Biol. *147*, 102–115.

Benson, D.A., Karsch-Mizrachi, I., Lipman, D.J., Ostell, J., and Wheeler, D.L. (2005). GenBank. Nucleic Acids Res. 33, D34–38.

Berman, S.A., Wilson, N.F., Haas, N.A., and Lefebvre, P.A. (2003). A novel MAP kinase regulates flagellar length in *Chlamydomonas*. Curr. Biol. *13*, 1145–1149.

Blatch, G.L., and Lassle, M. (1999). The tetratricopeptide repeat: a structural motif mediating protein–protein interactions. Bioessays *21*, 932–939.

Bouzid, M., Steverding, D., and Tyler, K.M. (2008). Detection and surveillance of waterborne protozoan parasites. Curr. Opin. Biotechnol. *19*, 302–306.

Bretscher, A., and Weber, K. (1980). Fimbrin, a new microfilament-associated protein present in microvilli and other cell surface structures. J. Cell Biol. *86*, 335–340.

Brugerolle, G. (1991). Flagellar and cytoskeletal systems in amitochondrial flagellates: Archamoeba, Metamonada and Parabasala. Protoplasma *164*, 70–90.

Brugerolle, G. (2006). The symbiotic fauna of the African termite *Hodotermes mossambicus* identification of four flagellate species of the genera *Spironympha*, *Trichomonoides* and *Retortamonas*. Parasitol. Res. *98*, 257–263.

Brugerolle, G., Bricheux, G., and Coffe, G. (1996). Actin cytoskeleton demonstration in *Trichomonas vaginalis* and in other trichomonads. Biol. Cell *88*, 29–36.

Brugerolle, G., Bricheux, G., and Coffe, G. (2000). Centrin protein and genes in *Trichomonas vaginalis* and close relatives. J. Eukaryot. Microbiol. *47*, 129–138.

Burkhard, P., Stetefeld, J., and Strelkov, S.V. (2001). Coiled coils: a highly versatile protein folding motif. Trends Cell Biol. *11*, 82–88.

Campanati, L., Holloschi, A., Troster, H., Spring, H., de Souza, W., and Monteiro-Leal, L.H. (2002). Video-microscopy observations of fast dynamic processes in the protozoon *Giardia lamblia*. Cell. Motil. Cytoskeleton *51*, 213–224.

Campanati, L., Troester, H., Monteiro-Leal, L.H., Spring, H., Trendelenburg, M.F., and de Souza, W. (2003). Tubulin diversity in trophozoites of *Giardia lamblia*. Histochem. Cell Biol. *119*, 323–331.

Carlton, J.M., Hirt, R.P., Silva, J.C., Delcher, A.L., Schatz, M., Zhao, Q., Wortman, J.R., Bidwell, S.L., Alsmark, U.C., Besteiro, S., Sicheritz-Ponten, T., *et al.* (2007). Draft genome sequence of the sexually transmitted pathogen *Trichomonas vaginalis*. Science *315*, 207–212.

Carvalho, P., Tirnauer, J.S., and Pellman, D. (2003). Surfing on microtubule ends. Trends Cell Biol. *13*, 229–237.

Casella, J.F., Maack, D.J., and Lin, S. (1986). Purification and initial characterization of a protein from skeletal muscle that caps the barbed ends of actin filaments. J. Biol. Chem. *261*, 10915–10921.

Chavez-Munguia, B., Tsutsumi, V., and Martinez-Palomo, A. (2006). *Entamoeba histolytica*: ultrastructure of the chromosomes and the mitotic spindle. Exp. Parasitol. *114*, 235–239.

Chen, M., and Shen, X. (2007). Nuclear actin and actin-related proteins in chromatin dynamics. Curr. Opin. Cell Biol. *19*, 326–330.

Cleveland, D.W. (1999). Introduction to tubulin and associated proteins, In Guidebook to the Cytoskeletal and Motor Proteins, T. Kreis, and R. Vale, eds. (Oxford: Oxford University Press), pp. 189–196.

Correa, G., Morgado-Diaz, J.A., and Benchimol, M. (2004). Centrin in *Giardia lamblia* – ultrastructural localization. FEMS Microbiol. Lett. *233*, 91–96.

Crossley, R., and Holberton, D. (1985). Assembly of 2.5 nm filaments from giardin, a protein associated with cytoskeletal microtubules in *Giardia*. J. Cell Sci. *78*, 205–231.

Crossley, R., and Holberton, D.V. (1983a). Characterization of proteins from the cytoskeleton of *Giardia lamblia*. J. Cell Sci. *59*, 81–103.

Crossley, R., and Holberton, D.V. (1983b). Selective extraction with Sarkosyl and repolymerization *in vitro* of cytoskeleton proteins from *Giardia*. J. Cell Sci. *62*, 419–438.

Crossley, R., Marshall, J., Clark, J.T., and Holberton, D.V. (1986). Immunocytochemical differentiation of microtubules in the cytoskeleton of *Giardia lamblia* using monoclonal antibodies to alpha-tubulin and polyclonal antibodies to associated low molecular weight proteins. J. Cell Sci. *80*, 233–252.

Davids, B.J., Williams, S., Lauwaet, T., Palanca, T., and Gillin, F.D. (2008). *Giardia lamblia* aurora kinase: a regulator of mitosis in a binucleate parasite. Int. J. Parasitol. *38*, 353–369.

Davis, L.H., Otto, E., and Bennett, V. (1991). Specific 33-residue repeat(s) of erythrocyte ankyrin associate with the anion exchanger. J. Biol. Chem. *266*, 11163–11169.

Dawson, S.C., Sagolla, M.S., Mancuso, J.J., Woessner, D.J., House, S.A., Fritz-Laylin, L., and Cande, W.Z. (2007). Kinesin-13 regulates flagellar, interphase, and mitotic microtubule dynamics in *Giardia intestinalis*. Eukaryot. Cell *6*, 2354–2364.

De Souza, C.P., and Osmani, S.A. (2007). Mitosis, not just open or closed. Eukaryot. Cell *6*, 1521–1527.

DeBoer, S.R., You, Y., Szodorai, A., Kaminska, A., Pigino, G., Nwabuisi, E., Wang, B., Estrada-Hernandez, T., Kins, S., Brady, S.T., and Morfini, G. (2008). Conventional kinesin holoenzymes are composed of heavy and light chain homodimers. Biochemistry *47*, 4535–4543.

Delgado-Viscogliosi, P., Brugerolle, G., and Viscogliosi, E. (1996). Tubulin post-translational modifications in the primitive protist *Trichomonas vaginalis*. Cell. Motil. Cytoskeleton 33, 288–297.

Delgado-Viscogliosi, P., Viscogliosi, E., Gerbod, D., Kulda, J., Sogin, M.L., and Edgcomb, V.P. (2000). Molecular phylogeny of parabasalids based on small subunit rRNA sequences, with emphasis on the Trichomonadinae subfamily. J. Eukaryot. Microbiol. *47*, 70–75.

Didry, D., Carlier, M.F., and Pantaloni, D. (1998). Synergy between actin depolymerizing factor/cofilin and profilin in increasing actin filament turnover. J. Biol. Chem. *273*, 25602–25611.

dos Remedios, C.G., Chhabra, D., Kekic, M., Dedova, I.V., Tsubakihara, M., Berry, D.A., and Nosworthy, N.J. (2003). Actin binding proteins: regulation of cytoskeletal microfilaments. Physiol. Rev. *83*, 433–473.

Drouin, G., Moniz de Sa, M., and Zuker, M. (1995). The *Giardia lamblia* actin gene and the phylogeny of eukaryotes. J. Mol. Evol. *41*, 841–849.

Dutcher, S.K. (2003). Long-lost relatives reappear: identification of new members of the tubulin superfamily. Curr. Opin. Microbiol. *6*, 634–640.

Dwight, S.S., Balakrishnan, R., Christie, K.R., Costanzo, M.C., Dolinski, K., Engel, S.R., Feierbach, B., Fisk, D.G., Hirschman, J., Hong, E.L., Issel-Tarver, L., *et al.* (2004). *Saccharomyces* genome database: underlying principles and organisation. Brief. Bioinform. *5*, 9–22.

Dyall, S.D., Brown, M.T., and Johnson, P.J. (2004). Ancient invasions: from endosymbionts to organelles. Science *304*, 253–257.

El-Sayed, N.M., Myler, P.J., Bartholomeu, D.C., Nilsson, D., Aggarwal, G., Tran, A.N., Ghedin, E., Worthey, E.A., Delcher, A.L., Blandin, G., Westenberger, S.J., *et al.* (2005). The genome sequence of *Trypanosoma cruzi*, etiologic agent of Chagas disease. Science *309*, 409–415.

Elmendorf, H.G., Dawson, S.C., and McCaffery, J.M. (2003). The cytoskeleton of *Giardia lamblia*. Int. J. Parasitol. 33, 3–28.

Elmendorf, H.G., Rohrer, S.C., Khoury, R.S., Bouttenot, R.E., and Nash, T.E. (2005). Examination of a novel head-stalk protein family in *Giardia lamblia* characterised by the pairing of ankyrin repeats and coiled-coil domains. Int. J. Parasitol. 35, 1001–1011.

Embley, T.M., and Martin, W. (2006). Eukaryotic evolution, changes and challenges. Nature *440*, 623–630.

Erickson, H.P. (2007). Evolution of the cytoskeleton. Bioassays 29, 668–677.

Erlandsen, S.L., and Feely, D.E. (1984). Trophozoite motility and the mechanism of attachment, In *Giardia* and Giardiasis, S.L. Erlandsen, and E.A. Meyer, eds. (New York: Plenum Press), pp. 33–63.

Erlandsen, S.L., Russo, A.P., and Turner, J.N. (2004). Evidence for adhesive activity of the ventrolateral flange in *Giardia lamblia*. J. Eukaryot. Microbiol. *51*, 73–80.

Ersfeld, K., Melville, S.E., and Gull, K. (1999). Nuclear and genome organization of *Trypanosoma brucei*. Parasitol. Today *15*, 58–63.

Evangelista, M., Zigmond, S., and Boone, C. (2003). Formins: signaling effectors for assembly and polarization of actin filaments. J. Cell Sci. *116*, 2603–2611.

Feely, D.E., and Erlandsen, S.L. (1982). Effect of cytochalasin-B, low Ca++ concentration, iodoacetic acid, and quinacrine-HCl on the attachment of *Giardia* trophozoites *in vitro*. J. Parasitol. *68*, 869–873.

Feely, D.E., Erlandsen, S.L., and Chase, D.G. (1984). Structure of the Trophozoite and Cyst, In *Giardia* and Giardiasis: Biology, Pathogenesis and Epidemiology, S.L. Erlandsen and E.A. Meyer, eds. (New York: Plenum Press), pp. 3–32.

Feely, D.E., Erlandsen, S.L., and Chase, D.G. (1990). The biology of *Giardia*, In Giardiasis, E.A. Meyer, ed. (Amsterdam: Elsevier), pp. 11–49.

Filice, F.P. (1952). Studies on the cytology and life history of a *Giardia* from the laboratory rat. Univ. Calif. Publ. Zool. *57*, 53–146.

Finn, R.D., Tate, J., Mistry, J., Coggill, P.C., Sammut, J.S., Hotz, H.R., Ceric, G., Forslund, K., Eddy, S.R., Sonnhammer, E.L., and Bateman, A. (2008). The Pfam protein families database. Nucleic Acids Res. *36*, D281–D288.

Fiori, P.L., Rappelli, P., and Addis, M.F. (1999). The flagellated parasite *Trichomonas vaginalis*: new insights into cytopathogenicity mechanisms. Microbes Infect. *1*, 149–156.

Friend, D.S. (1966). The fine structure of *Giardia muris*. J. Cell Biol. *29*, 317–332.

Furness, B.W., Beach, M.J., and Roberts, J.M. (2000). Giardiasis surveillance: United States, 1992–1997. MMWR CDC Surveill. Summ. *49*, 1–13.

Furtado, M.B., and Benchimol, M. (1998). Observation of membrane fusion on the interaction of *Trichomonas vaginalis* with human vaginal epithelial cells. Parasitol. Res. *84*, 213–220.

Garbarino, J.E., and Gibbons, I.R. (2002). Expression and genomic analysis of midasin, a novel and highly conserved AAA protein distantly related to dynein. BMC Genomics 3, 18.

Gerke, V., and Moss, S.E. (2002). Annexins: from structure to function. Physiol. Rev. *82*, 331–371.

Ghosh, S., Frisardi, M., Rogers, R., and Samuelson, J. (2001). How *Giardia* swim and divide. Infect. Immun. *69*, 7866–7872.

Gibson, C., Schanen, B., Chakrabarti, D., and Chakrabarti, R. (2006). Functional characterisation of the regulatory subunit of cyclic AMP-dependent protein kinase A homologue of *Giardia lamblia*: Differential expression of the regulatory and catalytic subunits during encystation. Int. J. Parasitol. *36*, 791–799.

Gillin, F.D., Reiner, D.S., and McCaffery, J.M. (1996). Cell biology of the primitive eukaryote *Giardia lamblia*. Annu. Rev. Microbiol. *50*, 679–705.

Goldman, R.D., and Chao, Y.-H. (1999). Introduction to intermediate filaments and associated proteins, In Guidebook to the Cytoskeletal and Motor Proteins, T. Kreis, and R. Vale, eds. (Oxford: Oxford University Press), pp. 281–284.

Gonzalez-Robles, A., Arguello, C., Chavez, B., Cedillo-Rivera, R., Ortega-Pierres, G., and Martinez-Palomo, A. (1989). *Giardia lamblia*: surface charge of human isolates in culture. Trans. R. Soc. Trop. Med. Hyg. *83*, 642–643.

Gonzalez-Robles, A., Lazaro-Haller, A., Espinosa-Cantellano, M., Anaya-Velazquez, F., and Martinez-Palomo, A. (1995). *Trichomonas vaginalis*: ultrastructural bases of the cytopathic effect. J. Eukaryot. Microbiol. *42*, 641–651.

Granger, B.L., Warwood, S.J., Benchimol, M., and De Souza, W. (2000). Transient invagination of flagella by *Tritrichomonas foetus*. Parasitol. Res. *86*, 699–709.

Gray, M.W. (2005). Evolutionary biology: the hydrogenosome's murky past. Nature *434*, 29–31.

Halpain, S., and Dehmelt, L. (2006). The MAP1 family of microtubule-associated proteins. Genome Biol. 7, 224.

Hampl, V., Cepicka, I., Flegr, J., Tachezy, J., and Kulda, J. (2007). Morphological and molecular diversity of the monocercomonadid genera *Monocercomonas*, *Hexamastix*, and *Honigbergiella* gen. nov. Protist *158*, 365–383.

Hansen, W.R., and Fletcher, D.A. (2008). Tonic shock induces detachment of *Giardia lamblia*. PLoS Negl. Trop. Dis. 2, e169.

Hansen, W.R., Tulyathan, O., Dawson, S.C., Cande, W.Z., and Fletcher, D.A. (2006). *Giardia lamblia* attachment force is insensitive to surface treatments. Eukaryot. Cell 5, 781–783.

Hao, L., and Scholey, J.M. (2009). Intraflagellar transport at a glance. J. Cell Sci. *122*, 889–892.

Hellman, M., Paavilainen, V.O., Naumanen, P., Lappalainen, P., Annila, A., and Permi, P. (2004). Solution structure of coactosin reveals structural homology to ADF/cofilin family proteins. FEBS Lett. *576*, 91–96.

Hernandez, Y., Castillo, C., Roychowdhury, S., Hehl, A., Aley, S.B., and Das, S. (2007). Clathrin-dependent pathways and the cytoskeleton network are involved in ceramide endocytosis by a parasitic protozoan, *Giardia lamblia*. Int. J. Parasitol. 37, 21–32.

Hertz-Fowler, C., Peacock, C.S., Wood, V., Aslett, M., Kerhornou, A., Mooney, P., Tivey, A., Berriman, M., Hall, N., Rutherford, K., Parkhill, J., *et al.* (2004). GeneDB: a resource for prokaryotic and eukaryotic organisms. Nucleic Acids Res. 32, D339–343.

Hill, K.L. (2003). Biology and mechanism of trypanosome cell motility. Eukaryot. Cell 2, 200–208.

Hirokawa, N. (1998). Kinesin and dynein superfamily proteins and the mechanism of organelle transport. Science 279, 519–526.

Hlavsa, M.C., Watson, J.C., and Beach, M.J. (2005). Giardiasis surveillance: United States, 1998-2002. MMWR Surveill. Summ. 54, 9–16.

Hoeng, J.C., Dawson, S.C., House, S.A., Sagolla, M.S., Pham, J.K., Mancuso, J.J., Lowe, J., and Cande, W.Z. (2008). High-resolution crystal structure and *in vivo* function of a kinesin-2 homologue in *Giardia intestinalis*. Mol. Biol. Cell *19*, 3124–3137.

Holberton, D.V. (1973a). Fine structure of the ventral disk apparatus and the mechanism of attachment in the flagellate *Giardia* muris. J. Cell Sci. *13*, 11–41.

Holberton, D.V. (1973b). Mechanism of attachment of *Giardia* to the wall of the small intestine. Trans. R. Soc. Trop. Med. Hyg. *67*, 29–30.

Holberton, D.V. (1974). Attachment of *Giardia*-a hydrodynamic model based on flagellar activity. J. Exp. Biol. *60*, 207–221.

Holberton, D.V. (1981). Arrangement of subunits in microribbons from *Giardia*. J. Cell Sci. *47*, 167–185.

Holberton, D.V., and Ward, A.P. (1981). Isolation of the cytoskeleton from *Giardia*. Tubulin and a low-molecular-weight protein associated with microribbon structures. J. Cell Sci. *47*, 139–166.

Holberton, D., Baker, D.A., and Marshall, J. (1988). Segmented alpha-helical coiled-coil structure of the protein giardin from the *Giardia* cytoskeleton. J. Mol. Biol. *204*, 789–795.

Honigberg, B.M. (1963). Evolutionary and systematic relationships in the flagellate order Trichomonadida Kirby. J. Protozool. *10*, 20–63.

Honigberg, B.M., and Brugerolle, G. (1990). Structure, In Trichomonads Parasitic in Humans, B.M. Honigberg, ed. (New York: Springer-Verlag), pp. 5–35.

Honigberg, B.M., and King, V.M. (1964). Structure of *Trichomonas vaginalis* Donné. J. Parasitol. *50*, 345–364.

Ikegami, K., Horigome, D., Mukai, M., Livnat, I., MacGregor, G.R., and Setou, M. (2008). TTLL10 is a protein polyglycylase that can modify nucleosome assembly protein 1. FEBS Lett. *582*, 1129–1134.

Iwabe, N., and Miyata, T. (2002). Kinesin-related genes from diplomonad, sponge, amphioxus, and cyclostomes: divergence pattern of kinesin family and evolution of giardial membrane-bounded organella. Mol. Biol. Evol. *19*, 1524–1533.

Janke, C., Rogowski, K., Wloga, D., Regnard, C., Kajava, A.V., Strub, J.M., Temurak, N., van Dijk, J., Boucher, D., van Dorsselaer, A., Suryavanshi, S., *et al.* (2005). Tubulin polyglutamylase enzymes are members of the TTL domain protein family. Science *308*, 1758–1762.

Jemilohun, P.F. (1998). Isolation and characterization of flagella from *Trichomonas vaginalis*. Parasitol. Res. *84*, 800–805.

Jemilohun, P.F., Garnett, T., and Burgess, D.E. (2005). Isolation, partial purification, and immunogenicity of flagella from *Tritrichomonas foetus*. J. Eukaryot. Microbiol. *52*, 245–249.

Jenkins, M., O'Brien, C.N., Murphy, C., Schwarz, R., Miska, K.B., Rosenthal, B.M., and James, T. (2009). Antibodies to the ventral disc protein delta-giardin prevent *in vitro* binding of *Giardia lamblia* trophozoites. J. Parasitol. *95*, 895–899.

Khoshnan, A., and Alderete, J.F. (1993). Multiple double-stranded RNA segments are associated with virus particles infecting *Trichomonas vaginalis*. J. Virol. *67*, 6950–6955.

Kirk-Mason, K.E., Turner, M.J., and Chakraborty, P.R. (1988). Cloning and sequence of beta tubulin cDNA from *Giardia lamblia*. Nucleic Acids Res. *16*, 2733.

Kirk-Mason, K.E., Turner, M.J., and Chakraborty, P.R. (1989). Evidence for unusually short tubulin mRNA leaders and characterization of tubulin genes in *Giardia lamblia*. Mol. Biochem. Parasitol. *36*, 87–99.

Krieger, J.N., Jenny, C., Verdon, M., Siegel, N., Springwater, R., Critchlow, C.W., and Holmes, K.K. (1993). Clinical manifestations of trichomoniasis in men. Ann. Intern. Med. *118*, 844–849.

Kull, F.J., Vale, R.D., and Fletterick, R.J. (1998). The case for a common ancestor: kinesin and myosin motor proteins and G proteins. J. Muscle Res. Cell Motil. *19*, 877–886.

Landfear, S.M., and Ignatushchenko, M. (2001). The flagellum and flagellar pocket of trypanosomatids. Mol. Biochem. Parasitol. *115*, 1–17.

Lauwaet, T., Davids, B.J., Reiner, D.S., and Gillin, F.D. (2007). Encystation of *Giardia lamblia*: a model for other parasites. Curr. Opin. Microbiol. *10*, 554–559.

Lawrence, C.J., Malmberg, R.L., Muszynski, M.G., and Dawe, R.K. (2002). Maximum likelihood methods reveal conservation of function among closely related Kinesin families. J. Mol. Evol. *54*, 42–53.

Lecke, S.B., Tasca, T., Souto, A.A., and De Carli, G.A. (2002). *Trichomonas vaginalis*: microtubule cytoskeleton distribution using fluorescent taxoid. Exp. Parasitol. *102*, 113–116.

Li, J., Mahajan, A., and Tsai, M.D. (2006). Ankyrin repeat: a unique motif mediating protein–protein interactions. Biochemistry 45, 15168–15178.

Linck, R.W., and Norrander, J.M. (2003). Protofilament ribbon compartments of ciliary and flagellar microtubules. Protist *154*, 299–311.

Lindmark, D.G., and Müller, M. (1973). Hydrogenosome, a cytoplasmic organelle of the anaerobic flagellate *Tritrichomonas foetus*, and its role in pyruvate metabolism. J. Biol. Chem. *248*, 7724–7728.

Liu, H.W., Chu, Y.D., and Tai, J.H. (1998). Characterization of *Trichomonas vaginalis* virus proteins in the pathogenic protozoan *T. vaginalis*. Arch. Virol. *143*, 963–970.

Lloyd, G.L., Case, J.R., De Frias, D., and Brannigan, R.E. (2003). *Trichomonas vaginalis* orchitis with associated severe oligoasthenoteratospermia and hypogonadism. J. Urol. *170*, 924.

Lopes, L.C., Ribeiro, K.C., and Benchimol, M. (2001). Immunolocalization of tubulin isoforms and post-translational modifications in the protists *Tritrichomonas foetus* and *Trichomonas vaginalis*. Histochem. Cell Biol. *116*, 17–29.

Lowe, J., and Amos, L.A. (1998). Crystal structure of the bacterial cell-division protein FtsZ. Nature *391*, 203–206.

Lujan, H.D., Mowatt, M.R., and Nash, T.E. (1997). Mechanisms of *Giardia lamblia* differentiation into cysts. Microbiol. Mol. Biol. Rev. *61*, 294–304.

Machesky, L.M., and Gould, K.L. (1999). The Arp2/3 complex: a multifunctional actin organizer. Curr. Opin. Cell Biol. *11*, 117–121.

Machesky, L.M., Mullins, R.D., Higgs, H.N., Kaiser, D.A., Blanchoin, L., May, R.C., Hall, M.E., and Pollard, T.D. (1999). Scar, a WASp-related protein, activates nucleation of actin filaments by the Arp2/3 complex. Proc. Natl. Acad. Sci. U.S.A. *96*, 3739–3744.

MacRae, T.H. (1997). Tubulin post-translational modifications. Enzymes and their mechanisms of action. Eur. J. Biochem. *244*, 265–278.

Maiato, H., Khodjakov, A., and Rieder, C.L. (2005). *Drosophila* CLASP is required for the incorporation of microtubule subunits into fluxing kinetochore fibres. Nat. Cell Biol. 7, 42–47.

Marshall, J., and Holberton, D.V. (1993). Sequence and structure of a new coiled coil protein from a microtubule bundle in *Giardia*. J. Mol. Biol. *231*, 521–530.

Marshall, J., and Holberton, D.V. (1995). *Giardia* gene predicts a 183 kDa nucleotide-binding head-stalk protein. J. Cell Sci. *108*, 2683–2692.

Marszalek, J.R., and Goldstein, L.S. (2000). Understanding the functions of kinesin-II. Biochim. Biophys. Acta *1496*, 142–150.

Mattern, C.F., Honigberg, B.M., and Daniel, W.A. (1967). The mastigont system of *Trichomonas gallinae* (Rivolta) as revealed by electron microscopy. J. Protozool. *14*, 320–339.

McDonald, H.B., Stewart, R.J., and Goldstein, L.S. (1990). The kinesin-like ncd protein of *Drosophila* is a minus end-directed microtubule motor. Cell *63*, 1159–1165.

McKean, P.G., Vaughan, S., and Gull, K. (2001). The extended tubulin superfamily. J. Cell Sci. *114*, 2723–2733.

McNally, F.J., and Vale, R.D. (1993). Identification of katanin, an ATPase that severs and disassembles stable microtubules. Cell 75, 419–429.

Melkonian, M., Reize, I.B., and Preisig, H.R. (1987). Maturation of a flagellum/basal body requires more than one cell cycle in algal flagellates: studies on *Nephroselmis olivacea* (Prasinophyceae), In Algal Development: Molecular and Cellular Aspects, W. Wiessner, D.G. Robinson, and R.C. Starr, eds. (Springer-Verlag, Berlin), pp. 102–113.

Meng, T.C., Aley, S.B., Svärd, S.G., Smith, M.W., Huang, B., Kim, J., and Gillin, F.D. (1996). Immunolocalization and sequence of caltractin/centrin from the early branching eukaryote *Giardia lamblia*. Mol. Biochem. Parasitol. 79, 103–108.

Miki, H., Okada, Y., and Hirokawa, N. (2005). Analysis of the kinesin superfamily: insights into structure and function. Trends Cell Biol. *15*, 467–476.

Monteiro-Leal, L.H., da Cunha-e-Silva, N.L., Benchimol, M., and de Souza, W. (1993). Isolation and biochemical characterization of the costa of *Tritrichomonas foetus*. Eur. J. Cell Biol. *60*, 235–242.

Monteiro-Leal, L.H., Farina, M., Benchimol, M., Kachar, B., and De Souza, W. (1995). Coordinated flagellar and ciliary beating in the protozoon *Tritrichomonas foetus*. J. Eukaryot. Microbiol. *42*, 709–714.

Monteiro-Leal, L.H., Farina, M., and de Souza, W. (1996). Free movement of *Tritrichomonas foetus* in a liquid medium: a video-microscopy study. Cell. Motil. Cytoskeleton 34, 206–214.

Morgan, R.O., and Fernandez, M.P. (1995). Molecular phylogeny of annexins and identification of a primitive homologue in *Giardia lamblia*. Mol. Biol. Evol. *12*, 967–979.

Morrison, H.G., McArthur, A.G., Gillin, F.D., Aley, S.B., Adam, R.D., Olsen, G.J., Best, A.A., Cande, W.Z., Chen, F., Cipriano, M.J., Davids, B.J., *et al.* (2007). Genomic minimalism in the early diverging intestinal parasite *Giardia lamblia*. Science *317*, 1921–1926.

Mueller, J.C., Jones, A.L., and Brandborg, L.L. (1974). Scanning electron microscope observations in human giardiasis, In Scanning Electron Microscopy, 1973, O. Johari, ed. (Chicago: IIT Research Institute), pp. 557–564.

Müller, M. (1993). The hydrogenosome. J. Gen. Microbiol. *139*, 2879–2889.

Narcisi, E.M., Paulin, J.J., and Fechheimer, M. (1994). Presence and localization of vinculin in *Giardia*. J. Parasitol. *80*, 468–473.

Narcisi, E.M., and Secor, W.E. (1996). *In vitro* effect of tinidazole and furazolidone on metronidazole-resistant *Trichomonas vaginalis*. Antimicrob. Agents Chemother. *40*, 1121–1125.

Noël, C., Gerbod, D., Fast, N.M., Wintjens, R., Delgado-Viscogliosi, P., Doolittle, W.F., and Viscogliosi, E. (2001). Tubulins in *Trichomonas vaginalis*: molecular characterization of alpha-tubulin genes, posttranslational modifications, and homology modeling of the tubulin dimer. J. Eukaryot. Microbiol. *48*, 647–654.

Nohria, A., Alonso, R.A., and Peattie, D.A. (1992). Identification and characterization of gamma-giardin and the gamma-giardin gene from *Giardia lamblia*. Mol. Biochem. Parasitol. *56*, 27–37.

Nohynkova, E., Draber, P., Reischig, J., and Kulda, J. (2000). Localization of gamma-tubulin in interphase and mitotic cells of a unicellular eukaryote, *Giardia intestinalis*. Eur. J. Cell Biol. *79*, 438–445.

Nohynkova, E., Tumova, P., and Kulda, J. (2006). Cell division of *Giardia intestinalis*: flagellar developmental cycle involves transformation and exchange of flagella between mastigonts of a diplomonad cell. Eukaryot. Cell *5*, 753–761.

Palm, D., Weiland, M., McArthur, A.G., Winiecka-Krusnell, J., Cipriano, M.J., Birkeland, S.R., Pacocha, S.E., Davids, B., Gillin, F., Linder, E., and Svärd, S. (2005). Developmental changes in the adhesive disk during *Giardia* differentiation. Mol. Biochem. Parasitol. *141*, 199–207.

Parry, D.A., Fraser, R.D., and Squire, J.M. (2008). Fifty years of coiled-coils and alpha-helical bundles: a close relationship between sequence and structure. J. Struct. Biol. *163*, 258–269.

Pathuri, P., Nguyen, E.T., and Luecke, H. (2006). Expression, purification, crystallization and preliminary X-ray diffraction analysis of alpha-11 giardin from *Giardia lamblia*. Acta Crystallogr. Sect. F Struct. Biol. Cryst. Commun. *62*, 1108–1112.

Pathuri, P., Nguyen, E.T., Ozorowski, G., Svärd, S.G., and Luecke, H. (2009). Apo and calcium-bound crystal structures of cytoskeletal protein alpha-14 giardin (annexin E1) from the intestinal protozoan parasite *Giardia lamblia*. J. Mol. Biol. *385*, 1098–1112.

Pathuri, P., Nguyen, E.T., Svärd, S.G., and Luecke, H. (2007). Apo and calcium-bound crystal structures of Alpha-11 giardin, an unusual annexin from *Giardia lamblia*. J. Mol. Biol. *368*, 493–508.

Peattie, D.A., Alonso, R.A., Hein, A., and Caulfield, J.P. (1989). Ultrastructural localization of giardins to the edges of disk microribbons of *Giardia lamblia* and the nucleotide and deduced protein sequence of alpha giardin. J. Cell Biol. *109*, 2323–2335.

Pereira, A.J., Dalby, B., Stewart, R.J., Doxsey, S.J., and Goldstein, L.S. (1997). Mitochondrial association of a plus end-directed microtubule motor expressed during mitosis in *Drosophila*. J. Cell Biol. *136*, 1081–1090.

Pereira-Neves, A., Ribeiro, K.C., and Benchimol, M. (2003). Pseudocysts in trichomonads – new insights. Protist *154*, 313–329.

Piva, B., and Benchimol, M. (2004). The median body of *Giardia lamblia*: an ultrastructural study. Biol. Cell *96*, 735–746.

Pollard, T.D. (2001). Genomics, the cytoskeleton and motility. Nature *409*, 842–843.

Pollard, T.D., and Beltzner, C.C. (2002). Structure and function of the Arp2/3 complex. Curr. Opin. Struct. Biol. *12*, 768–774.

Pruitt, K.D., Tatusova, T., and Maglott, D.R. (2005). NCBI Reference Sequence (RefSeq): a curated non-redundant sequence database of genomes, transcripts and proteins. Nucleic Acids Res. *33*, D501–504.

Ralston, K.S., and Hill, K.L. (2006). Trypanin, a component of the flagellar Dynein regulatory complex, is essential in bloodstream form African trypanosomes. PLoS Pathog. *2*, e101.

Raynaud-Messina, B., and Merdes, A. (2007). Gamma-tubulin complexes and microtubule organization. Curr. Opin. Cell Biol. *19*, 24–30.

Rendtorff, R.C. (1954). The experimental transmission of human intestinal protozoan parasites. II. *Giardia lamblia* cysts given in capsules. Am. J. Hyg. *59*, 209–220.

Ribeiro, K.C., Monteiro-Leal, L.H., and Benchimol, M. (2000). Contributions of the axostyle and flagella to closed mitosis in the protists *Tritrichomonas foetus* and *Trichomonas vaginalis*. J. Eukaryot. Microbiol. *47*, 481–492.

Ribeiro, K.C., Pereira-Neves, A., and Benchimol, M. (2002). The mitotic spindle and associated membranes in the closed mitosis of trichomonads. Biol. Cell *94*, 157–172.

Roeben, A., Kofler, C., Nagy, I., Nickell, S., Hartl, F.U., and Bracher, A. (2006). Crystal structure of an archaeal actin homolog. J. Mol. Biol. *358*, 145–156.

Rohrig, U., Gerisch, G., Morozova, L., Schleicher, M., and Wegner, A. (1995). Coactosin interferes with the capping of actin filaments. FEBS Lett. *374*, 284–286.

Roskens, H., and Erlandsen, S.L. (2002). Inhibition of *in vitro* attachment of *Giardia* trophozoites by mucin. J. Parasitol. *88*, 869–873.

Rupp, G., and Porter, M.E. (2003). A subunit of the dynein regulatory complex in *Chlamydomonas* is a homologue of a growth arrest-specific gene product. J. Cell Biol. *162*, 47–57.

Sagolla, M.S., Dawson, S.C., Mancuso, J.J., and Cande, W.Z. (2006). Three-dimensional analysis of mitosis and cytokinesis in the binucleate parasite *Giardia intestinalis*. J. Cell Sci. *119*, 4889–4900.

Sant'Anna, C., Campanati, L., Gadelha, C., Lourenco, D., Labati-Terra, L., Bittencourt-Silvestre, J., Benchimol, M., Cunha-e-Silva, N.L., and De Souza, W. (2005). Improvement on the visualization of cytoskeletal structures of protozoan parasites using high-resolution field emission scanning electron microscopy (FESEM). Histochem. Cell Biol. *124*, 87–95.

Scholey, J.M. (2003). Intraflagellar transport. Annu. Rev. Cell Dev. Biol. *19*, 423–443.

Schwebke, J.R., and Burgess, D. (2004). Trichomoniasis. Clin. Microbiol. Rev. *17*, 794–803.

Sekine, Y., Okada, Y., Noda, Y., Kondo, S., Aizawa, H., Takemura, R., and Hirokawa, N. (1994). A novel microtubule-based motor protein (KIF4) for organelle transports, whose expression is regulated developmentally. J. Cell Biol. *127*, 187–201.

Sellers, J.R. (2000). Myosins: a diverse superfamily. Biochim. Biophys. Acta *1496*, 3–22.

Sheffield, H.G., and Bjorvat, B. (1977). Ultrastructure of the cyst of *Giardia lamblia*. Am. J. Trop. Med. Hyg. *26*, 23–30.

Sheterline, P., Clayton, J., and Sparrow, J.C. (1998). Actin (Oxford: Oxford University Press).

Shih, Y.L., and Rothfield, L. (2006). The bacterial cytoskeleton. Microbiol. Mol. Biol. Rev. *70*, 729–754.

Silacci, P., Mazzolai, L., Gauci, C., Stergiopulos, N., Yin, H.L., and Hayoz, D. (2004). Gelsolin superfamily proteins: key regulators of cellular functions. Cell. Mol. Life Sci. *61*, 2614–2623.

Silflow, C.D., and Lefebvre, P.A. (2001). Assembly and motility of eukaryotic cilia and flagella. Lessons from *Chlamydomonas reinhardtii*. Plant Physiol. *127*, 1500–1507.

Silva, J.C., Bastida, F., Bidwell, S.L., Johnson, P.J., and Carlton, J.M. (2005). A potentially functional mariner transposable element in the protist *Trichomonas vaginalis*. Mol. Biol. Evol. *22*, 126–134.

Smertenko, A., Saleh, N., Igarashi, H., Mori, H., Hauser-Hahn, I., Jiang, C.J., Sonobe, S., Lloyd, C.W., and Hussey, P.J. (2000). A new class of microtubule-associated proteins in plants. Nat. Cell Biol. *2*, 750–753.

Smith, B.F., and Stewart, B.T. (1966). Fine structure of *Trichomonas vaginalis*. Exp. Parasitol. *19*, 52–63.

Smith, E.F., and Lefebvre, P.A. (1997). The role of central apparatus components in flagellar motility and microtubule assembly. Cell. Motil. Cytoskeleton *38*, 1–8.

Smith, H.E., and Arnott, H.J. (1974). Axostyle structure in the termite protozoon *Pyrsonympha vertens*. Tissue Cell *6*, 193–207.

Solari, A.J., Rahn, M.I., Saura, A., and Lujan, H.D. (2003). A unique mechanism of nuclear division in *Giardia lamblia* involves components of the ventral disk and the nuclear envelope. Biocell *27*, 329–346.

Soloviev, M.M. (1968). Morphological and biological peculiarities of *Giardia lamblia* in connection with their habitat on the brush border of intestinal epithelium. Acta Protozool. *6*, 365–373.

Sousa, M.C., Goncalves, C.A., Bairos, V.A., and Poiares-Da-Silva, J. (2001). Adherence of *Giardia lamblia* trophozoites to Int-407 human intestinal cells. Clin. Diagn. Lab. Immunol. *8*, 258–265.

Susalka, S.J., Nikulina, K., Salata, M.W., Vaughan, P.S., King, S.M., Vaughan, K.T., and Pfister, K.K. (2002). The roadblock light chain binds a novel region of the cytoplasmic dynein intermediate chain. J. Biol. Chem. *277*, 32939–32946.

Sutcliffe, S., Giovannucci, E., Alderete, J.F., Chang, T.H., Gaydos, C.A., Zenilman, J.M., De Marzo, A.M., Willett, W.C., and Platz, E.A. (2006). Plasma antibodies against *Trichomonas vaginalis* and subsequent risk of prostate cancer. Cancer Epidemiol. Biomarkers Prev. *15*, 939–945.

Torian, B.E., Barnes, R.C., Stephens, R.S., and Stibbs, H.H. (1984). Tubulin and high-molecular-weight polypeptides as *Giardia lamblia* antigens. Infect. Immun. *46*, 152–158.

Tovar, J., Leon-Avila, G., Sanchez, L.B., Sutak, R., Tachezy, J., van der Giezen, M., Müller, M., and Lucocq, J.M. (2003). Mitochondrial remnant organelles of *Giardia* function in iron-sulphur protein maturation. Nature *426*, 172–176.

Tumova, P., Kulda, J., and Nohynkova, E. (2007). Cell division of *Giardia intestinalis*: assembly and disassembly of the adhesive disc, and the cytokinesis. Cell. Motil. Cytoskeleton *64*, 288–298.

Uetrecht, A.C., and Bear, J.E. (2006). Coronins: the return of the crown. Trends Cell Biol. *16*, 421–426.

Vahrmann, A., Saric, M., Koebsch, I., and Scholze, H. (2008). alpha14-Giardin (annexin E1) is associated with tubulin in trophozoites of *Giardia lamblia* and forms local slubs in the flagella. Parasitol. Res. *102*, 321–326.

Vale, R.D. (2003). The molecular motor toolbox for intracellular transport. Cell *112*, 467–480.

Vale, R.D., and Milligan, R.A. (2000). The way things move: looking under the hood of molecular motor proteins. Science *288*, 88–95.

Vaughan, S., Wickstead, B., Gull, K., and Addinall, S.G. (2004). Molecular evolution of FtsZ protein sequences encoded within the genomes of archaea, bacteria, and eukaryota. J. Mol. Evol. *58*, 19–29.

Vieira, P.B., Borges, F.P., Gottardi, B., Stuepp, C., Larre, A.B., Tasca, T., and De Carli, G.A. (2008). Analysis of microtubule cytoskeleton distribution using a fluorescent taxoid in two trichomonadid protozoa: *Trichomonas gallinae* and *Tritrichomonas foetus*. Exp. Parasitol. *119*, 186–191.

Viscogliosi, E., and Brugerolle, G. (1994a). Cytoskeleton in trichomonads. III Study of the morphogenesis during division by using monoclonal antibodies against cytoskeletal structures. Eur. J. Protistol. *30*, 129–138.

Viscogliosi, E., and Brugerolle, G. (1994b). Striated fibers in trichomonads: costa proteins represent a new class of proteins forming striated roots. Cell. Motil. Cytoskeleton *29*, 82–93.

Viscogliosi, E., Philippe, H., Baroin, A., Perasso, R., and Brugerolle, G. (1993). Phylogeny of trichomonads based on partial sequences of large subunit rRNA and on cladistic analysis of morphological data. J. Eukaryot. Microbiol. *40*, 411–421.

Wang, A., Wang, C.C., and Alderete, J.F. (1987). *Trichomonas vaginalis* phenotypic variation occurs only among trichomonads infected with the double-stranded RNA virus. J. Exp. Med. *166*, 142–150.

Warton, A., and Honigberg, B.M. (1979). Structure of trichomonads as revealed by scanning electron microscopy. J. Protozool. *26*, 56–62.

Weber, K., Geisler, N., Plessmann, U., Bremerich, A., Lechtreck, K.F., and Melkonian, M. (1993). SF-assemblin, the structural protein of the 2-nm filaments from striated microtubule associated fibers of algal flagellar roots, forms a segmented coiled coil. J. Cell Biol. *121*, 837–845.

Weber, K., Schneider, A., Westermann, S., Müller, N., and Plessmann, U. (1997). Posttranslational modifications of α- and β-tubulin in *Giardia lamblia*, an ancient eukaryote. FEBS Lett. *419*, 87–91.

Weiland, M.E., McArthur, A.G., Morrison, H.G., Sogin, M.L., and Svärd, S.G. (2005). Annexin-like alpha giardins: a new cytoskeletal gene family in *Giardia lamblia*. Int. J. Parasitol. *35*, 617–626.

Weiland, M.E., Palm, J.E., Griffiths, W.J., McCaffery, J.M., and Svärd, S.G. (2003). Characterisation of alpha-1 giardin: an immunodominant *Giardia lamblia* annexin with glycosaminoglycan-binding activity. Int. J. Parasitol. *33*, 1341–1351.

Weinstock, H., Berman, S., and Cates, W., Jr. (2004). Sexually transmitted diseases among American youth: incidence and prevalence estimates, 2000. Perspect. Sex. Reprod. Health *36*, 6–10.

WHO/UNICEF (2000). Global Water Supply and Sanitation Assessment 2000 Report (Geneva: WHO/UNICEF Joint Monitoring Programme for Water Supply and Sanitation).

WHO (2001). Global Prevalence and Incidence of Selected Curable Sexually Transmitted Infections. (Geneva: World Health Organization).

Wickstead, B., and Gull, K. (2006). A 'holistic' kinesin phylogeny reveals new kinesin families and predicts protein functions. Mol. Biol. Cell *17*, 1734–1743.

Wickstead, B., and Gull, K. (2007). Dyneins across eukaryotes: a comparative genomic analysis. Traffic *8*, 1708–1721.

Yang, P., Diener, D.R., Yang, C., Kohno, T., Pazour, G.J., Dienes, J.M., Agrin, N.S., King, S.M., Sale, W.S., Kamiya, R., Rosenbaum, J.L., and Witman, G.B. (2006). Radial spoke proteins of *Chlamydomonas* flagella. J. Cell Sci. *119*, 1165–1174.

Yang, X.J., and Gregoire, S. (2007). Metabolism, cytoskeleton and cellular signalling in the grip of protein N^{ε} – and O-acetylation. EMBO Rep. *8*, 556–562.

Yokoyama, R., O'Toole, E., Ghosh, S., and Mitchell, D.R. (2004). Regulation of flagellar dynein activity by a central pair kinesin. Proc. Natl. Acad. Sci. U.S.A. *101*, 17398–17403.

Yu, L.Z., Birky, C.W., Jr., and Adam, R.D. (2002). The two nuclei of *Giardia* each have complete copies of the genome and are partitioned equationally at cytokinesis. Eukaryot. Cell *1*, 191–199.

Zhang, P., Knowles, B.A., Goldstein, L.S., and Hawley, R.S. (1990). A kinesin-like protein required for distributive chromosome segregation in *Drosophila*. Cell *62*, 1053–1062.

Entamoeba histolytica Genomic Analyses

6

Upinder Singh and Gretchen M. Ehrenkaufer

Abstract

The sequencing of the *Entamoeba histolytica* genome, as well as the development of genome-wide techniques to allow study of gene expression, has led to many advances in our understanding of the biology of this parasite. In this chapter, we examine recent work using genomic technologies (DNA microarrays and proteomics) to study important aspects of amoebic biology, including pathogenesis, host interaction, phagocytosis, stage conversion, responses to stressful stimuli, and mechanisms of gene regulation. The studies described in this chapter identify pathways involved in regulating major functions in this important human pathogen.

Genomic analyses of *E. histolytica*

Study of *E. histolytica* entered a 'post-genome' stage with the initiation of the genome sequencing effort (Loftus *et al.*, 2005). Subsequently, a number of large-scale genomic technologies have been applied successfully to the study of amoebic biology, including transcriptional profiling with microarray technology and measurement of protein abundance with proteomics. In this chapter we will summarize the recent work in *E. histolytica* using these two approaches.

Microarray and proteomics technologies applied to the study of *E. histolytica*

Transcriptional profiling using microarray technology has emerged as a powerful technique and has been applied widely to the study of biological systems (Chu *et al.*, 1998; Cleary *et al.*, 2002; Hayward *et al.*, 2000; Lockhart *et al.*, 1996; Saxena *et al.*, 2007).

Microarray experiments provide snapshots of mRNA abundance at a genome-wide level. This approach has been used to gain insights into regulation of gene expression under biologically relevant conditions. In parasite systems, this approach has been applied successfully to identify an organism's transcriptional response during host invasion, stage conversion, and drug exposure.

A number of studies have indicated that gene expression in *E. histolytica* is regulated at the level of mRNA abundance; thus, microarray technology can be applied to studying aspects of amoebic biology. Various microarray platforms have been developed for *E. histolytica*, including both sequenced and non-sequenced genomic DNA microarrays (Debnath *et al.*, 2004; Shah *et al.*, 2005), oligonucleotide microarrays of selected amoebic genes (Beck *et al.*, 2005; Tillack *et al.*, 2007; Weber *et al.*, 2006), and whole-genome microarrays using long and short oligonucleotides (Davis *et al.*, 2007; Gilchrist *et al.*, 2006). Each array format has unique advantages.

Arrays made from genomic DNA can be used for comparison of divergent *Entamoeba* strains or species as the spotted target sequences (fragments of genomic DNA) are longer and can be bound by labelled probe even in the presence of some sequence divergence. Additionally, as intergenic regions are included on the array, differences between non-coding regions of DNA can be observed. Two genomic microarray platforms were developed for *E. histolytica* (one using non-sequenced and one using sequenced genomic DNA fragments) and used for expression profiling of *Entamoeba* species and strains (Debnath *et al.*, 2004; MacFarlane and Singh, 2006), as well as comparative genomic hybridizations (Shah *et al.*, 2005). These approaches could be developed even before genome assembly and annotation was complete and the entire process was relatively cost-effective. With improved genome annotation, oligonucleotide microarrays were developed, which allowed significantly greater sensitivity for low abundance transcripts and allowed discrimination between gene family members. Microarrays representing the extensive repertoires of amoebic transmembrane kinases and cysteine proteases were developed (Beck *et al.*, 2005; Tillack *et al.*, 2007). Subsequently, a targeted approach using probes for selected amoebic genes (Weber *et al.*, 2006) as well as whole-genome arrays have emerged (Davis *et al.*, 2007; Gilchrist *et al.*, 2006). Greater signal resolution is a significant advantage but increased cost and technology limitations have hindered the use of these arrays by some members of the scientific community. One additional advantage of the short oligonucleotide arrays based on Affymetrix technology is that extremely small amounts of RNA can be used, by utilizing the technology of RNA amplification, allowing an unprecedented analysis of amoebic gene expression under conditions in which the starting material typically has been a limiting factor.

Genome-wide analysis of the protein composition of an organism is a powerful approach to study the functional repertoire of a system (James, 1997). Whereas microarray technology is limited to RNA expression analysis, proteomic analysis identifies the stable protein makeup of an organism and is more likely to identify functional components of an organism, including identifying post-translational modifications that are often crucial to protein function (e.g. phosphorylation or ubiquitination). A number of proteomics approaches can be used, including two-dimensional gel electrophoresis, chromatography, capillary electrophoresis and mass spectrometry. The extremely high numbers and expression of proteolytic enzymes in *E. histolytica* have provided challenges to the use of proteomic approaches but some investigators have used this approach successfully to study aspects of amoebic biology (Davis *et al.*, 2006; Marion *et al.*, 2005; Okada *et al.*, 2005; Okada *et al.*, 2006; Tazreiter *et al.*, 2008; Tolstrup *et al.*, 2007).

Host invasion

Amoebic colitis

The most common clinical manifestation of amoebic disease is dysentery (Haque *et al.*, 2003) and animal models for this disease process have been developed (Houpt *et al.*, 2004; Zhang *et al.*, 1994). Understanding the molecular mechanisms used by the parasite as it invades the human host is important for an understanding of the virulence arsenal used by the parasite. Transcriptional profiling of both the parasite and host response during amoebic colitis has been accomplished, providing important insights into this manifestation of amoebic disease.

Parasite transcriptional networks associated with colonic invasion

Using a mouse model of colitis in which chronic inflammation and tissue ulcers form, Gilchrist *et al.* determined the transcriptome of *E. histolytica* in colonic disease (Gilchrist *et al.*, 2006). Parasites were harvested from the colitis model (at days 1 and 29 after implantation into a mouse cecum) and compared to 'input' parasites (i.e. from parallel *in vitro* laboratory cultures harvested at time of injection into mice). The parasites obtained from animal infections were used directly for array experiments with no *in vitro* culture steps to increase the parasite number. Because of the extremely small numbers of parasites and amounts of RNA obtained, the authors used methods of signal amplification that have been shown not to affect the results adversely (Dumur *et al.*, 2004). By comparing the transcriptome of parasites under *in vitro* tissue culture conditions to those harvested from a mouse, a molecular signature associated with amoebic colonization and/or invasion was identified.

An Affymetrix microarray containing probes for 9435 of the 9938 annotated genes was utilized for this study. The overall transcriptional modulation was surprisingly limited with 326 genes showing changes in transcript abundance at day 1 of mouse colonization, 109 genes showing changes in transcript abundance at day 29 of mouse colonization, and only 88 genes with changes in transcript abundance at both time points (Fig. 6.1). Genes with altered transcript levels were found in many functional categories, including energy metabolism, vesicle transport and membrane transport. Interestingly, transcripts encoding genes involved in oxygen detoxification were downregulated, including ferredoxin and the enzymes that synthesize iron-sulfur clusters, possibly indicating a reduced oxygen stress in the colonic environment compared to *in vitro* growth. Also regulated were multiple genes encoding intracellular signalling molecules such as kinases, phosphatases, and the *E. histolytica* calcium binding protein (EhCaBP1), all of which could function to initiate the parasite's responses to the colonic environment. Importantly, known virulence determinants such as the Gal/GalNAc lectin, cysteine proteases 1, 2 and 5, and amoebapore were not significantly upregulated during mouse colonization. This may be because these genes have very high basal expression during axenic tissue culture conditions.

In this analysis, parasites from axenic tissue culture conditions were compared to parasites from a mouse cecum, where parasites are exposed to mucin, host epithelial cells, bacterial flora and immune inflammatory cells. Considering the vast differences in the environment of the parasites, the extent of transcriptional modulation was relatively modest: 3.5% of genes differentially expressed at day 1; 1.2% of genes differentially expressed at day 29; and 0.93% of genes differentially expressed at both time points. Multiple explanations may account for this observation. One option is that the parasites under *in vitro* culture conditions are already primed for tissue invasion and are thus expressing high levels of the proteins involved in tissue invasion. An analogous situation occurs with macrophages, which become activated by attaching to the plastic of tissue culture flasks (Gagliardi *et al.*, 1995). This

	Liver abscess[a]	Colitis[b]	Cyst[c]
Array used	Long oligonucleotide ~3,000 genes	Affymetrix ~9,435 genes	Affymetrix ~9,435 genes
Experimental details	Liver abscess parasites Cultured for 7 days c/w long-term cultures	Mouse colitis parasites No in vitro culture c/w input parasites	Parasites from clinical isolates Cultured for ≤ 8 weeks c/w long-term cultured parasites
Gene regulation	29 genes upregulated 21 genes downregulated	379 genes upregulated 143 genes downregulated	672 genes upregulated 767 genes downregulated
Overlap	7 genes (colitis) 11 genes (cyst)	7 genes (liver abscess) 140 genes (cyst)	11 genes (liver abscess) 140 genes (colitis)
Further analysis	KERP1[d]	HMG protein[e] URE3-binding protein[f]	Myb protein[g]
Future questions	•Whole genome analysis •No tissue culture step •Which genes cause disease	•Parasites deeply invading •Variability in parasite response •Single parasite analysis	•Synchronized *in vitro* encystation and excystation

Figure 6.1 Summary of microarray data from amoebic liver abscess, amoebic colitis, and trophozoite to cyst stage conversion. The format of the array used, experimental details, extent of gene regulation, overlap between the conditions listed, gene-specific analysis, and questions for the future are outlined. (a) Adapted from Santi-Rocca *et al.*, 2008. (b) Adapted from Gilchrist *et al.*, 2006. (c) Adapted from Ehrenkaufer *et al.*, 2007. (d) Adapted from Santi-Rocca *et al.*, 2008. (e) Adapted from Abhyankar *et al.*, 2008. (f) Adapted from Gilchrist *et al.*, 2008. (g) Adapted from Ehrenkaufer *et al.*, 2009.

hypothesis is concordant with the observation that the majority of known virulence factors are expressed at high levels under axenic culture conditions. In fact, it has been shown previously that *E. histolytica* trophozoites in tissue culture are able to lyse Caco-2 monolayers with minimal modulation of the parasite transcriptome (MacFarlane *et al.*, 2005). Since that work was performed with genomic DNA arrays representing only 1971 genes it was hypothesized that technical issues (the limited number of genes analysed, low-level transcriptional modulation, and cross-hybridization among gene family members leading to poor data resolution) might have accounted for the low level of transcriptional changes observed. However, given the results of Gilchrist *et al.*, who used a whole genome microarray, a highly sensitive Affymetrix technology, and parasites exposed to the full spectrum of host factors, it appears that axenic *E. histolytica* trophozoites do not need to undergo significant transcriptional modulation in order to colonize/invade the host colon. The second possibility is that the methods of parasite retrieval from the mouse colon missed parasites that had invaded deep into the host epithelium and thus any transcriptional modulation in those parasites was not detected. In fact, because the parasite material was harvested from cecal material without tissue dissection, the authors were careful to discuss the microarray results in the context of parasite colonization and/or invasion. Future studies using microdissection of parasites invading deep into host tissue will be useful to address this issue. A further possibility is that amoebae respond during tissue invasion by post-transcriptional changes, which would not be detected by microarray analysis.

Pathways regulating colonic disease: gene-specific analysis

Further analysis of genes regulated during amoebic establishment of colonic disease has been performed for two transcription factors, the Upstream Regulatory Element 3 Binding Protein (URE3-BP) and a High Mobility Group (HMG) protein (Abhyankar *et al.*, 2008; Gilchrist *et al.*, 2008). URE3-BP is a calcium-binding protein initially identified as binding to an upstream regulatory element (URE3) found in the promoters of a gene encoding a lectin family member (*hgl5*) and ferredoxin (*fdx1*) (Gilchrist *et al.*, 2001; Gilchrist *et al.*, 1998). The transcript encoding URE3-BP was down regulated *in vivo* during host invasion. When a calcium insensitive version of this gene, which was predicted to act as a dominant positive regulator, was overexpressed in *E. histolytica* it resulted in a phenotype of increased migration of trophozoites though a transwell membrane. In order to identify potential URE3-BP target genes, gene expression in these parasites was monitored using the whole genome Affymetrix array. Altered expression was observed for 50 transcripts (42 decreased and 8 increased), compared to parasites expressing a control plasmid. The URE3-BP consensus site was present in the promoters of 27 URE3-BP regulated genes, and was significantly enriched compared promoters of all genes ($p < 0.0001$), providing evidence that URE3-BP may regulate these genes directly. The majority of proteins encoded by the genes fell into four categories: proteins containing a transmembrane domain, those involved in cellular metabolic processes, cytoskeletal proteins, and transcription/translation proteins. In contrast to URE3-BP, the gene encoding a putative High-Mobility-Group Box protein (EhHMGB1), a member of a group of conformation-specific DNA binding proteins, was found to be upregulated during colonic colonization. Recombinant EhHMGB1 was demonstrated to localize to the nucleus, have DNA bending activity and to enhance binding of the p53 transcription factor to its target sequence, establishing that it can function as a HMGB protein (Abhyankar *et al.*, 2008). When EhHMGB1 was overexpressed in *E. histolytica* trophozoites, Abhyankar *et al.* found 33 genes that showed differential expression. Significantly, 20 of these transcripts were also modulated *in vivo*, implying that EhHMGB1 may in fact help regulate the parasite response to the colonic environment. Additionally, four genes related to virulence, including lectin light chain and a cysteine protease (EhCP-A7) showed altered transcript levels. The studies with URE 3-BP and EhHMGB1 proteins, as follow-up analyses to the host invasion expression profiling, indicate that such data can be mined sequentially to characterize further the important factors in specific aspects of amoebic biology.

Host response during parasite invasion and colitis

The host response to *E. histolytica* colonization and invasion during amoebic colitis has also been investigated in detail. By using a mouse model of amoebic colitis and human microarrays, Zhang and Stanley (2004) characterized the response of human tissue to amoebic infection. Human colonic xenografts implanted in SCID-HU-INT mice were infected with *E. histolytica* trophozoites for 4 and 24 hours, and RNA from the human tissue was compared to uninfected xenografts using a two-colour human cDNA microarray. Colonic tissue was also infected with the bacterium *Shigella flexneri* in order to determine which portions of the host response to infection were shared by the two systems. Interestingly, significant overlap was found between the host responses to the two pathogens, including the induction of proteins involved in the inflammatory response (IL-1B, IL-8 and TNF-α pathway members), tissue repair (matrix metalloproteases, growth factors), and response to hypoxia (lactate dehydrogenase, enolase). However, some changes in gene expression were pathogen-specific. Overall, fewer genes were induced specifically in *E. histolytica* infection than in *S. flexneri* infection, perhaps due to the extracellular nature of infection by *Entamoeba*. Genes that showed increased expression only in *E. histolytica* infection include integrin and laminin family members as well as beta-catenin, which may be induced as the epithelial layer attempts to maintain integrity upon exposure to *Entamoeba* proteases and cytotoxic factors.

Amoebic liver abscess

The most common extraintestinal clinical manifestation of amoebic disease is liver abscess (Haque *et al.*, 2003) and animal models for this disease process have been developed (Zhang *et al.*, 1994). Transcriptional profiling of both the parasite and host response during amoebic liver abscess (ALA) formation has been accomplished.

Parasite transcriptional networks associated with hepatic invasion

The transcriptome of *E. histolytica* parasites during liver abscess formation has been examined (Fig. 6.1) (Santi-Rocca *et al.*, 2008). *E. histolytica* trophozoites from the virulent strain HM-1:IMSS were injected into the portal vein of a male Syrian Gold hamster, a liver abscess allowed to develop for 7 days, and the parasites harvested from the resulting abscess. Because of the low number of organisms recovered from the liver abscess material, parasites were first grown in culture for 16 hours, then a transcriptome analysis was performed using an array containing 1131 unique genes. Results were compared to *E. histolytica* trophozoites of the same strain that had been attenuated by years of axenic culture and were no longer capable of causing liver abscesses. Data analysis revealed 29 genes with higher and 21 genes with lower expression in parasites from liver abscess compared to the attenuated axenically cultured parasites. Among the genes that had higher expression in the liver-derived parasites were many that encoded stress-related proteins, such as heat shock proteins and peroxiredoxin. Genes with lower expression in liver-derived parasites included several cytoskeletal genes (spectrin and an unconventional myosin).

Compared to the data from the amoebic colitis model, few known virulence factors showed significant changes in expression. One exception was the surface protein Ariel. The exact function of this protein is not known, but it has significant similarity to the surface antigen Serine Rich *E. histolytica* Protein (SREHP) and is absent from the genome of the non-pathogenic species *E. dispar*, suggesting a potential role in pathogenesis. This study also identified numerous novel proteins, some of which could play a role in virulence. However, the overall impact of the work is limited by use of an array that did not cover the majority of the genome, making it likely that genes important in hepatic invasion may have been missed.

Pathways regulating hepatic disease: gene-specific analysis

Further analysis of a lysine and glutamic acid-rich protein (KERP1) identified as being upregulated during hepatic invasion was undertaken subsequently (Santi-Rocca *et al.*, 2008). This gene has been linked to the parasite's ability to cause hepatic disease. KERP1 was found to localize to the plasma membrane of trophozoites and was able to bind to cultured human epithelial cells, both as purified protein (bound to beads) and in the context of an *E. histolytica*-Caco2 cell co-culture experiment. Specifically, KERP1 was found to associate with brush border microvilli, the major point of attachment of *E. histolytica* to epithelial cells, indicating that it may help to mediate adhesion, along with the Gal/GalNac lectin and other proteins (Seigneur *et al.*, 2005). Significantly, parasites with reduced levels of KERP1 protein had an impaired ability to form liver abscesses in a hamster model, providing further evidence of a role for this protein in amoebic hepatic abscess formation (Santi-Rocca *et al.*, 2008).

Host response during amoebic liver abscess formation

The host response during amoebic liver abscess formation has also been investigated (Pelosof *et al.*, 2006). Using a SCID mouse model of amoebic liver infection, Pelosof *et al.* looked for changes in host gene expression at 4, 12 and 24 hours post-infection. RNA was isolated from infected and uninfected livers, hybridized to an Affymetrix oligonucleotide array, and genes with significantly altered transcript levels were identified. As in the colonic infection model, inflammatory pathways controlled by IL-1, NF-κB and TNF-α were induced, along with other acute response genes such as hypoxia-inducible factor and Jun kinase. However, significant differences exist between the liver abscess and colitis transcriptomes. Significantly, genes encoding pro-apoptotic proteins, such as Noxa and Ask1 had increased expression in the infected liver but not in the infected colon, perhaps helping to explain why apoptosis is a significant feature of amoebic liver abscess but not of amoebic colitis.

Directed analysis of gene families with roles in virulence

A number of other studies have analysed directly genes thought to be important in amoebic virulence. One unexpected finding of the *E. histolytica* genome sequencing effort was the identification of large families of transmembrane kinases (TMK), a feature previously found only in complex eukaryotic systems (Loftus *et al.*, 2005). At least some of these TMKs have been shown to localize to the *E. histolytica* plasma membrane, suggesting that they are involved in the reception of extracellular signals (Beck *et al.*, 2005) and some of these gene family members have proven roles in phagocytosis (Boettner *et al.*, 2008). Using a long oligonucleotide microarray at least 20 TMKs were found to be significantly expressed during log-phase growth (Beck *et al.*, 2005). Importantly, some TMKs that had low expression in normal axenic culture had modulated transcription under other conditions, including stage conversion (Ehrenkaufer *et al.*, 2007b), tissue invasion (Gilchrist *et al.*, 2006; Santi-Rocca *et al.*, 2008), and stress (Vicente *et al.*, 2009). Overall, the data indicate that regulation of TMK expression may be important in the response of the parasite to its environment.

Cysteine proteases and their roles in amoebic pathogenesis have been studied extensively (Ankri *et al.*, 1998; Bruchhaus *et al.*, 2003; Hellberg *et al.*, 2001). However, the contribution of proteases to amoebic biology was not fully appreciated as the focus of most studies was the few highly expressed cysteine protease genes. The extensive network of protease genes encoded in the *E. histolytica* genome was identified and catalogued by Tillack *et al.* (2007). A total of 86 genes coding for putative proteases were identified in the genome, including 50 cysteine proteases (the majority belonging to the C1 papain family), 22 metallo-proteases, 10 serine proteases, and 4 aspartic proteases. Using long oligonucleotide microarrays, it was determined that despite their large number the majority of protease genes were not expressed under *in vitro* culture conditions or during heat shock. Furthermore, these genes were not differentially expressed in amoebic strains with variable virulence potential. Other studies have identified some of these proteases as being developmentally regulated (Ehrenkaufer *et al.*, 2007b) or regulated by oxidative or nitrosative stress (Vicente *et al.*, 2009). However, the full impact of this extensive protease arsenal in *E. histolytica* is not fully understood at present and future investigations to delineate their function in amoebic biology will be necessary.

Amoebic stage conversion: trophozoites and cysts

E. histolytica has a simple life cycle in which it interconverts between trophozoites and cysts. The cyst is environmentally resistant and survives outside the human host, thus allowing transmission of disease. Upon ingestion excystation occurs in the human small intestine and the parasite converts to a motile trophozoite form that is capable of causing disease symptoms (Haque *et al.*, 2003). Since the cyst stage is required to propagate infection, it is essential to understand the molecular basis of stage conversion. Furthermore, understanding amoebic development may allow identification of novel ways to prevent pathogen transmission and novel treatment strategies. There is currently no method for generating large numbers of *E. histolytica* cysts *in vitro*. Thus, until recently, very little was known about the molecular basis of *E. histolytica* stage conversion.

Earlier studies had demonstrated that parasites recently isolated from a human host and maintained in complex xenic medium retained encystation competence (Balamuth, 1951). Under these conditions amoebic cysts with a chitin containing cyst wall could be identified. As parasite culture methods developed to allow axenic cultivation, the growth of amoebae became easier, but the ability to generate cysts *in vitro* appears to have been lost (Eichinger, 1997). Using multiple stimuli (including addition of metal ions, hypo-osmotic shock, and high pH) a number of groups have attempted to generate *E. histolytica* cysts *in vitro*, but these attempts have failed to generate mature cysts (Barron-Gonzalez *et al.*, 2008; Chayen *et al.*, 1988).

The *E. histolytica* cyst transcriptome

Re-exploring the concept that clinical isolates of *E. histolytica* maintain encystation competence in complex media, Ehrenkaufer *et al.* generated an expression profile of recent clinical isolates grown in Robinson's medium (Fig. 6.1) (Ehrenkaufer *et al.*, 2007b). Three cultures from two separate clinical isolates of *E. histolytica* were obtained (isolated from patients and in culture for less than 2 months) and cysts were identified in the cultures. Expression data from these isolates were compared to expression profiles of a large number of *E. histolytica* trophozoite samples (including virulent and non-virulent strains, parasites exposed to stress conditions, and parasites from a mouse model of colitis). Consistent with previous observations and with the identification of cysts in the cultured material, data from the transcriptional profiling indicated that the parasites grown in complex medium had a substantially different expression profile to parasites from other conditions. The cultures used in this study contained a mixed population of trophozoites and cysts and were likely to include cells that were encysting or excysting. Therefore, genes identified by this study as being developmentally regulated may have varied expression patterns over the entire *E. histolytica* life cycle. Overall, 1439 genes (~15% of the genome) were identified as being developmentally regulated with 672 cyst-specific genes and 767 trophozoite-specific genes. This level of transcriptional change is similar to that seen in yeast in different life cycle stages (Chu *et al.*, 1998). Importantly, a number of the developmentally regulated genes identified in *E. histolytica* were also identified as being developmentally regulated in *Entamoeba invadens*, a related reptile parasite used as a model for *Entamoeba* development. This indicates that some aspects of development are conserved among *Entamoeba* species. Genes confirmed as being stage-specific in *E. invadens* had varying temporal regulation, with some genes upregulated as early as 24h after transfer into encystation medium while the expression of other genes did not change until 48–72 hours after encystation induction.

Examples of stage-specific genes include several cysteine proteases (CP7, CP18) and signalling molecules, including TMKs, that could be involved in sensing developmental signals from the colonic environment. Other developmentally regulated genes with potential roles in the regulation of encystation were identified. These include signal transduction pathway members such as cytoplasmic kinases, calcium binding proteins and phospholipase D, which helps to regulate the protein kinase C pathway. Additionally, likely transcriptional regulator genes, including genes encoding Myb domain-containing proteins and a leucine zipper protein of the Dopey family, showed stage-regulated expression levels in cysts and trophozoites. These developmentally regulated proteins may directly regulate transcription of stage-specific genes. Interestingly, many known virulence factors, such as Gal/GalNac lectin light chain and cysteine protease CP1, are significantly downregulated in cysts. Since cysts are a non-invasive form of the parasite, the lower expression of virulence genes is not unexpected. A substantial number of stage-specific genes were identified as 'hypothetical proteins' restricted to *Entamoeba*. These deserve further investigation as they may represent novel targets for diagnostic or therapeutic approaches.

In addition to contributing to our understanding of gene regulation during encystation, this work also provided insight into conditions that stimulate encystation in *E. histolytica*. The clinical isolates used were grown in a complex medium containing bacteria and rice starch, significantly different from the axenic media most often used in laboratory culture of trophozoites. The observation of spontaneous encystation under these conditions is consistent with earlier observations that *E. histolytica* grown in complex media maintain encystation competence, while axenized strains do not (Eichinger, 1997; Phillips, 1973). It has also been observed that the specific bacterial flora co-cultured with xenic parasites can influence the ability to encyst (Singh and Ehrenkaufer, 2009). The cyst transcriptome provided the first insights into this developmental pathway in *E. histolytica*. However, in order to truly understand the changing patterns of gene expression through the *Entamoeba* life cycle a method for inducing regulated expression in *E. histolytica* eventually will be required.

Pathways regulating *Entamoeba* stage conversion

Gene-specific analysis

Developmentally regulated genes can be controlled by a number of pathways that include transcription factors. In searching for genes that initiate the encystation cascade, the timing of gene expression is an important consideration. Genes involved in the regulation of stage conversion should be expressed only while the cell is differentiating. Recently, an *E. histolytica* Myb gene has been identified that is itself developmentally regulated (EhMyb-dr) and which regulates expression of a number of stage-specific genes (Ehrenkaufer *et al.*, 2009). Overexpression of EhMyb-dr in *E. histolytica* trophozoites resulted in parasites with a transcriptional profile that overlapped significantly with amoebic cysts. However, no cyst wall structures were noted in these cultures indicating that although the transcriptional pathway to encystation was initiated, no mature cysts were formed. Whether this indicates that other factors are necessary to generate mature cysts or that the parasite strain used is not encystation competent is not known. Analysis of the promoter regions of genes regulated by EhMyb-dr identified a conserved promoter motif (CCCCCC) to which amoebic nuclear proteins bound. Using chromatin immunoprecipitation, promoters of EhMyb-dr regulated genes containing the C-rich motif were identified as interacting directly with the EhMyb-dr protein. The EhMyb-dr belongs to the SHAKQY family of Myb genes and the identification of this first developmentally regulated transcription factor in *E. histolytica* serves as an important advance in understanding the molecular framework that regulates stage conversion in this system.

Histone acetylation

In addition to the identification of stage-specific genes, the study by Ehrenkaufer *et al.* (2007b) helped provide insights into possible methods of stage-specific regulation. The researchers found that some developmentally regulated genes (both cyst and trophozoites-specific) were clustered together in the same genomic region; up to six genes on a single scaffold were co-regulated (Ehrenkaufer *et al.*, 2007b). The observation that developmentally regulated genes are clustered suggests that the regulation of chromatin structure may be a factor in controlling gene expression during stage conversion. In order to investigate the role of histone modification, *E. histolytica* strain 200:NIH was exposed to high doses of the histone deacetylase inhibitor Trichostatin A (TSA). Significant decreases in parasite growth rate and transcriptional changes reminiscent of stage conversion were observed (Ehrenkaufer *et al.*, 2007a). A total of 163 genes were regulated under these conditions (122 genes up-regulated and 41 genes down-regulated) with a substantial overlap of this gene set with the developmentally regulated genes previously identified, including some genes that are in genomic clusters. Decreased expression of a number of virulence genes was also detected. Overall, the decreased parasite growth rate, increased expression of cyst-specific genes, and decreased expression of virulence genes was consistent with the data linking effects of TSA exposure to stage conversion. However, contrasting results were obtained using strain HM-1:IMSS and exposure to TSA (Isakov *et al.*, 2008). In this strain, TSA exposure resulted in an increase in parasite virulence and no induction of stage-specific genes. Whether the differences in the amoebic strains or doses of TSA contributed to these disparate data sets is not known and will need further investigation. Additionally, whether effects on gene expression are a direct or indirect effect of histone acetylation remains to be determined. Interestingly, when *E. invadens* cultures that had been induced to encyst were exposed to TSA, cyst formation was inhibited (Byers *et al.*, 2005). These apparently conflicting results between the effects of TSA in *E. histolytica* and *E. invadens* may be a result of differences between regulatory pathways in the two species, or could indicate that both cyst and trophozoite specific genes are regulated by histone acetylation. Overall, the work has only just begun to unravel the molecular mechanisms by which developmentally regulated genes are controlled in *E. histolytica*.

Molecular signatures of virulent and non-virulent *Entamoeba*

The majority of infections by *E. histolytica* are asymptomatic, yet in some individuals infection progresses to severe, life-threatening disease (Haque *et al.*, 2003). The host and parasite factors responsible for this variable clinical outcome are not known but outcome is likely to be multifactorial. One aspect that has received recent attention is characterization of the parasite subtype and identification of virulence determinants by comparing virulent and non-virulent strains of *E. histolytica*. The two strains most useful in this regard are *E. histolytica* HM-1:IMSS, a virulent strain that was isolated from a patient with amoebic colitis, and *E. histolytica* Rahman, which was isolated from an asymptomatic cyst passer and exhibits no cytotoxic ability in either *in vitro* assays or in animal models (Davis *et al.*, 2006). *E. histolytica* HM-1:IMSS is the canonical lab strain and was used to determine the amoeba genome sequence (Loftus *et al.*, 2005), and thus it is the strain upon which all the microarrays are based. Another *Entamoeba* species, *E. dispar*, is capable of colonizing the human intestine but appears to have no potential to invade tissue and cause disease, even in immunocompromised individuals, and treatment of colonization is not recommended (WHO, 1997). Previous attempts to analyse the differences between *E. histolytica* and *E. dispar* have been productive avenues for identifying virulence pathways in *E. histolytica* (Ankri *et al.*, 1999; Bruchhaus *et al.*, 1996). A number of groups have used microarray technology to compare virulent and non-virulent strains in order to identify novel virulence determinants (Fig. 6.2).

Comparative genomic hybridization

With the development of a genomic DNA microarray the comparative genomic hybridization (CGH) approach could be undertaken as a means to identify differences between virulent and non-virulent *Entamoeba* at the level of DNA sequence (Shah *et al.*, 2005). Genomic DNA isolated from two strains of *E. dispar* (SAW760 and SAW1734), three virulent strains of *E. histolytica* (HM-1:IMSS, 200:NIH and HK-9) and one non-virulent *E. histolytica* strain (Rahman) were hybridized to a sequenced genomic clone microarray. Numerous clones that were divergent in both *E. dispar* strains but conserved in *E. histolytica* were identified, including a gene encoding a putative protein kinase and a peroxiredoxin gene, as well as many novel genes. In addition, 9 genes were identified that were conserved in all virulent strains, but not conserved in either *E. dispar* or *E. histolytica* Rahman. This work not only revealed potential new virulence determinants, but also DNA sequences that could be used to develop diagnostic tests to distinguish *E. histolytica* and *E. dispar* infections.

Transcriptional analyses of strains with variable virulence potential

Genome wide analysis of transcript (Davis *et al.*, 2007; MacFarlane and Singh, 2006) and protein (Davis *et al.*, 2006) levels in different *Entamoeba* strains has also been used to identify molecular signatures of virulence. Using the genomic microarray mentioned above, MacFarlane and Singh (2006) compared expression profiles of *E. dispar* SAW760, *E. histolytica* HM-1:IMSS and *E. histolytica* Rahman. The data indicated lower

	Non-virulent vs Virulent		
Array used	Genomic clone ~2,200 genes	Long oligonucleotide[a] ~9,000 genes	Affymetrix[b] ~9,435 genes
Experimental details	Expression profiling[c]; Comparative genomic hybridization[d]		
Gene regulation	32 genes with higher expression in virulent strains	152 genes with higher expression in virulent strains 201 genes with lower expression in virulent strains	254 genes with higher expression in virulent strains 358 genes with lower expression in virulent strains
Further analysis	STIRP[e]	Peroxiredoxin[f]	
Future questions	Clinical isolates from different geographic locations and disease outcomes Identification of novel virulence genes Identification of novel drug targets		

Figure 6.2 Summary of microarray data from comparison of non-virulent and virulent *Entamoeba* species and strains. The format of the array used, experimental details, extent of gene regulation, gene-specific analysis, and questions for the future are outlined. (a) Adapted from Davis *et al.*, 2007. (b) Adapted from Ehrenkaufer *et al.*, 2007. (c) Adapted from MacFarlane *et al.*, 2006. (d) Adapted from Shah *et al.*, 2005. (e) Adapted from MacFarlane *et al.*, 2007. (f) Adapted from Davis *et al.*, 2008.

expression in non-virulent strains of genes linked to oxidative stress, including genes encoding Fe-hydrogenase, peroxiredoxin, and a type A flavoprotein. Davis *et al.* (2006) found similar results, using differential 2-D gel electrophoresis to demonstrate reduced levels of peroxiredoxin as well as superoxide dismutase (SOD) protein. Trophozoites are not exposed to oxygen during colonic lumen colonization, but upon tissue invasion they encounter an oxygen-rich environment. Thus, increased expression of these proteins may enable survival of virulent parasites in tissue. The peroxiredoxin gene product has been linked to virulence, as overexpression of this gene in *E. histolytica* Rahman was shown to increase parasite survival under oxidative stress conditions (Davis *et al.*, 2006). Other interesting variations between *Entamoeba* strains uncovered by these studies include differences in expression of several small GTPases, lysozyme, calmodulin, and retrotransposon elements.

Gene-specific analysis: identification of novel virulence genes

To prove that genes with differential expression between virulent and non-virulent strains of *Entamoeba* have the potential to be virulence factors, MacFarlane *et al.* (2007) used a genetic approach with the E. histolytica serine-, threonine-, isoleucine- rich protein (*Eh*STIRP). EhSTIRP contains a transmembrane domain and a large putative extracellular domain, and is expressed at high levels in virulent strains of *E. histolytica*. However, it has no detectable expression in *E. histolytica* Rahman. Knockdown of the transcript of EhSTIRP in a virulent strain using dsRNA decreased cytotoxicity and adhesion. This work illustrates that identification of molecular differences between virulent and non-virulent *Entamoeba* is a productive avenue of investigation for identification of novel virulence genes and pathways relevant to amoebic pathogenesis.

Transcriptional networks associated with stress

How the parasite modulates its genetic repertoire during tissue invasion is an important area of investigation. A number of stress conditions have been tested *in vitro*, and transcriptomes of the parasite response obtained (Fig. 6.3).

Heat shock

E. histolytica must adapt to increased temperatures during tissue invasion, especially during hepatic invasion and within hepatic abscesses. Recent investigations have begun to shed light on the molecular changes that occur during heat stress in *Entamoeba* (Hackney *et al.*, 2007; Weber *et al.*, 2006). Significant downregulation of gene expression was seen in both studies, with Hackney *et al.* finding ~13% of genes downregulated ≥ 2-fold and Weber *et al.* finding even more drastic changes with over 40% of genes showing decreased expression. This difference may reflect the platform used (Weber *et al.* used a 1131 gene 70-mer oligonucleotide array, while Hackney *et al.* used a full genome Affymetrix-based array) or the conditions used for heat shock (4 hr at 42 °C vs. 1 hr at 42 °C). In both studies, however, the number of genes downregulated was much larger than the number upregulated, indicating that downregulation of transcription is probably a real biological response. Furthermore, this response was very rapid, with significant changes observed after only one hour of heat shock, indicating that either the regulated genes are rapidly turned off with heat stress, and/or their transcripts have a short half life and are rapidly degraded. Significant upregulation of genes known in other organisms to participate in the response to heat shock and the accumulation of unfolded proteins, including Hsp family proteins, chaperones, ubiquitin, and protein disulfide isomerase, were also observed, as expected. Additionally, members of a large family of stress sensitive proteins apparently restricted to *Entamoeba* were also upregulated by heat shock, and identified as having a conserved promoter motif architecture (Hackney *et al.*, 2007).

	Heat shock		Oxidative and nitrosative stress
Array used	Long oligonucleotide[a] ~3,000 genes	Affymetrix[b] ~9,435 genes	Affymetrix[c] ~9,435 genes
Experimental details	42 °C, 4h	42 °C, 1h	H_2O_2 1mM, 60min DPTA-NONOate 200μM, 60min
Gene regulation	604 genes upregulated 1507 genes downregulated		H_2O_2 • 184 genes upregulated • 102 genes downregulated DPTA • 443 genes upregulated • 593 genes downregulated
Overlap	111 genes regulated in H_2O_2 322 genes regulated in DPTA		164 genes regulated in both H_2O_2 and DPTA
Future questions	Relevance to host invasion and adaptation Virulence determinants		

Figure 6.3 Summary of microarray data from heat shock and oxidative and nitrosative stress conditions. The format of the array used, experimental details, extent of gene regulation, overlap between the conditions listed, and questions for the future are outlined. (a) Adapted from Weber *et al.*, 2006. (b) Adapted from Hackney *et al.*, 2007. (c) Adapted from Vicente *et al.*, 2009.

Oxidative and nitrosative stress

During tissue invasion *E. histolytica* must adapt to changing oxygen tensions; the colonic lumen is largely anaerobic whereas the colonic tissue is oxygen-rich (Stanley, 2003). In this environment, the parasite copes with cytotoxic reactive oxygen species (ROS) and reactive nitrogen species (RNS) that are produced and released by the activated phagocytes attracted to the site of infection. Therefore, one aspect of *E. histolytica*'s pathogenic potential is its ability to cope with oxidative and nitrosative stresses generated during tissue invasion. In order to investigate the transcriptional response to oxidative and nitrosative stress, virulent *E. histolytica* HM-1:IMSS trophozoites were exposed to H_2O_2 (1mM) or a NO donor (DPTA-NONOate, 200μM) for 1 hour and expression profiling performed using whole-genome DNA microarrays (Vicente *et al.*, 2009). In both cases, significant changes in gene expression were noted, with 1036 genes differentially regulated by DPTA-NONOate, and 286 genes differentially regulated by H_2O_2 treatment. There was substantial overlap in the transcriptional response to the two compounds, indicating that there are some general responses to these stresses. As expected, a number of the transcriptionally regulated genes were in signalling, regulatory, repair and metabolic pathways. However, a substantial number of genes regulated under these conditions encode proteins of unknown function, suggesting potentially novel response pathways in *E. histolytica*. In contrast to the substantial expression changes noted in the virulent *E. histolytica* strain, trophozoites of the non-virulent *E. histolytica* strain Rahman had a significantly muted transcriptional response to oxidative stress. Most genes induced by H_2O_2 treatment in *E. histolytica* HM-1:IMSS were not induced by the same treatment in *E. histolytica* Rahman. Additionally, even the limited number of genes that had similar response patterns between the virulent and non-virulent strains had significantly lower levels of transcriptional change in the latter. The substantial difference hints that differential response to oxidative stress may be one factor that contributes to the pathogenic potential of *E. histolytica* and could help to explain the decreased virulence of *E. histolytica* Rahman. Further studies with specific genes will be useful in dissecting the transcriptional networks associated with response to oxidative and nitrosative stresses.

Epigenetic control of transcriptional regulatory networks

Investigations of gene expression regulation have demonstrated that epigenetic modifications such as DNA methylation and histone acetylation can influence gene expression in *Entamoeba*. For example, a DNA methyltransferase 2 family member appears to be active in the control of *E. histolytica* Hsp70 gene expression (Fisher *et al.*, 2004). The effects of DNA methylation on amoebic gene expression and parasite biology was investigated by treatment of *E. histolytica* trophozoites with 5-Azacytidine, which inhibits cytosine methylation (Fig. 6.4). This treatment reduces the cytotoxic ability of *E. histolytica* trophozoites as well as causing changes in genes expression (Ali *et al.*, 2007; Fisher *et al.*, 2004). Surprisingly, the effect on parasite gene expression was relatively limited, with only 199 genes (~2% of annotated genes) showing significant changes. Given the substantial effects on cytotoxicity the limited transcriptional response was unexpected. The effect of histone acetylation was discussed previously.

Identification of complex transcriptional regulatory networks in *E. histolytica*

Bioinformatic analyses of large-scale genomic datasets allow identification of transcriptional regulatory networks. In systems for which extremely large microarray datasets are publicly available, this approach has been used successfully (Shen-Orr *et al.*, 2002). Such efforts facilitated identification of genetic regulatory networks and promoter motifs that control coordinately regulated genes. Canonical promoter elements have been identified as regulating amoebic gene expression. Thus, this bioinformatic approach was applied to microarray data from *E. histolytica* to help identify regulatory motifs controlling coordinately regulated amoebic genes. Using microarray data from two arrays, Hackney *et al.* (2007) identified promoter motifs that were over-represented in genes with either low or high gene expression in *E. histolytica* trophozoites. A number of promoter

	5AzaC[a]	TSA	
Array used	Affymetrix ~9,435 genes	Long oligonucleotide[b] ~3,000 genes	Affymetrix[c] ~9,435 genes
Experimental details		HM-1:IMSS 50nM TSA 48h Increased virulence	200:NIH 150nM TSA 48h Decreased virulence
Gene regulation	68 genes upregulated 131 genes downregulated	13 genes upregulated 31 genes downregulated	122 genes upregulated 41 genes downregulated
Overlap	Negligible	Negligible	Developmentally regulated genes
Future questions	Mechanism of action Effectors of methylation pathway	Causes of variable TSA effects on virulence (parasite strain, dose effect?) Biological consequences	

Figure 6.4 Summary of microarray data from 5-azacytidine treatment and Trichostatin exposure. The format of the array used, experimental details, extent of gene regulation, overlap between the conditions listed, and questions for the future are outlined. (a) Adapted from Ali *et al.*, 2007. (b) Adapted from Isakov *et al.*, 2008. (c) Adapted from Ehrenkaufer *et al.*, 2007a.

motifs were shown to bind amoebic nuclear proteins specifically, validating the ability of bioinformatic approaches to identify functional promoter motifs. One motif ($^{A}/_{T}$AAACCCT) was highly enriched in the promoters of ribosomal protein and tRNA synthetase genes, suggesting that these genes are co-regulated.

In the promoters of genes with low levels of gene expression, three co-associated promoter motifs (GAATGATGC; AACTATTTAAACAT/C; TGAACTTATAAACATC) were identified. Overall 182 genes in the *Entamoeba* genome have either two or three of these motifs in their promoters. Genes with at least two promoter motifs, in any combination, were found to have significantly lower than average expression. Additionally, this set of three promoter motifs was highly enriched in a family of stress responsive genes, Ehssp. This gene family initially was identified as responsive to heat and oxidative stress (Satish *et al.*, 2003). Many members of this gene family were responsive to heat shock and regulated during stage conversion (Ehrenkaufer *et al.*, 2007b). Specific binding of amoebic nuclear protein(s) to one of these motifs under heat shock conditions was detected. This highly complex regulatory circuit was identified using relatively limited transcriptional datasets. Thus, a similar bioinformatics approach could be utilized in the future to identify further regulatory networks relevant to amoebic pathogenesis, invasion, stress response, and stage conversion.

Proteomic studies of *Entamoeba*

Proteomic analyses of *Entamoeba* have investigated the steady-state repertoire of amoebic proteins (Tolstrup *et al.*, 2007), phagosome composition and differences among *E. histolytica* strains (Marion *et al.*, 2005; Okada *et al.*, 2005; Okada *et al.*, 2006), differences between virulent and non-virulent amoebic strains (Davis *et al.*, 2006), and responses to drug challenge (exposure to metronidazole) (Tazreiter *et al.*, 2008).

Using differential in-gel 2D electrophoresis (DIGE) with two colour images, Davis *et al.* (2006) determined the protein repertoire of virulent *E. histolytica* HM-1:IMSS compared to non-virulent *E. histolytica* Rahman. On a given slide approximately 2754 distinct fluorescent spots were observed computationally with ~20 spots identified as being fivefold higher in HM-1:IMSS compared with Rahman and an equal number being fivefold lower in HM-1:IMSS compared with Rahman. However, when the criteria were made more stringent (a minimum of fivefold difference in intensity, reproducibly identified in two separate DIGE runs with biological replicates), only six proteins were identified that showed differential expression between Rahman and HM-1:IMSS. Two of these proteins (peroxiredoxin and superoxide dismutase) have known antioxidant properties while three proteins were of unknown function (grainin 1, grainin 2 and a protein containing a LIM-domain). Alcohol dehydrogenase was also differentially expressed between the virulent and non-virulent amoebic strains. The function of peroxiredoxin in amoebic virulence was demonstrated by generating an *E. histolytica* Rahman strain that overexpressed the peroxiredoxin protein and showing that these parasites have increased resistance to oxidative stress and greater virulence properties.

E. histolytica is professionally phagocytic and this ability is considered essential to the parasite's virulence arsenal. In order to determine the protein makeup of the phagosome, two groups generated proteomic profiles of the early amoebic phagosomes (Marion *et al.*, 2005; Okada *et al.*, 2005). Marion *et al.* used magnetic beads covered with proteins from human serum to study the early events involved in phagocytosis by *E. histolytica*. Purified early phagosomes were isolated and proteomic analysis by liquid chromatography and tandem mass spectroscopy (LC-MS/MS) undertaken. Proteins identified in this manner included proteins involved in cytoskeleton activity, signalling, endocytosis, lytic activity and cell surface proteins. A number of signalling proteins present in the phagosome were localized by immunolabelling during erythrophagocytosis and were identified at the tips of the extending phagocytic pseudopods. This localization suggests an active role for these proteins during erythrophagocytosis and confirmed the feasibility of a proteomic approach to identify proteins in the phagosome. Okada *et al.* (2005) used carboxylated latex beads that were internalized by centrifugation and the resulting phagosomes purified. Proteomic analysis of phagosomes was by liquid chromatography and mass spectrometry and this identified 85 proteins involved in surface recognition, actin cytoskeleton rearrangement, vesicular trafficking, and degradation. Phagosome localization of representative proteins was verified by immunofluorescence assays. A follow-up study analysed the kinetics of phagocytosis and differences in phagocytosis among *E. histolytica* strains (Okada *et al.*, 2006). Intact phagosomes were isolated at various intervals after internalization of latex beads and analysed by reversed-phase capillary liquid chromatography and ion trap tandem mass spectrometry. A total of 159 proteins were identified from the reference strain at different stages of phagosome maturation of which ~70% were detected in a time-dependent fashion, suggesting dynamic aspects of phagosome biogenesis. The kinetics of representative proteins were verified by immunoblots and video microscopy. When phagosome profiles of recent clinical isolates were determined, significant strain-specific differences were observed. A total of ~229 proteins were detected in at least one of these three strains with ~60% identified only in one strain and only 20% of the proteins detected in all three strains.

Conclusions and future directions

Identifying the transcriptome and proteome of *E. histolytica* parasites has been a productive avenue of investigation. The work has taken the study of amoebic biology into the post genomic arena. Multiple novel aspects of amoebic biology have been elucidated including identification of genes regulated specifically during host invasion and stage conversion. Future investigations of the roles of transcriptionally modulated genes, especially the large numbers of genes restricted to *Entamoeba*, will be an important avenue of investigation.

References

Abhyankar, M.M., Hochreiter, A.E., Hershey, J., Evans, C., Zhang, Y., Crasta, O., Sobral, B.W., Mann, B.J., Petri, W.A., Jr., and Gilchrist, C.A. (2008). Characterization of an *Entamoeba histolytica* high-mobility-group box protein induced during intestinal infection. Eukaryot. Cell 7, 1565–1572.

Ali, I.K.M., Ehrenkaufer, G.M., Hackney, J.A., and Singh, U. (2007). Growth of the protozoan parasite *Entamoeba histolytica* in 5-azacytidine has limited effects on parasite gene expression. BMC Genomics *8*, 7.

Ankri, S., Padilla-Vaca, F., Stolarsky, T., Koole, L., Katz, U., and Mirelman, D. (1999). Antisense inhibition of expression of the light subunit (35 kDa) of the Gal/GalNac lectin complex inhibits *Entamoeba histolytica* virulence. Mol. Microbiol. *33*, 327–337.

Ankri, S., Stolarsky, T., and Mirelman, D. (1998). Antisense inhibition of expression of cysteine proteinases does not affect *Entamoeba histolytica* cytopathic or haemolytic activity but inhibits phagocytosis. Mol. Microbiol. *28*, 777–785.

Balamuth, W. (1951). Biological studies on *Entamoeba histolytica*. III. Induced encystation in several mediums, including an account of a new procedure. J. Infect. Dis. *88*, 230–236.

Barron-Gonzalez, M.P., Villarreal-Trevino, L., Resendez-Perez, D., Mata-Cardenas, B.D., and Morales-Vallarta, M.R. (2008). *Entamoeba histolytica*: cyst-like structures *in vitro* induction. Exp. Parasitol. *118*, 600–603.

Beck, D.L., Boettner, D.R., Dragulev, B., Ready, K., Nozaki, T., and Petri, W.A., Jr. (2005). Identification and gene expression analysis of a large family of transmembrane kinases related to the Gal/GalNAc lectin in *Entamoeba histolytica*. Eukaryot. Cell *4*, 722–732.

Boettner, D.R., Huston, C.D., Linford, A.S., Buss, S.N., Houpt, E., Sherman, N.E., and Petri, W.A., Jr. (2008). *Entamoeba histolytica* phagocytosis of human erythrocytes involves PATMK, a member of the transmembrane kinase family. PLoS Pathog. *4*, e8.

Bruchhaus, I., Jacobs, T., Leippe, M., and Tannich, E. (1996). *Entamoeba histolytica* and *Entamoeba dispar*: differences in numbers and expression of cysteine proteinase genes. Mol. Microbiol. *22*, 255–263.

Bruchhaus, I., Loftus, B.J., Hall, N., and Tannich, E. (2003). The intestinal protozoan parasite *Entamoeba histolytica* contains 20 cysteine protease genes, of which only a small subset is expressed during *in vitro* cultivation. Eukaryot. Cell *2*, 501–509.

Byers, J., Faigle, W., and Eichinger, D. (2005). Colonic short-chain fatty acids inhibit encystation of *Entamoeba invadens*. Cell. Microbiol. *7*, 269–279.

Chayen, A., Avron, B., Nuchamowitz, Y., and Mirelman, D. (1988). Appearance of sialoglycoproteins in encysting cells of *Entamoeba histolytica*. Infect. Immun. *56*, 673–681.

Chu, S., DeRisi, J., Eisen, M., Mulholland, J., Botstein, D., Brown, P.O., and Herskowitz, I. (1998). The transcriptional program of sporulation in budding yeast. Science *282*, 699–705.

Cleary, M.D., Singh, U., Blader, I.J., Brewer, J.L., and Boothroyd, J.C. (2002). *Toxoplasma gondii* asexual development: identification of developmentally regulated genes and distinct patterns of gene expression. Eukaryot. Cell *1*, 329–340.

Davis, P.H., Schulze, J., and Stanley, S.L., Jr. (2007). Transcriptomic comparison of two *Entamoeba histolytica* strains with defined virulence phenotypes identifies new virulence factor candidates and key differences in the expression patterns of cysteine proteases, lectin light chains, and calmodulin. Mol. Biochem. Parasitol. *151*, 118–128.

Davis, P.H., Zhang, X., Guo, J., Townsend, R.R., and Stanley, S.L., Jr. (2006). Comparative proteomic analysis of two *Entamoeba histolytica* strains with different virulence phenotypes identifies peroxiredoxin as an important component of amoebic virulence. Mol. Microbiol. *61*, 1523–1532.

Debnath, A., Das, P., Sajid, M., and McKerrow, J.H. (2004). Identification of genomic responses to collagen binding by trophozoites of *Entamoeba histolytica*. J. Infect. Dis. *190*, 448–457.

Dumur, C.I., Garrett, C.T., Archer, K.J., Nasim, S., Wilkinson, D.S., and Ferreira Gonzalez, A. (2004). Evaluation of a linear amplification method for small samples used on high-density oligonucleotide microarray analysis. Anal. Biochem. *331*, 314–321.

Ehrenkaufer, G., Hackney, J.A., and Singh, U. (2009). A developmentally regulated Myb domain protein regulates expression of a subset of stage-specific genes in *Entamoeba histolytica*. Cell. Microbiol. *11*, 898–910.

Ehrenkaufer, G.M., Eichinger, D.J., and Singh, U. (2007a). Trichostatin A effects on gene expression in the protozoan parasite *Entamoeba histolytica*. BMC Genomics *8*, 216.

Ehrenkaufer, G.M., Haque, R., Hackney, J.A., Eichinger, D.J., and Singh, U. (2007b). Identification of developmentally regulated genes in *Entamoeba histolytica*: insights into mechanisms of stage conversion in a protozoan parasite. Cell. Microbiol. *9*, 1426–1444.

Eichinger, D. (1997). Encystation of *Entamoeba* parasites. Bioessays *19*, 633–639.

Fisher, O., Siman-Tov, R., and Ankri, S. (2004). Characterization of cytosine methylated regions and 5-cytosine DNA methyltransferase (Ehmeth) in the protozoan parasite *Entamoeba histolytica*. Nucleic Acids Res. *32*, 287–297.

Gagliardi, M.C., De Petrillo, G., Salemi, S., Boffa, L., Longobardi, M.G., Dellabona, P., Casorati, G., Tanigaki, N., Harris, R., Lanzavecchia, A., *et al.* (1995). Presentation of peptides by cultured monocytes or activated T cells allows specific priming of human cytotoxic T lymphocytes *in vitro*. Int. Immunol. *7*, 1741–1752.

Gilchrist, C.A., Baba, D.J., Zhang, Y., Crasta, O., Evans, C., Caler, E., Sobral, B.W., Bousquet, C.B., Leo, M., Hochreiter, A., *et al.* (2008). Targets of the *Entamoeba histolytica* Transcription Factor URE3-BP. PLoS Negl. Trop. Dis. *2*, e282.

Gilchrist, C.A., Holm, C.F., Hughes, M.A., Schaenman, J.M., Mann, B.J., and Petri, W.A., Jr. (2001). Identification and characterization of an *Entamoeba histolytica* upstream regulatory element 3 sequence-specific DNA-binding protein containing EF-hand motifs. J. Biol. Chem. *276*, 11838–11843.

Gilchrist, C.A., Houpt, E., Trapaidze, N., Fei, Z., Crasta, O., Asgharpour, A., Evans, C., Martino-Catt, S., Baba, D.J., Stroup, S., *et al.* (2006). Impact of intestinal colonization and invasion on the *Entamoeba histolytica* transcriptome. Mol. Biochem. Parasitol. *147*, 163–176.

Gilchrist, C.A., Mann, B.J., and Petri, W.A., Jr. (1998). Control of ferredoxin and Gal/GalNAc lectin gene expression in *Entamoeba histolytica* by a cis-acting DNA sequence. Infect. Immun. *66*, 2383–2386.

Hackney, J.A., Ehrenkaufer, G.M., and Singh, U. (2007). Identification of putative transcriptional regulatory networks in *Entamoeba histolytica* using Bayesian inference. Nucleic Acids Res. *35*, 2141–2152.

Haque, R., Huston, C.D., Hughes, M., Houpt, E., and Petri, W.A., Jr. (2003). Amebiasis. N. Engl. J. Med. *348*, 1565–1573.

Hayward, R.E., Derisi, J.L., Alfadhli, S., Kaslow, D.C., Brown, P.O., and Rathod, P.K. (2000). Shotgun DNA microarrays and stage-specific gene expression in *Plasmodium falciparum* malaria. Mol. Microbiol. *35*, 6–14.

Hellberg, A., Nickel, R., Lotter, H., Tannich, E., and Bruchhaus, I. (2001). Overexpression of cysteine proteinase 2 in *Entamoeba histolytica* or *Entamoeba dispar* increases amoeba-induced monolayer destruction *in vitro* but does not augment amoebic liver abscess formation in gerbils. Cell. Microbiol. *3*, 13–20.

Houpt, E., Barroso, L., Lockhart, L., Wright, R., Cramer, C., Lyerly, D., and Petri, W.A. (2004). Prevention of intestinal amebiasis by vaccination with the *Entamoeba histolytica* Gal/GalNac lectin. Vaccine *22*, 611–617.

Isakov, E., Siman-Tov, R., Weber, C., Guillen, N., and Ankri, S. (2008). Trichostatin A regulates peroxiredoxin expression and virulence of the parasite *Entamoeba histolytica*. Mol. Biochem. Parasitol. *158*, 82–94.

James, P. (1997). Protein identification in the post-genome era: the rapid rise of proteomics. Q. Rev. Biophys. *30*, 279–331.

Lockhart, D.J., Dong, H., Byrne, M.C., Follettie, M.T., Gallo, M.V., Chee, M.S., Mittmann, M., Wang, C., Kobayashi, M., Horton, H., *et al.* (1996). Expression monitoring by hybridization to high-density oligonucleotide arrays. Nat. Biotechnol. *14*, 1675–1680.

Loftus, B., Anderson, I., Davies, R., Alsmark, U.C., Samuelson, J., Amedeo, P., Roncaglia, P., Berriman, M., Hirt, R.P., Mann, B.J., *et al.* (2005). The genome of the protist parasite *Entamoeba histolytica*. Nature *433*, 865–868.

MacFarlane, R.C., Shah, P.H., and Singh, U. (2005). Transcriptional profiling of *Entamoeba histolytica* trophozoites. Int. J. Parasitol. *35*, 533–542.

MacFarlane, R.C., and Singh, U. (2006). Identification of differentially expressed genes in virulent and non-virulent *Entamoeba* species: potential implications for amebic pathogenesis. Infect. Immun. *74*, 340–351.

MacFarlane, R.C., and Singh, U. (2007). Identification of an *Entamoeba histolytica* serine-, threonine-, and isoleucine-rich protein with roles in adhesion and cytotoxicity. Eukaryot. Cell *6*, 2139–2146.

Marion, S., Laurent, C., and Guillen, N. (2005). Signalization and cytoskeleton activity through myosin IB during the early steps of phagocytosis in *Entamoeba histolytica*: a proteomic approach. Cell. Microbiol. *7*, 1504–1518.

Okada, M., Huston, C.D., Mann, B.J., Petri, W.A., Jr., Kita, K., and Nozaki, T. (2005). Proteomic analysis of phagocytosis in the enteric protozoan parasite *Entamoeba histolytica*. Eukaryot. Cell *4*, 827–831.

Okada, M., Huston, C.D., Oue, M., Mann, B.J., Petri, W.A., Jr., Kita, K., and Nozaki, T. (2006). Kinetics and strain variation of phagosome proteins of *Entamoeba histolytica* by proteomic analysis. Mol. Biochem. Parasitol. *145*, 171–183.

Pelosof, L.C., Davis, P.H., Zhang, Z., Zhang, X., and Stanley, S.L., Jr. (2006). Co-ordinate but disproportionate activation of apoptotic, regenerative and inflammatory pathways characterizes the liver response to acute amebic infection. Cell. Microbiol. *8*, 508–522.

Phillips, B.P. (1973). *Entamoeba histolytica*: concurrent irreversible loss of infectivity-pathogenicity and encystment potential after prolonged maintenance in axenic culture *in vitro*. Exp. Parasitol. *34*, 163–167.

Santi-Rocca, J., Weber, C., Guigon, G., Sismeiro, O., Coppee, J.Y., and Guillen, N. (2008). The lysine- and glutamic acid-rich protein KERP1 plays a role in *Entamoeba histolytica* liver abscess pathogenesis. Cell. Microbiol. *10*, 202–217.

Satish, S., Bakre, A.A., Bhattacharya, S., and Bhattacharya, A. (2003). Stress-dependent expression of a polymorphic, charged antigen in the protozoan parasite *Entamoeba histolytica*. Infect. Immun. *71*, 4472–4486.

Saxena, A., Lahav, T., Holland, N., Aggarwal, G., Anupama, A., Huang, Y., Volpin, H., Myler, P.J., and Zilberstein, D. (2007). Analysis of the *Leishmania donovani* transcriptome reveals an ordered progression of transient and permanent changes in gene expression during differentiation. Mol. Biochem. Parasitol. *152*, 53–65.

Seigneur, M., Mounier, J., Prevost, M.C., and Guillen, N. (2005). A lysine- and glutamic acid-rich protein, KERP1, from *Entamoeba histolytica* binds to human enterocytes. Cell. Microbiol. *7*, 569–579.

Shah, P.H., MacFarlane, R.C., Bhattacharya, D., Matese, J.C., Demeter, J., Stroup, S.E., and Singh, U. (2005). Comparative genomic hybridizations of *Entamoeba* strains reveal unique genetic fingerprints that correlate with virulence. Eukaryot. Cell *4*, 504–515.

Shen-Orr, S.S., Milo, R., Mangan, S., and Alon, U. (2002). Network motifs in the transcriptional regulation network of *Escherichia coli*. Nat. Genet. *31*, 64–68.

Singh, U., and Ehrenkaufer, G.M. (2009). Recent insights into *Entamoeba* development: identification of transcriptional networks associated with stage conversion. Int. J. Parasitol. *39*, 41–47.

Stanley, S.L., Jr. (2003). Amoebiasis. Lancet *361*, 1025–1034.

Tazreiter, M., Leitsch, D., Hatzenbichler, E., Mair-Scorpio, G.E., Steinborn, R., Schreiber, M., and Duchene, M. (2008). *Entamoeba histolytica*: response of the parasite to metronidazole challenge on the levels of mRNA and protein expression. Exp. Parasitol. *120*, 403–410.

Tillack, M., Biller, L., Irmer, H., Freitas, M., Gomes, M.A., Tannich, E., and Bruchhaus, I. (2007). The *Entamoeba histolytica* genome: primary structure and expression of proteolytic enzymes. BMC Genomics *8*, 170.

Tolstrup, J., Krause, E., Tannich, E., and Bruchhaus, I. (2007). Proteomic analysis of *Entamoeba histolytica*. Parasitology *134*, 289–298.

Vicente, J.B., Ehrenkaufer, G.M., Saraiva, L.M., Teixeira, M., and Singh, U. (2009). *Entamoeba histolytica* modulates a complex repertoire of novel genes in response to oxidative and nitrosative stresses: implications for amebic pathogenesis. Cell. Microbiol. *11*, 51–69.

Weber, C., Guigon, G., Bouchier, C., Frangeul, L., Moreira, S., Sismeiro, O., Gouyette, C., Mirelman, D., Coppee, J.Y., and Guillen, N. (2006). Stress by heat shock induces massive down regulation of genes and allows differential allelic expression of the Gal/GalNAc lectin in *Entamoeba histolytica*. Eukaryot. Cell *5*, 871–875.

WHO (1997). Amoebiasis. Wkly. Epidemiol. Rec. *72*, 97–99.

Zhang, T., Cieslak, P.R., and Stanley, S.L., Jr. (1994). Protection of gerbils from amebic liver abscess by immunization with a recombinant *Entamoeba histolytica* antigen. Infect. Immun. *62*, 1166–1170.

Zhang, Z., and Stanley, S.L., Jr. (2004). Stereotypic and specific elements of the human colonic response to *Entamoeba histolytica* and *Shigella flexneri*. Cell. Microbiol. *6*, 535–554.

Manipulation of Gene Expression in *Entamoeba histolytica*

7

Rivka Bracha, Nancy Guillén and David Mirelman

Abstract

Our knowledge of the functions of different structural proteins and virulence factors in the cellular organization and pathogenesis of *Entamoeba histolytica* has significantly increased following the introduction of various molecular techniques that enable the manipulation of gene expression. Unfortunately, to date, all the attempts to integrate exogenous DNA into the parasite's genome have failed and most methods for up- and down-regulation of gene expression have been based on the transfection of stably maintained plasmids. Down-regulation has been achieved by plasmids encoding: (i) antisense RNA, (ii) truncated or mutated genes that exert dominant-negative effects, and (iii) inverted loops that generate double stranded RNA molecules. Small interfering RNA oligonucleotides incorporated directly from the culture medium also cause the down-regulation of specific genes. Recently, epigenetic silencing of the amoebapore gene, which encodes a toxic polypeptide that disrupts the membranes of bacteria and mammalian cells, was induced following transfections with a plasmid construct containing an upstream segment of the amoebapore gene. A number of additional genes have been silenced subsequently in a similar way. Amoebae in which the expression of virulence genes has been down-regulated have in many cases an altered phenotype which sheds light on the specific role of that gene.

Introduction

Entamoeba histolytica is not amenable to many standard genetic techniques as it has no known sexual cycle, the genome is thought to be tetraploid and many virulence determinants are members of large gene families. Methods of homologous recombination have not yet been developed for *E. histolytica*, therefore the only ways to study gene function are by the expression of mutant genes or by gene inactivation. In view of these limitations most of the attempts to manipulate the expression of genes in this organism have focused on stable episomal transfection, for both the over-expression and down-regulation of genes of interest (Alon *et al.*, 1997; Hamann *et al.*, 1995; Moshitch-Moshkovitch *et al.*, 1996; Vines *et al.*, 1995). These transfections require the use of strong promoters of the parasite to drive the expression of the desired transcript, whether in sense, anti-sense or dominant negative gene constructs. Inducible gene expression vectors based on the tetracycline (Tet) repressor system were also developed for *E. histolytica*. The first Tet-inducible plasmid constructs were prepared by Hamann (Hamann *et al.*, 1997) and Ramakrishnan (Ramakrishnan *et al.*, 1997) and were used for the short term induction of over-expression of genes coding for virulence factors, so as to prevent the toxicity associated with accumulation of the up-regulated molecules in the parasite. Recently two additional possibilities for manipulation of gene expression have been introduced: (i) selective inhibition of gene expression by either direct administration or episomal generation of short interfering RNA (siRNA) molecules (Kaur and Lohia, 2004; Vayssié *et al.*, 2004), and (ii) epigenetic gene silencing (Bracha *et al.*, 2003). Each of the above mentioned techniques

(Fig. 7.1) has advantages and disadvantages, as will be discussed in the respective sections below.

Episomal over-expression of selected genes

The development of transfection vectors and methods was not a simple task in *E. histolytica.* The first reported successful transient transfections (Nickel and Tannich, 1994; Purdy *et al.*, 1994) were achieved following the exposure of trophozoites to high voltage electrical pulses. Transient expression was observed from plasmid constructs that had reporter genes under the control of the 5′ and 3′ regulating elements of *E. histolytica* genes such as the galactose/N-acetylgalactosamine (Gal/GalNAc) lectin heavy subunit (*Ehhgl*) or actin (*Ehact*) genes. Other promoters that are widely used for expression in eukaryotic cells, such as viral promoters, failed to express transfected genes in the parasite. The relatively low efficiency of transfection required the development of homogenous and stable cultures. *E. histolytica* trophozoites are very resistant to most antibiotics and thus for the preparation of plasmid constructs that would enable selection of stable transfectants the two possibilities were to introduce either the Neomycin or Hygromycin resistance genes. These resistance genes were cloned into plasmids under the regulatory elements of the *Ehact* or *Ehhgl* genes (Hamann *et al.*, 1995; Ramakrishnan *et al.*, 1997; Vines *et al.*, 1995). In order to increase the efficiency of

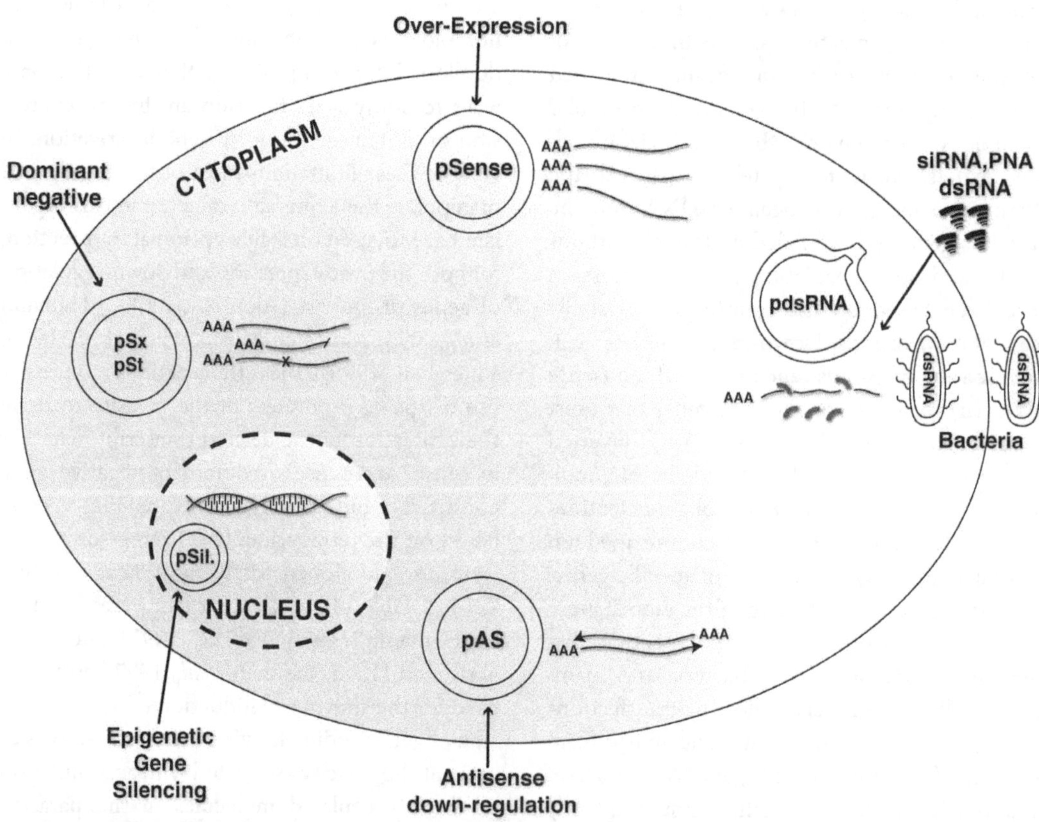

Figure 7.1 Different methods and techniques that have been used for the manipulation of gene expression in *E. histolytica.* Transfection with: (i) pSense – plasmids for over-expression of genes, (ii) pSx and pSt – plasmids that over-express mutated or truncated transcripts for dominant-negative effects, (iii) pAS – plasmids that express anti-sense transcripts for down-regulation, (iv) pdsRNA – plasmids that contain an inverted repeat which expresses dsRNA, (v) pSil. – plasmids that cause epigenetic gene silencing. Delivery of dsRNA has also been accomplished by phagocytosis of bacteria that harbour plasmids which generate dsRNA. Down-regulation of gene expression by PNA or siRNA involves small molecules that are taken up by the parasites when added to the culture medium. Plasmids are depicted in the location of their effect but all are presumed to be located in the nucleus.

replication, an element with autonomous replication sequences (ARS) (Grodberg *et al.*, 1990; Lohia *et al.*, 1990) was also incorporated and a number of plasmid constructs became available for over-expression of genes of choice.

The over-expression of a number of genes has been reported to affect cellular functions of *E. histolytica*. Over-expression of a P-glycoprotein gene was shown to confer resistance to the anti-amoebic drug emetine (Ghosh *et al.*, 1996). Over-expression of myosin IB in *E. histolytica* trophozoites enhanced viscosity of the cytoplasm and reduced phagocytosis (Marion *et al.*, 2004). Following the over-expression of *EhGEF1*, a Rho/Rac guanine nucleotide exchange factor, a decrease in F-actin and a reduction in cell migration and in the amoeba's capacity to damage mammalian cells was observed (Aguilar-Rojas *et al.*, 2005). Over-expression was also used to identify enhancer binding proteins (EhEBPs) that recognize the URE4 promoter motifs (Schaenman *et al.*, 2001). Over-expression of *EhEBP1* resulted in a 7-fold drop in promoter activity in transiently transfected reporter gene constructs that included the URE4 motif, demonstrating the DNA sequence binding specificity of EBP1.

To test the role of the various cysteine proteinases (CPs) in amoeba-induced pathology, the three major proteinases of the parasite, namely EhCP1, EhCP2 and EhCP5, which together account for about 90% of total proteinase activity, were over-expressed by stable episomal transfection. Total CP activity of the transfected amoebae increased by three- to six-fold depending on the gene transfected (Tillack *et al.*, 2006). Interestingly, over-expression of *EhCP1* or *EhCP2* increased the activity of the corresponding enzyme only, whereas over-expression of *EhCP5* increased the activity of all three enzymes, which suggested that EhCP5 also has an enzyme-converting activity. Cytopathic activity, measured by *in vitro* monolayer disruption, was dramatically increased in *EhCP5* transfectants (five-fold) but showed only a modest increase in *EhCP1*- or *EhCP2*-transfectants. In addition, overexpression of *EhCP5* but not of *EhCP1* or *EhCP2* significantly increased the size of amoebic liver abscesses formed in laboratory animals.

Another important aspect of metabolism investigated by up-regulation of gene expression was the mechanism of resistance of *E. histolytica* to metronidazole. Episomal transfection and simultaneous overexpression of two antioxidant enzymes, the iron-containing superoxide dismutase Fe-SOD and peroxiredoxin, revealed a significant reduction in the susceptibility of parasites to metronidazole (Wassmann *et al.*, 1999). Results, however, are not always as expected. An example where over-expression yielded an unexpected result was in the case of amoebapore A (*Ehap-a*) where most of the over-produced protein was dispersed in the cytoplasm rather than localized in the small granules, its normal location, causing a reduction in trophozoite virulence instead of the anticipated increase (Bracha *et al.*, 2002).

Down-regulation of gene expression by anti-sense transcripts

The anti-sense method has been used in many organisms to down-regulate the expression of genes of interest by disrupting the endogenous gene transcript via an anti-sense transcript produced from a stable episomal construct (Scherczinger *et al.*, 1992) The first successful use of the anti-sense method in *E. histolytica* was based on the finding that the parasite possesses two genes encoding ribosomal protein L21 (named *EhRP-L21-gLE3* and *EhRP-L21-g34*) (Petter *et al.*, 1994). Although the open reading frame (ORF) sequences of the two genes are almost identical, they were found to differ significantly in their levels of translation and in the sequences of their 5′ untranslated (UTR) regions. The *gLE3* gene copy contains a short 5′ UTR [15 nucleotides (nt)] and was found to be transcribed and translated, whereas the *g34* gene copy, which contains a longer 5′ UTR (34 nt), was transcribed but not translated in *E. histolytica*. The inverse phenomenon was found to occur in *E. dispar* where the gene copy homologous to *g34* was the one that was actively translated while the *gLE3* copy was only transcribed (Moshitch-Moshkovitch *et al.*, 1998). A high accumulation of the chloroamphenicol acetyl transferase (*CAT*) reporter gene mRNA was observed in transfectants of strain HM-1:IMSS in which the *CAT* gene was placed in a plasmid under the regulation of the *EhRP-L21-g34* gene promoter sequences.

The high levels of *CAT* transcripts observed gave us the idea to exploit this vector for the expression and accumulation of high levels of anti-sense transcripts of a gene of choice so as to increase the blocking of the sense transcript. The first *Entamoeba* gene to be down-regulated using the above concept was that encoding a lysine-rich 30 kDa surface antigen that is normally expressed in *E. dispar* but not in *E. histolytica* (Alon *et al.*, 1997; Moshitch-Moshkovitch *et al.*, 1997). The cloning of the gene in the anti-sense orientation under the *gLE3* promoter (the gene that is not translated in *E. dispar*) and transfection of *E. dispar* trophozoites resulted in a significant decrease in the expression of the 30 kDa antigen (Alon *et al.*, 1997). This anti-sense principle also proved to be effective for down-regulating gene expression in *E. histolytica*. Three genes coding for different virulence factors, *EhCP5* (Ankri *et al.*, 1998), the light subunit (35 kDa) of the Gal/GalNac lectin (*Ehlgl*) (Ankri *et al.*, 1999a) and *Ehap-a* (Bracha *et al.*, 1999), were each cloned in the anti-sense orientation under the promoter of the *EhRP-L21-g34* gene, and after transfection into *E. histolytica* trophozoites each of the three genes showed a very significant reduction in the level of both the sense transcript and the corresponding protein of the endogenous gene. The level of the reduction varied from 60% for the amoebapore protein to 90% in the case of the cysteine proteinase inhibition.

The reduction in the expression of each of these three important virulence factors enabled investigations that have shed light on their role in the parasite's pathogenicity (Ankri *et al.*, 1999b; Bracha *et al.*, 1999). In a model system that uses human intestinal xenografts in SCID mice, the cysteine proteinase-deficient amoebae failed to induce intestinal epithelial cell production of the inflammatory cytokines interleukin (IL)–1B and IL-8, and caused significantly less gut inflammation and damage to the intestinal permeability barrier (Zhang *et al.*, 2000b). These amoebae also failed to produce liver abscesses in hamsters (Ankri *et al.*, 1999b) and had a reduced ability to degrade human intestinal mucus (Moncada *et al.*, 2006). Anti-sense inhibition of *Ehlgl* resulted in reduction of the trophozoites' cytopathic activity, cytotoxic activity, and ability to induce the formation of liver lesions in hamsters (Ankri *et al.*, 1999a). Anti-sense down-regulation of the *Ehap-a* gene proved to be more problematic, since it required the inclusion of the 5′ and 3′ UTR regions of the gene in the anti-sense plasmid construct (Bracha *et al.*, 2000). However, once the transfectant was produced and the level of amoebapore became reduced, the amoebae were found to be significantly less virulent. Cytopathic and cytolytic activities as well as lysis of red blood cells were markedly inhibited (Bracha *et al.*, 1999). Moreover, trophozoite extracts of those transfectants displayed lower pore-forming activity and were less potent in inhibiting bacterial growth compared to controls. In addition, their ability to induce the formation of liver abscesses in hamsters was totally impaired. These experiments provided the first direct evidence for the crucial role that amoebapores play in parasite virulence.

Using the same anti-sense principle, expression of the gene coding for aldehyde dehydrogenase (*EhADH2*) was also reduced (Chen *et al.*, 2004). The NAD-dependent ADH activity of the transfected trophozoites was 60–70% lower and this was accompanied by a marked decrease in amoebic growth and viability. In addition, the down-regulation of *EhMLBP*, a protein that specifically binds to methylated long interspersed elements (LINEs), resulted in trophozoites with impaired growth and cytopathic activity (Lavi *et al.*, 2008).

Tetracycline-induced plasmids to down-regulate gene expression

The Tet-inducible system has also been successfully adapted to express anti-sense transcripts (Hamann *et al.*, 1997). This system was first introduced to inhibit the expression of *EhCaBP1*, one of the calcium binding protein genes of *E. histolytica* (Sahoo *et al.*, 2004). A 50% reduction in the level of the protein was obtained under tetracycline induction and resulted in inhibition of cellular proliferation. The investigators also found that *EhCaBP1* is involved in dynamic changes of the actin cytoskeleton and the down-regulation of *EhCaBP1* resulted in severely impaired endocytosis and phagocytosis. The Tet anti-sense system has also been used to down-regulate the expression of the *E. histolytica*

GlcNAc-phosphatidylinositol deacetylase, which participates in the biosynthesis of glycosylphosphatidylinositol (GPI) anchors (Vats *et al.*, 2005). Induction of the expression of the anti-sense transcript resulted in reduced cell proliferation, endocytosis, and adhesion to target cells. Further evidence for the importance of GPI anchors for amoebic pathogenesis was obtained by the down regulation of the expression of mannosyltransferase (*EhPIG-M1*), which is another gene required for synthesis of GPI. Lower amounts of GPI and proteophosphoglycan (PPG) (Weber *et al.*, 2008) were found on the cell surfaces and the transfected trophozoites had higher sensitivity to complement and displayed a dramatic reduction in their ability to induce liver abscesses in hamsters.

More recently, inhibition of gene expression by Tet-induced antisense RNA technology was used to down-regulate (by 55%) the expression of the 29 kDa thiol-dependent peroxidase Eh29. The transfected trophozoites showed increased reactive oxygen species (ROS) accumulation and had a significantly lower viability during oxidative stress. Cytopathic and cytotoxic activities were also found to decrease in the Eh29 down-regulated trophozoites (Sen *et al.*, 2007). An intricate regulation of gene expression was discovered following anti-sense expression of the gene encoding KERP1, the lysine and glutamic acid rich protein 1 (Santi-Rocca *et al.*, 2008). Induction of anti-sense *EhKERP1* mRNA by Tet led to the transcription of a short RNA that had homology to the 5′ end of *EhKERP1* mRNA and caused, under stress conditions, a reduction in the levels of KERP1. These amoebae were particularly sensitive to tissue responses during early stages of liver abscess formation.

Peptide nucleic acids (PNA)

A different way to down-regulate the expression of genes is by using synthetic anti-sense peptide nucleic acids (PNAs) (Ray and Norden, 2000). PNA oligomers have a polyamide backbone in place of the sugar–phosphate DNA backbone, are uncharged, and are stable to cleavage by proteases and nucleases. A significant inhibition of episomally expressed neomycin phosphorotransferase (NPT) was observed following the addition of micromolar concentrations of PNA anti-sense oligomers specific for NPT to the growth medium of growing trophozoite cultures. PNA oligomers also succeeded in inhibiting the expression of *EhErd2*, a chromosomal gene that codes for a Golgi membrane receptor homologue (Stock *et al.*, 2001). In both cases the amount of the relevant protein detected was reduced by around 45% and this caused a reduction in trophozoite proliferation. Another gene whose expression was significantly inhibited following treatment of amoebae (96h) with 20 μM of a specific PNA oligomer was *EhSec61*, a conserved essential integral component of the endoplasmic reticulum (Sánchez *et al.*, 2005). The decrease in Sec61 protein levels was 80% and this also correlated with a decrease in proliferation of 50%, but the dramatic decrease did not affect the viability of the surviving trophozoites and did not significantly inhibit their ability to induce liver damage.

Expression of genes that produce a dominant-negative effect

Down-regulation of the activity of certain proteins was demonstrated in several cases by over-expressing a mutated or truncated form of the gene product, which then interfered in a dominant-negative form with the role of the native protein. This approach was used successfully in *E. histolytica* in order to understand the role of Myosin II in the amoebic capping process (Coudrier *et al.*, 2005). Over-expression of the light meromyosin domain (LMM), which is essential for myosin II filament formation, resulted in dominant-negative transfectants that had properties similar to a myosin II null phenotype. The transfectants grew at a slower rate, poly-nucleation was observed in numerous cells, they had slower motility and shorter pseudopods, and the formation of caps and uroids was reduced dramatically. LMM parasites were unable to form liver abscesses and were highly sensitive to the immune response (Tavares *et al.*, 2005).

The heavy subunit of the Gal/GalNAc lectin, which is an important virulence factor, was also subjected to a dominant-negative investigation by the over-expression of its cytoplasmic tail (Hgl1 Lectin-GFP fusion 224). The transfectants were found to have a decreased extracellular adhesion activity (Vines *et al.*, 1998). The introduction of a mutation in the sequence of the over-expressed cytoplasmatic tail resulted in loss of the dominant-negative phenotype. Interestingly, the altered phenotype in the 224-expressing amoebae specifically resulted in inhibition of the adherence function of the lectin but not in cytolysis (independent of adherence), phagocytosis of human erythrocytes or resistance to complement lysis. A second type of dominant negative amoeba (HGL2) over-expressing the cytoplasmic domain of the *E. histolytica* Gal/GalNAc lectin heavy subunit gene *hgl2* caused reduced adhesion to enterocytes (Tavares *et al.*, 2005). The trophozoites no longer exhibited chemotaxis towards tumour necrosis factor (TNF) and were very deficient in inducing liver lesions (Blazquez *et al.*, 2008; Blazquez *et al.*, 2007). HGL2 parasites were impaired in their motility, especially when living in a 3D organized hepatic tissue (Coudrier *et al.*, 2005).

Dominant-negative techniques have also been used to try to understand the role of the light subunit of the Gal/GalNAc lectin (Katz *et al.*, 2002). Over-expression of a truncated form of the gene in which the 55 N-terminal amino acids had been removed had a dominant-negative effect. The over-produced 31kDa truncated protein replaced most of the native 35-kDa light subunits in the disulfide linked Gal/GalNAc lectin heterodimeric complex and exerted a dominant-negative effect that included a significant decrease in the ability of the parasites to adhere to and kill mammalian cells, as well as in their capacity to form rosettes with and to phagocytose erythrocytes. The transfected dominant-negative trophozoites were also impaired in their ability to cap the heavy subunit of the Gal/GalNAc lectin to the uroid region following the binding of anti-lectin antibodies (Katz *et al.*, 2002). The *E. histolytica* bifunctional alcohol dehydrogenase 2 (EhADH2) protein is necessary for amoebic growth and survival. *EhADH2* appears to encode a fusion protein, with a separate N-terminal acetaldehyde dehydrogenase activity (EhALDH) and a C- terminal alcohol dehydrogenase (EhADH) domain. In this case, transfection used a plasmid containing a copy of the *EhADH2* gene in which a single histidine in the putative active site of the ADH domain was mutated; EhADH2 enzymatic activity was completely eliminated, affecting both the ADH and the ALDH activities (Chen *et al.*, 2004).

Transfection and over-expression of a mutated *EhRacG-V12* gene, which encodes an *E. histolytica* protein that is homologous to the human Rac1 small GTPases, resulted in a dominant-negative phenotype. Cells had an altered polarization, were inhibited in cytokinesis, and had alterations in the actin cytoskeleton (Guillén *et al.*, 1998). Several other virulence functions were found to be defective, including the ability to adhere to and destroy host cells and to release cysteine proteinases. The role of the p21-activated kinases (PAKs) as regulators of *E. histolytica* cell motility and morphology was also investigated using the dominant-negative approach. Labruyère *et al.* (Labruyère *et al.*, 2003) showed that over-production of the C-terminal domain of *EhPAK1*, which contains a constitutive kinase activity, caused a significant reduction in amoeboid migration as measured by dynamic image analysis. These amoebae, however, displayed increased erythrophagocytosis.

Rab7 small GTPase plays a crucial role in the regulation of trafficking to late endosomes, lysosomes and phagosomes. The roles of the two representative Rab7 isotypes, EhRab7A and EhRab7B, have been investigated (Saito-Nakano *et al.*, 2007). EhRab7A localizes to a non-acidic compartment while EhRab7B localizes exclusively to acidic vacuoles. Over-expression of *EhRab7A* and *EhRab7B* both resulted in increased late endosome/lysosome acidification. In contrast, expression of a permanently activated *EhRab7B-GTP* mutant gave rise to a dominant-negative phenotype that included a decrease in late endosome/lysosome acidification and defective secretion of lysosomal proteins, while expression of an *EhRab7A-GTP* mutant enhanced acidification but had no effect on cysteine proteinase activity, either intracellular or secreted. Expression of *EhRab7B* and the *EhRab7B-GTP* mutant both caused defects in phagocytosis as well as affecting formation and disassembly of pre-phagosomal vacuoles, a compartment known to be linked to efficient ingestion. Together, these data indicate that both Rab7 isotypes are involved in lysosome and phagosome biogenesis but have distinct, although coordinated, roles (Saito-Nakano *et al.*, 2007).

In summary, the use of anti-sense techniques to down-regulate the expression of *E. histolytica* genes and the over-expression of dominant-negative mutated forms have contributed significantly to clarifying the roles that many genes play, especially in pathogenicity. All of these methods have, however, a number of limitations (Petri and Ramakrishnan, 1999). First, in most of the cases, the inhibition of expression that was attained was incomplete. Second, in order to achieve optimal and stable episomal expression of anti-sense or mutated transcripts there is a need to grow the transfected parasites at high concentrations of the selective antibiotic (Geneticin [G418] or Hygromycin, with or without Tetracycline). Prolonged growth of the parasites under such conditions causes some toxicity and affects their growth, and proper controls with similar but sham selective plasmids have to be designed and included in the studies. In the case of the PNA anti-sense oligomers, the addition of high concentrations of such synthetic molecules to the culture medium can be very costly and precludes their long-term use.

Gene silencing by siRNA

Gene silencing by interference with messenger RNA (RNAi) is a process in which a double-stranded RNA (dsRNA) molecule is cleaved into short RNAs that interfere with endogenous RNAs (siRNAs), and these guide the RNase-mediated cleavage of the homologous target RNA. The RNase III-family enzyme Dicer catalyses cleavage of the dsRNA and generates 21 to 25 nt dsRNA with characteristic 5′-phosphate groups and 2-nt 3′-overhangs. The small dsRNA is then unwound by an as-yet unidentified helicase (Peters and Meister, 2007). One of the strands of the siRNA is recruited to the RNAi silencing protein complex (RISC) and binds to a RISC component, Argonaute (Ago), where the siRNA guides the cleavage of the target RNA (Peters and Meister, 2007). The second strand is degraded by nucleases (Song *et al.*, 2004). Seminal contributions to this field have been made by Andrew Fire and Craig C. Mello, for which they were awarded the Nobel Prize for Medicine in 2006 (Fire, 2007). The production of dsRNA in eukaryotic cells generally results from viral replication or transcription of transposable elements and repetitive DNA sequences. *In vivo* production of dsRNA involves a RNA-dependent RNA polymerase (RdRP) (Iyer *et al.*, 2003) that differs from the RdRPs of viruses. Eukaryotic RdRP enzymes produce dsRNA from aberrant single-stranded RNA (presumably lacking normal processing signals such as a 5′-cap or a poly(A) tail). In *E. histolytica*, genes encoding orthologues of RdRP and of Ago (containing a PAZ and a PIWI domain) have been reported to exist (Table 7.1) (Abed and Ankri, 2005). Orthologues of Dicer have not been found in this parasite although an enzyme bearing a RNase III domain is present and active in degrading RNA (Abed and Ankri, 2005; Abed and Ankri, 2008).

Eukaryotic cells also have a second mechanism for RNAi: the micro RNA (miRNA) of 22 nt generated by a Dicer pathway (Guarnieri and

Table 7.1 Genes encoding proteins potentially involved in RNA interference in *E. histolytica*

Gene accession no.		
Pathema	NCBI	Protein
EHI_139420	XM_646217	RdRP
EHI_179800	XM_650495	RdRP
EHI_125650	XM_651422	Ago-like
EHI_188650	XM_651344	Ago-like
EHI_177170	XM_647727	Ago-like
EHI_068740	XM_645408	RNAse III

RdRP, RNA-dependent RNA Polymerase; Ago-like, Argonaute-like; RNaseIII, Ribonuclease III

DiLeone, 2008). The sequences of miRNAs are encoded in the genome, transcribed as primary miRNAs (pri-miRNAs) by RNA polymerase II and cleaved by the RNase III Drosha within the nuclear compartment. The cleavage of pri-miRNA produces a hairpin RNA of 65 nt known as pre-miRNA, which is then transported into the cytoplasm by exportin 5 and further processed by Dicer before entering into the RISC complex. Ago1 and Ago4 are involved in incorporation of miRNA in the RISC complex but no mRNA cleavage has been observed; basically, gene expression is blocked by repression of mRNA translation. Micro RNAs have been found in all metazoans and are involved in regulation of development, cell death, cell proliferation and many other processes. More than 700 miRNAs have been identified in mammals. Using bioinformatics, 32 potential miRNAs have been proposed for *E. histolytica* (De *et al.*, 2006) but as yet no experimental data have supported their potential role in regulation of amoebic gene expression. An interesting finding concerns the description of 27-nt small RNAs in *E. histolytica* that have 5′-polyphosphate termini, an anti-sense orientation to certain genes, and associate with a PIWI-related protein (Zhang *et al.*, 2008). Similar 5′-polyphosphate moiety-bearing small RNAs have been observed only in *Caenorhabditis elegans* (Faehnle and Joshua-Tor, 2007; Pak and Fire, 2007) where they amplify gene silencing. These small RNAs seem to arise from a mechanism independent of Dicer (Sijen *et al.*, 2007), making the observation highly relevant for *E. histolytica,* since this organism is also devoid of Dicer homologues.

RNAi in *E. histolytica*

RNAi has been shown to be effective in *E. histolytica* following the production of dsRNA within the trophozoites. The 5′ end of a Formin gene (Diaphanous) was cloned into a plasmid head to head with an intervening non-specific stuffer DNA and expressed, following transfection in trophozoites, as a dsRNA with a hairpin loop (Kaur and Lohia, 2004). Synthesis of Formin in parasites carrying the dsRNA-encoding construct was inhibited. This method has been used to reduce the expression of several genes, including the transmembrane kinase PATMK (Boettner *et al.*, 2008) and the STIRP gene family (MacFarlane and Singh, 2007). However, after one-year of growth, the strain carrying the STIRP dsRNA construct recovered the wild-type phenotype (MacFarlane and Singh, 2008). The mechanism by which *E. histolytica* became resistant to dsRNA interference is completely unknown. Successful targeting of *E. histolytica* γ-tubulin mRNA has been observed following the soaking of parasites in growth medium containing chemically synthesized siRNA (Solis and Guillén, 2008; Vayssié *et al.*, 2004). The siRNA interference method is reversible since, upon depletion of siRNA from the culture medium, the γ-tubulin mRNA levels returned to normal. Recently, based on the observation that *E. histolytica* ingests bacteria, bacteria expressing dsRNAs were fed to trophozoites and shown to confer specific interference effects (N. Guillén, unpublished data). Feeding with RNAi-producing bacteria has the advantage of being considerably less expensive than soaking, which requires the *in vitro* or chemical synthesis of siRNA. However, the production of RNAi by feeding bacteria requires significant preparatory molecular biology work to generate plasmids that produce dsRNA for the gene of interest in the bacterial strain.

Another function of RNAi that deserves comment is transposon inactivation, since the *E. histolytica* genome is littered with transposable elements. There are two major types of autonomous LINEs of which there are three subtypes (*EhLINE* 1, 2 and 3), and there are three types of short interspersed elements (SINEs) (*EhSINE* 1, 2 and 3) (Clark *et al.*, 2007). These elements are inactive, although ESTs corresponding to *EhLINE*s and *EhSINE*s have been found by cDNA sequencing projects, indicating that transcription of these elements in *E. histolytica* still occurs and contributes significantly to the polyadenylated RNA of the cells.

Transposons are a major source of endogenous siRNAs in *Drosophila* (Obbard and Finnegan, 2008). Because of the mutagenic consequences of mobile genetic elements like transposons, elaborate defences have evolved to restrict their activity. Recently, an important mechanism of RNAi that suppresses transposon activity has been reported and uses Piwi-interacting RNAs (piRNA) of 24–30 nt (Chung *et al.*, 2008). These RNAs are generated from sense or anti-sense single-stranded RNAs. Generation of these endogenous siRNAs from transposon sequences depends on an Argonaute family in the germline of the flies. In adult flies, small RNAs derived from transposons seem to rely on the participation of Dicer2 and Argonaute2 but these conclusions have not yet been fully proven (Kawamura *et al.*, 2008). Also, it is not evident that endogenous siRNAs affect transposition *per se* but may be involved in the maintenance of heterochromatin. RNAi represents a breakthrough technology for functional genomics research in *E. histolytica*, although the molecular mechanisms are still a mystery and not all the components have been characterized. The enzymatic activity that appears to degrade dsRNA has not yet been characterized but the small RNAs that are produced appear to induce silencing in *E. histolytica* by RNAi, following the pathways of siRNA, miRNA or both. In *E. histolytica*, an epigenetic mechanism of transcriptional gene silencing has been observed (next section) and the involvement of non-coding RNA molecules generated from the numerous *SINE* or *LINE* repetitive elements is still a possibility.

Transcriptional gene silencing

Silencing of gene expression can proceed by one of two pathways. The most common type, as discussed above, is at the post-transcriptional level (PTGS) (Grewal and Jia, 2007; Hannon, 2002; Locke and Martienssen, 2006). Silencing at the transcriptional level (TGS) (Cogoni, 2001; Moazed, 2001) has been shown in a number of systems, including plants, following the chromosomal insertion of homologous repeats of a transgene promoter region without the transcribed sequences normally associated with it in the genome (Muskens *et al.*, 2000; Vaucheret and Fagard, 2001). Such silencing is inheritable in the progeny and it can persist even when the silencer sequence is subsequently excised (Zhang *et al.*, 2000a). Epigenetic changes can be induced by DNA methylation (Bird, 2002), as well as through induction of a repressed chromatin structure via post-translational modifications

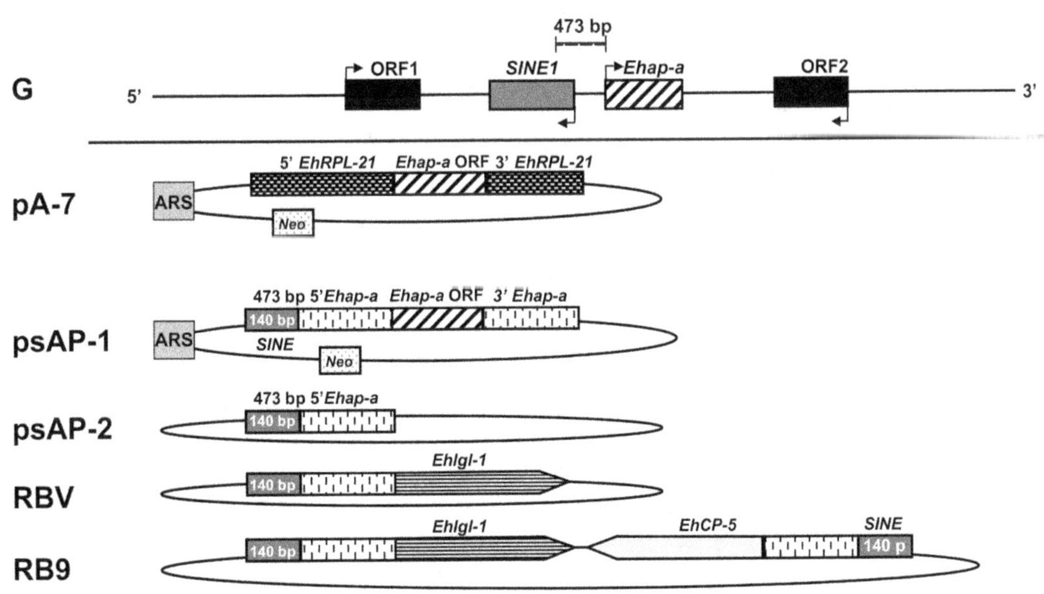

Figure 7.2 The genomic organization (G) around the *Ehap-a* gene depicting the location of the 473 bp 5′ upstream region and the different plasmids constructed for the transfections. pA-7 was the plasmid constructed for over-expression of the *Ehap-a* gene using the ribosomal protein (*EhRPL-21*) 3′ and 5′ regulatory elements. All the other constructs (psAP-1-RB9) included the same 473 bp segment (see for example psAP-2). ARS – an AT rich region containing autonomous replicating sequences from *E. histolytica*. *Neo* – neomycin phosphotransferase gene encoding resistance to Geneticin (G418). 140 bp – the truncated 5′ SINE1 segment upstream of the *Ehap-a* gene promoter. *Ehlgl1* – gene encoding a light subunit of the Gal/GalNAc lectin. *EhCP5* – gene encoding cysteine proteinase 5.

of the amino terminal tails of core histones by acetylation, deacetylation, methylation, phosphorylation, ADP-ribosylation, ubiquitination or α-sumoylation (Grewal and Jia, 2007; Ruthenburg *et al.*, 2007).

Epigenetic gene silencing in *E. histolytica*

The discovery that transcriptional gene silencing can occur in *E. histolytica* was made by coincidence (Bracha *et al.*, 2003; Mirelman *et al.*, 2008). Gene silencing happened during an attempt to over-express an amoebapore gene (*Ehap-a*) by transfecting trophozoites of the virulent strain HM-1:IMSS with a plasmid (psAP-1, Fig. 7.2) that contained the genomic sequence of the *Ehap-a* gene, including segments of its 5′ and 3′ flanking elements. Surprisingly, instead of over-expressing the amoebapore A (AP-A) protein, the transfectants were found to be completely devoid of this protein. Over-expression of AP-A protein was observed, however, when transfection was done with a second plasmid (pA-7, Fig. 7.2) in which the ORF of the *Ehap-a* gene was flanked by the 5′ and 3′ regulating elements of another gene (*EhRP-L21gLE-3*) (Bracha *et al.*, 2002). The ability to silence the *Ehap-a* gene was found to reside in the 5′ upstream flanking segment (473 bp) of the *Ehap-a* gene. Plasmid psAP-2 (Fig. 7.2), containing only this sequence without the ORF and the 3′ regulating region, was able to induce silencing even faster (in 2 vs 3 weeks). Sequence analysis of the 473 bp 5′ flanking element revealed that it contained, upstream of the promoter motifs, a truncated segment (140 bp) of a SINE1 retroposon element that is transcribed in the opposite orientation and is preceded by a unique, T- rich stretch of 48 bp (Willhoeft *et al.*, 2002). Removal of the *Ehap-a* gene promoter sequences from the 473 bp 5′ flanking element or deletion of more than 60 bp of the SINE1 segment abolished the ability of the plasmid to induce silencing of the *Ehap-a* gene. Furthermore, inclusion of the entire SINE1 element (550 bp) together with its 3′ poly-adenylation signal sequences also prevented the silencing of the *Ehap-a* gene (Anbar *et al.*, 2005).

Removal of plasmid psAP-2 from silenced cells by discontinuing the selective antibiotic resulted in plasmid-less clones that did not resume transcription of the *Ehap-a* gene. One such clone, termed G3, was characterized and found to be incapable of expressing the AP-A protein even following re-transfection with plasmid pA-7 (Fig. 7.2), which contains the *Ehap-a* gene under the regulatory elements of a different gene. This inability was restricted to the *Ehap-a* gene, as transfection of G3 trophozoites with an analogous plasmid in which the *Ehap-a* gene was replaced by the CAT reporter gene, led to transcription and expression of CAT (Bracha *et al.*, 2003). Irmer and Tannich (Irmer *et al.*, 2009) have recently confirmed the silencing of the *Ehap-a* gene by using a somewhat similar plasmid construct. They have found that the plasmids that have produced silencing of the *Ehap-a* gene also possess part of a tRNA array (Clark *et al.*, 2006) encoding glutamic acid and tyrosine tRNAs, and that removal of this sequence also prevents silencing of the gene. It is still not clear why silencing of the *Ehap-a* gene following transfection with plasmid psAP-2 has only been obtained in one strain of *E. histolytica* (HM-1:IMSS); other strains were not silenced despite the sequence identity of their 5′ upstream regions. The failure in silencing maybe due to small variations in the genomic organization of the numerous SINE1 retroposon elements in the different strains of *E. histolytica* (Srivastava *et al.*, 2005).

It is, however, important to mention that trophozoites lacking the AP-A protein, such as the G3 trophozoites that were derived from the virulent strain HM-1:IMSS, are avirulent as they are incapable of producing liver lesions in hamsters or mice; however they are still able to induce amoebic colitis in the SCID-HU-mice animal model of intestinal amoebiasis (Bracha *et al.*, 2003; Bujanover *et al.*, 2003; Zhang *et al.*, 2004).

Silencing of additional genes in G3 trophozoites

Analysis of the expression of certain other members of the SAPLIP gene family (*Ehap-b* and *SAPLIP1*) in G3 trophozoites revealed that they were also down-regulated whereas other SAPLIPs, such as *Ehap-c,* continued to be expressed (Bracha *et al.*, 2006). This finding gave us the idea that we could perhaps direct the silencing of other genes of choice. Silencing of additional genes, such as the light subunit of the Gal/GalNAc lectin

(*Ehlgl1*) and cysteine proteinase 5 (*EhCP5*), was successfully induced in the G3 plasmid-less trophozoites using a hybrid plasmid construct that contained the second gene directly ligated to the 473 bp 5′ upstream region of the *Ehap-a* gene mentioned previously (Bracha *et al.*, 2006) (Fig. 7.2). In the parent strain HM-1:IMSS, transfection of such plasmids resulted in over-expression of the second gene. Interestingly, there was no need to insert all the ORF of the second gene or the 3′ regulatory sequences; as little as 120 bp of the 5′ end of the ORF were sufficient to induce silencing of the genomic copy. Plasmids in which the SINE element sequence (140 bp) was removed from the 473 bp 5′ upstream region (Fig. 7.2) were incapable of inducing the silencing of a second gene. Silencing of the *Ehlgl1* gene caused the simultaneous silencing of two additional genes from the same gene family (*Ehlgls 2–3*) (Bracha *et al.*, 2007), as was originally shown for the SAPLIPs. The two other members of this family (*Ehlgls 4–5*), which encode slightly shorter proteins, became over-expressed indicating that these two latter genes may compensate for the loss of function in *Ehlgls 1–3* (Bracha *et al.*, 2007). The opposite was found when G3 trophozoites were transfected with a silencing plasmid that contained the ORF of *Ehlgl5*; it caused the simultaneous silencing of *Ehlgls 4–5* but induced the over-expression of *Ehlgls 1–3*. Attempts to silence all five *Ehlgl* genes simultaneously have failed and the trophozoites died suggesting that total abrogation of expression of all the light subunit genes is lethal.

Using this technique we have managed to silence a number of additional genes, such as those coding for *EhCP5* (Bracha *et al.*, 2006), a Lim-like protein (*EhLimA*) (Wender *et al.*, 2007) and the arginine methyl transferase PRMT1 (*EhPRMT1*) (M. Persky, unpublished results). Recently we have been successful in the simultaneous silencing of three genes (Bracha *et al.*, 2007). Transfection of G3 trophozoites with a plasmid construct (RB9, Fig. 7.2) in which two genes (*Ehlgl1* and *EhCP5*) were each placed adjacent to the 5′ upstream 473 bp segment of the *Ehap-a* gene enabled the simultaneous silencing of both genes. This offspring of the G3 amoeba (RB9) failed to express three of its main virulence genes (*Ehap-a, Ehlgl1* and *EhCP5*). Using the same technique we have failed, however, to silence a number of other 'second' genes. This could be due to the importance of such second genes for the survival of the parasite.

In collaboration with Drs. C. Gilchrist and W.A. Petri (University of Virginia), a comparison was undertaken of the transcriptomes of the parent strain HM-1:IMSS and the G3 silenced trophozoites using their Affimetrix microarrays. A large number of genes was found to be significantly up-regulated and another group was markedly down-regulated, including the genes *Ehap-a* and *Ehap-b* which had earlier been detected as transcriptionally silenced. Validation of the changes in expression levels of several of these genes as well as the possible reasons for such significant transcription modulation are currently being investigated.

Investigating the gene silencing mechanism in G3 trophozoites

As mentioned above, sequencing of the *E. histolytica* genome made it possible to search for genes with homology to those of other organisms that are known to participate in the different mechanisms of gene silencing (Table 7.1). Although a classic Dicer enzyme has not been found, the presence of small RNA molecules was recently reported (Zhang *et al.*, 2008).

The mechanism of gene silencing in the G3 trophozoites of *E. histolytica* appears to be of the TGS type. Nuclear run-on experiments using G3 trophozoites indicated that there was no transcription initiation (< 0.1%) of the *Ehap-a* gene (Bracha *et al.*, 2003). In addition there was no detection of any DNA rearrangements or changes in the restriction pattern of genomic DNA and no DNA methylation was found in the promoter region of the *Ehap-a* gene in the G3 silenced trophozoites. The silencing of the *Ehap-a* gene persisted in the plasmid-less G3 trophzoites and remained stable even when cultures were grown for long periods in the presence of the DNA methylation inhibitor 5-azacytidine or the histone deacetylase inhibitor Trichostatin A. Growth in the presence of nicotinamide, which is an inhibitor of Sir2 histone deacetylase, caused a very limited resumption of transcription of the *Ehap-a* gene (R. Bracha, unpublished results).

Chromatin immunoprecipitation (ChIP) analysis with an antibody against methylated lysine 4 of histone H3 (H3mK4) showed demethylation of K4 at the *Ehap-a* gene chromatin domain in G3 trophozoites, indicating transcriptional inactivation (Anbar *et al.*, 2005). Unfortunately, due to the significant sequence divergence between the histones of *E. histolytica* and those of other eukaryotes (Födinger *et al.*, 1993), it has not yet been possible to study additional histone modifications in the silenced amoebae due to a lack of specific detection reagents. Supporting evidence for the maintenance of a heterochromatin structure on the silenced region was obtained by observing a significantly higher resistance to micrococcal nuclease digestion at the chromatin domain of the *Ehap-a* promoter region in G3 trophozoites versus the parent strain HM-1:IMSS (Huguenin *et al.*, 2009). The silenced region appears to be quite limited in length as the gene that is downstream of *Ehap-a* (EHI_159470) and the *SINE1* element that is upstream are both transcribed in G3 trophozoites.

Silencing of the *Ehap-a* gene appears not to involve siRNA. Exhaustive analysis of RNA extracts from G3 trophozoites failed to detect any siRNA molecules of 21–24nt that would hybridize with probes for the *Ehap-a* gene, but they did reveal, however, the presence of single stranded, non-coding RNA molecules of approximately 140nt with sequence homology to the 5′ end of SINE1 elements (Anbar *et al.*, 2005). These truncated non coding *SINE1* ssRNA molecules were found to accumulate in the nuclei of G3 trophozoites. They did not possess polyA tails, which are usually present on the normal SINE1 transcripts, and were not detected in nuclear RNA extracts of the *E. histolytica* parent strain HM-1:IMSS. The presence of the truncated *SINE1* ssRNA molecules in the nuclei may be linked to our finding that in G3 trophozoites there is an up-regulation of transcription of the SINE1 element located upstream of the *Ehap-a* gene (Anbar *et al.*, 2005). At present we do not know the role of these non-coding, truncated *SINE1* ssRNA transcripts. One possibility is that due to their nuclear accumulation, they may bind to the homologous SINE1 DNA region located upstream of the *Ehap-a* gene, triggering the binding of a silencing complex, which then spreads downstream and is responsible for the chromatin modification, as has been proposed for mammals (Wang *et al.*, 2006). The heterochromatin that is formed inhibits transcription of the downstream gene and becomes a stable inheritable structure that recognizes the promoter sequences of the *Ehap-a* gene in the progeny also. Transfection of G3 trophozoites with a new plasmid (RBV or RB9) in which a second gene is placed next to the 473 bp 5′ *Ehap-a* gene promoter region (Fig. 7.2) enables the heterochromatin structure, unique to the *Ehap-a* promoter sequences, to become bound to the same region present on the newly inserted plasmid. Once there, it continues to spread its chromatin modification downstream to the ligated second gene because there is no boundary or insulator region between them (Hannon, 2002; Wang *et al.*, 2006; Grewal and Moazed, 2003) (Fig. 7.3). Once the new heterochromatin structure is formed on the plasmid, it recognizes the sequence of the second gene in the genome, and inhibits its transcription. This inhibition also becomes inheritable.

Silencing of the *Ehap-a* gene, as mentioned above, also occurred following transfection with plasmid psAP-1 (Fig. 7.2), which contains the ORF as well as a 3′ flanking segment of the *Ehap-a* gene. Removal of the G418 selection pressure resulted in a plasmid-less amoeba, termed F, in which the *Ehap-a* gene remained inheritably silenced (Bracha *et al.*, 2006). However, in contrast to what was described above for the G3 amoeba, transfection of the F trophozoites with the hybrid plasmid containing the *Ehlgl1* gene directly ligated to the 473 bp 5′ upstream segment of the *Ehap-a* gene (RBV, Fig. 7.2), did not result in silencing of the second gene. The reason for this could be due to an inability of the silencing factors formed in the F trophozoites, which recognize the promoter and ORF sequences of the *Ehap-a* gene, to identify and bind to the newly transfected hybrid plasmid, which has a different sequence in the second ORF.

In conclusion, multiple gene silencing cannot provide much information about the individual contribution of each gene to the observed non-virulent phenotype due to the fact that they all lack EhAP-A protein. However, it can, in some cases, provide insights into their specific function, as was the case with our findings on the different

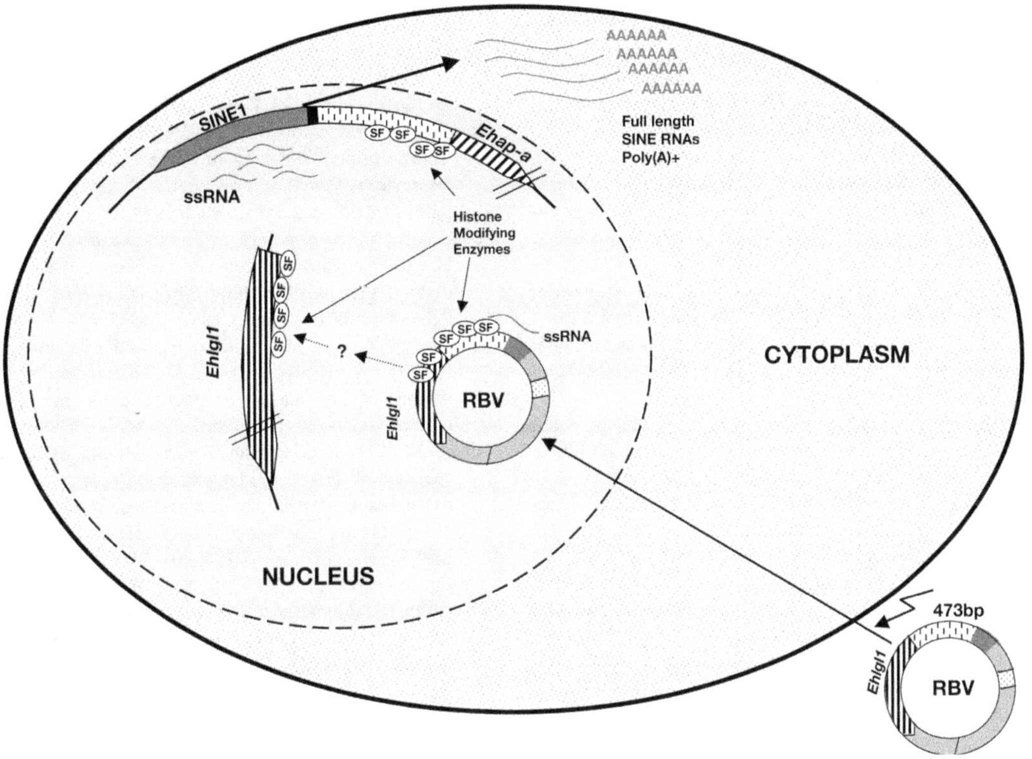

Figure 7.3 Proposed model for silencing of a second gene. Transfection of plasmid RBV (see Figure 7.2) into the G3 gene-silenced amoeba in which the silencing factors (SFs) recognize the promoter sequences of the *Ehap-a* gene, results in an accumulation and attachment of such SFs to the same *Ehap-a* sequences on the newly inserted RBV plasmid. The SFs then continue to spread to the *Ehlgl1* gene on the same plasmid. Once the SFs recognize the sequences of the *Ehlgl1* gene, they are transferred, in trans, to the same gene in the parasite genome and cause its silencing.

light subunits of the Gal/GalNAc-lectin, and also on the inter-relationship among the members of gene families. In order to understand the individual gene contribution to amoebic virulence it would be of great value if it were possible, as a first step, to induce the silencing of a gene that does not have an important role in parasite virulence. This, unfortunately, is yet to be achieved.

Last, but not least, the fact that the multiple epigenetic gene silencing process has resulted in a viable parasite with highly impaired virulence, but which still preserves the antigenic repertoire of the virulent strain, suggests that this amoeba may be a candidate for an attenuated vaccine strain. Trophozoites of *E. histolytica* evade the host immune system by capping and shedding of surface receptors that become bound to host antibodies or innate immunity components. The silencing of the light subunit of the Gal/GalNAc-lectin caused a defect in the capping of surface molecules (Bracha *et al.*, 2006; Bracha *et al.*, 2007) and thus such parasites may evoke a different response by the host immune system.

Acknowledgements

The research performed in the DM lab was supported by grants received from Mrs. Erica Drake, from Mr. Henry H. Meyer Jr. A joint Grant from the Conseil Pasteur-Weizmann has supported the laboratories of NG and DM.

References

Abed, M., and Ankri, S. (2005). Molecular characterization of *Entamoeba histolytica* RNase III and AGO2, two RNA interference hallmark proteins. Exp. Parasitol. *110*, 265–269.

Abed, M., and Ankri, S. (2008). Progress and prospects of gene inactivation in *Entamoeba histolytica*. Exp. Parasitol. *118*, 151–155.

Aguilar-Rojas, A., Almaraz-Barrera, M., Krzeminski, J., Robles-Flores, M., Hernández-Rivas, R., Guillén, N., Maroun, R.C., and Vargas, M. (2005). *Entamoeba histolytica*: inhibition of cellular functions by overexpression of EhGEF1, a novel Rho/Rac guanine nucleotide exchange factor. Exp. Parasitol. *109*, 150–162.

Alon, R.N., Bracha, R., and Mirelman, D. (1997). Inhibition of expression of the lysine-rich 30kDa surface antigen of *Entamoeba dispar* by the transcription of its antisense RNA. Mol. Biochem. Parasitol. *90*, 193–201.

Anbar, M., Bracha, R., Nuchamowitz, Y., Li, Y., Florentin, A., and Mirelman, D. (2005). Involvement of a short interspersed element in epigenetic transcriptional silencing of the amoebapore gene in *Entamoeba histolytica*. Eukaryot. Cell *4*, 1775–1784.

Ankri, S., Padilla-Vaca, F., Stolarsky, T., Koole, L., Katz, U., and Mirelman, D. (1999a). Antisense inhibition of expression of the light subunit (35 kDa) of the Gal/GalNac lectin complex inhibits *Entamoeba histolytica* virulence. Mol. Microbiol. *33*, 327–337.

Ankri, S., Stolarsky, T., Bracha, R., Padilla-Vaca, F., and Mirelman, D. (1999b). Antisense inhibition of expression of cysteine proteinases affects *Entamoeba histolytica*-induced formation of liver abscess in hamsters. Infect. Immun. *67*, 421–422.

Ankri, S., Stolarsky, T., and Mirelman, D. (1998). Antisense inhibition of expression of cysteine proteinases in *Entamoeba histolytica* does not affect cytopathic or hemolytic activity but inhibits phagocytosis. Mol. Microbiol. *28*, 777–785.

Bird, A. (2002). DNA methylation patterns and epigenetic memory. Genes Dev. *16*, 6–21.

Blazquez, S., Guigon, G., Weber, C., Syan, S., Sismeiro, O., Coppée, J.Y., Labruyère, E., and Guillén, N. (2008). Chemotaxis of *Entamoeba histolytica* towards the pro-inflammatory cytokine TNF is based on PI3K signalling, cytoskeleton reorganization and the Galactose/N-acetylgalactosamine lectin activity. Cell. Microbiol. *10*, 1676–1686.

Blazquez, S., Rigothier, M.C., Huerre, M., and Guillen, N. (2007). Initiation of inflammation and cell death during liver abscess formation by *Entamoeba histolytica* depends on activity of the galactose/N-acetyl-D-galactosamine lectin. Int. J. Parasitol. *37*, 425–433.

Boettner, D.R., Huston, C.D., Linford, A.S., Buss, S.N., Houpt, E., Sherman, N.E., and Petri, W.A., Jr. (2008). *Entamoeba histolytica* phagocytosis of human erythrocytes involves PATMK, a member of the transmembrane kinase family. PLoS Pathog. *4*, e8.

Bracha, R., Nuchamowitz, Y., Anbar, M., and Mirelman, D. (2006). Transcriptional silencing of multiple genes in trophozoites of *Entamoeba histolytica*. PLoS Pathog. *2*, e48.

Bracha, R., Nuchamowitz, Y., Leippe, M., and Mirelman, D. (1999). Antisense inhibition of amoebapore expression in *Entamoeba histolytica* causes a decrease in amoebic virulence. Mol. Microbiol. *34*, 463–472.

Bracha, R., Nuchamowitz, Y., and Mirelman, D. (2000). Inhibition of gene expression in *Entamoeba* by the transcription of antisense RNA: effect of 5′ and 3′ regulatory elements. Mol. Biochem. Parasitol. *107*, 81–90.

Bracha, R., Nuchamowitz, Y., and Mirelman, D. (2002). Amoebapore is an important virulence factor of *Entamoeba histolytica*. J. Biosci. *27*, 579–589.

Bracha, R., Nuchamowitz, Y., and Mirelman, D. (2003). Transcriptional silencing of an amoebapore gene in *Entamoeba histolytica:* Molecular analysis and effect on pathogenicity. Eukaryot. Cell *2*, 295–305.

Bracha, R., Nuchamowitz, Y., Wender, N., and Mirelman, D. (2007). Transcriptional gene silencing reveals two distinct groups of *E. histolytica* Gal/GalNAc lectin light subunits. Eukaryot. Cell *6*, 1758–1765.

Bujanover, S., Katz, U., Bracha, R., and Mirelman, D. (2003). A virulence attenuated amoebapore-less mutant of *Entamoeba histolytica* and its interaction with host cells. Int. J. Parasitol. *33*, 1655–1663.

Chen, M., Li, E., and Stanley, S.L., Jr. (2004). Structural analysis of the acetaldehyde dehydrogenase activity of *Entamoeba histolytica* alcohol dehydrogenase 2 (EhADH2), a member of the ADHE enzyme family. Mol. Biochem. Parasitol. *137*, 201–205.

Chung, W.J., Okamura, K., Martin, R., and Lai, E.C. (2008). Endogenous RNA interference provides a somatic defense against *Drosophila* transposons. Curr. Biol. *18*, 795–802.

Clark, C.G., Ali, I.K., Zaki, M., Loftus, B.J., and Hall, N. (2006). Unique organisation of tRNA genes in *Entamoeba histolytica*. Mol. Biochem. Parasitol. *146*, 24–29.

Clark, C.G., Alsmark, U.C., Tazreiter, M., Saito-Nakano, Y., Ali, V., Marion, S., Weber, C., Mukherjee, C., Bruchhaus, I., Tannich, E., *et al.* (2007). Structure and content of the *Entamoeba histolytica* genome. Adv. Parasitol. *65*, 51–190.

Cogoni, C. (2001). Homology-dependent gene silencing mechanisms in fungi. Annu. Rev. Microbiol. *55*, 381–406.

Coudrier, E., Amblard, F., Zimmer, C., Roux, P., Olivo-Marin, J.C., Rigothier, M.C., and Guillén, N. (2005). Myosin II and the Gal-GalNAc lectin play a crucial role in tissue invasion by *Entamoeba histolytica*. Cell. Microbiol. *7*, 19–27.

De, S., Pal, D., and Ghosh, S.K. (2006). *Entamoeba histolytica*: computational identification of putative microRNA candidates. Exp. Parasitol. *113*, 239–243.

Faehnle, C.R., and Joshua-Tor, L. (2007). Argonautes confront new small RNAs. Curr. Opin. Chem. Biol. *11*, 569–577.

Fire, A.Z. (2007). Gene silencing by double-stranded RNA (Nobel Lecture). Angew. Chem. Int. Ed. Engl. *46*, 6966–6984.

Födinger, M., Ortner, S., Plaimauer, B., Wiedermann, G., Scheiner, O., and Duchêne, M. (1993). Pathogenic *Entamoeba histolytica:* cDNA cloning of a histone H3 with a divergent primary structure. Mol. Biochem. Parasitol. *59*, 315–322.

Ghosh, S.K., Lohia, A., Kumar, A., and Samuelson, J. (1996). Overexpression of P-glycoprotein gene 1 by transfected *Entamoeba histolytica* confers emetine-resistance. Mol. Biochem. Parasitol. *82*, 257–260.

Grewal, S.I., and Moazed, D., (2003). Heterochromatin and epigenetic control of gene expression. Science *301*, 798–802.

Grewal, S.I., and Jia, S. (2007). Heterochromatin revisited. Nat. Rev. Genet. *8*, 35–46.

Grodberg, Salazar, N., Oren, R., and Mirelman, D. (1990). Autonomous replication sequences in an extrachromosomal element of a pathogenic *Entamoeba histolytica*. Nucleic Acids Res. *18*, 5515–5519.

Guarnieri, D.J., and DiLeone, R.J. (2008). MicroRNAs: a new class of gene regulators. Ann Med *40*, 197–208.

Guillén, N., Boquet, P., and Sansonetti, P. (1998). The small GTP-binding protein RacG regulates uroid formation in the protozoan parasite *Entamoeba histolytica*. J. Cell Sci. *111*, 1729–1739.

Hamann, L., Buss, H., and Tannich, E. (1997). Tetracycline-controlled gene expression in *Entamoeba histolytica*. Mol. Biochem. Parasitol. *84*, 83–91.

Hamann, L., Nickel, R., and Tannich, E. (1995). Transfection and continuous expression of heterologous genes in the protozoan parasite *Entamoeba histolytica*. Proc. Natl. Acad. Sci. U.S.A. *92*, 8975–8979.

Hannon, G.J. (2002). RNA interference. Nature *418*, 244–251.

Huguenin, M., Bracha, R., Chookajorn, T., and Mirelman, D. (2009). Epigenetic transcriptional gene silencing in *Entamoeba histolytica*: insight into histone and chromatin modifications. Parasitology *23*, 1–9.

Irmer, H., Tillack, M., Biller, L., Handal, G., Leippe, M., Roeder, T., Tannich, E., and Bruchhaus, I. (2009). Major cysteine peptidases of *Entamoeba histolytica* are required for aggregation and digestion of erythrocytes but are dispensable for phagocytosis and cytopathogenicity. Mol. Microbiol. *72*, 658–667.

Iyer, L.M., Koonin, E.V., and Aravind, L. (2003). Evolutionary connection between the catalytic subunits of DNA-dependent RNA polymerases and eukaryotic RNA-dependent RNA polymerases and the origin of RNA polymerases. BMC Struct. Biol. *3*, 1–23.

Katz, U., Ankri, S., Stolarsky, T., Nuchamowitz, Y., and Mirelman, D. (2002). *Entamoeba histolytica* expressing a dominant negative N-truncated light subunit of its gal-lectin are less virulent. Mol. Biol. Cell *13*, 4256–4265.

Kaur, G., and Lohia, A. (2004). Inhibition of gene expression with double strand RNA interference in *Entamoeba histolytica*. Biochem. Biophys. Res. Commun. *320*, 1118–1122.

Kawamura, Y., Saito, K., Kin, T., Ono, Y., Asai, K., Sunohara, T., Okada, T.N., Siomi, M.C., and Siomi, H. (2008). Drosophila endogenous small RNAs bind to Argonaute 2 in somatic cells. Nature *453*, 793–797.

Labruyère, E., Zimmer, C., Galy, V., Olivo-Marin, J.C., and Guillén, N. (2003). EhPAK, a member of the p21-activated kinase family, is involved in the control of *Entamoeba histolytica* migration and phagocytosis. J. Cell Sci. *116*, 61–71.

Lavi, T., Siman-Tov, R., and Ankri, S. (2008). EhMLBP is an essential constituent of the *Entamoeba histolytica* epigenetic machinery and a potential drug target. Mol. Microbiol. *69*, 55–66.

Locke, S.M., and Martienssen, R.A. (2006). Slicing and spreading of heterochromatic silencing by RNA interference. Cold Spring. Harb. Symp. Quant. Biol. *71*, 497–503.

Lohia, A., Haider, N., and Biswas, B.B. (1990). Characterization of a repetitive DNA family from *Entamoeba histolytica* containing *Saccharomyces cerevisiae* ARS consensus sequences. Gene *96*, 197–203.

MacFarlane, R.C., and Singh, U. (2007). Identification of an *Entamoeba histolytica* serine-, threonine-, and isoleucine-rich protein with roles in adhesion and cytotoxicity. Eukaryot. Cell *6*, 2139–2146.

MacFarlane, R.C., and Singh, U. (2008). Loss of dsRNA-based gene silencing in *Entamoeba histolytica*: implications for approaches of genetic analysis. Exp. Parasitol. *119*, 296–300.

Marion, S., Wilhelm, C., Voigt, H., Bacri, J.C., and Guillén, N. (2004). Overexpression of myosin IB in living *Entamoeba histolytica* enhances cytoplasm viscosity and reduces phagocytosis. J. Cell Sci. *117*, 3271–3279.

Mirelman, D., Anbar, M., and Bracha, R. (2008). Epigenetic transcriptional gene silencing in *Entamoeba histolytica*. IUBMB Life *60*, 598–604.

Moazed, D. (2001). Common themes in mechanisms of gene silencing. Mol. Cell *8*, 489–498.

Moncada, D., Keller, K., Ankri, S., Mirelman, D., and Chadee, K. (2006). Antisense inhibition of *Entamoeba histolytica* cysteine proteases inhibits colonic mucus degradation. Gastroenterology *130*, 721–730.

Moshitch-Moshkovitch, S., Petter, R., Levitan, A., Stolarsky, T., and Mirelman, D. (1998). Regulation of expression of ribosomal protein L-21 genes of *Entamoeba histolytica* and *E. dispar* is at the post-transcriptional level. Mol. Microbiol. *27*, 677–685.

Moshitch-Moshkovitch, S., Stolarsky, T., Mirelman, D., and Alon, R.N. (1996). Stable episomal transfection and gene expression in *Entamoeba dispar*. Mol. Biochem. Parasitol. *83*, 257–261.

Moshitch-Moshkovitch, S., Stolarsky, T., Petter, R., and Mirelman, D. (1997). Upstream regions of rp L21 genes play a role in regulation of expression at the post-transcriptional level in *E. histolytica* and *E. dispar*. Arch. Med. Res. *28 Spec No*, 36–38.

Muskens, M.W.M., Vissers, A.P.A., Mol, J.N.M., and Kooter, J.M. (2000). Role of inverted DNA repeats in transcriptional and post-transcriptional gene silencing. Plant Mol. Biol. *43*, 243–260.

Nickel, R., and Tannich, E. (1994). Transfection and transient expression of chloroamphenicol acetyl transferase gene in the protozoan parasite *Entamoeba histolytica*. Proc. Natl. Acad. Sci. U.S.A. *91*, 7095–7098.

Obbard, D.J., and Finnegan, D.J. (2008). RNA interference: endogenous siRNAs derived from transposable elements. Curr. Biol. *18*, R561–563.

Pak, J., and Fire, A. (2007). Distinct populations of primary and secondary effectors during RNAi in *C. elegans*. Science *315*, 241–244.

Peters, L., and Meister, G. (2007). Argonaute proteins: mediators of RNA silencing. Mol. Cell *26*, 611–623.

Petri, W.A., Jr., and Ramakrishnan, G. (1999). Applying antisense technology to the study of *Entamoeba histolytica* pathogenesis. Trends Microbiol. *7*, 471–474.

Petter, R., Moshitch, S., Rozenblatt, S., Nuchamowitz, Y., and Mirelman, D. (1994). Characterization of two distinct gene transcripts for ribosomal protein L21 from pathogenic and nonpathogenic strains of *Entamoeba histolytica*. Gene *150*, 181–186.

Purdy, J.E., Mann, B.J., Pho, L.T., and Petri, W.A., Jr. (1994). Transient transfection of the enteric parasite *Entamoeba histolytica* and expression of firefly luciferase. Proc. Natl. Acad. Sci. U.S.A. *91*, 7099–7103.

Ramakrishnan, G., Vines, R.R., Mann, B.J., and Petri, W.A., Jr. (1997). A tetracycline-inducible gene expression system in *Entamoeba histolytica*. Mol. Biochem. Parasitol. *84*, 93–100.

Ray, A., and Norden, B. (2000). Peptide nucleic acid (PNA): its medical and biochemical applications and promise for the future. FASEB Journal *14*, 1041–1060.

Ruthenburg, A.J., Li, H., Patel, D.J., and Allis, C.D. (2007). Multivalent engagement of chromatin modifications by linked binding modules. Nat. Rev. Mol. Cell Biol. *8*, 983–994.

Sahoo, N., Labruyère, E., Bhattacharya, S., Sen, P., Guillén, N., and Bhattacharya, A. (2004). Calcium binding protein 1 of the protozoan parasite *Entamoeba histolytica* interacts with actin and is involved in cytoskeleton dynamics. J. Cell Sci. *117*, 3625–3634.

Saito-Nakano, Y., Mitra, B.N., Nakada-Tsukui, K., Sato, D., and Nozaki, T. (2007). Two Rab7 isotypes, EhRab7A and EhRab7B, play distinct roles in biogenesis of lysosomes and phagosomes in the enteric protozoan parasite *Entamoeba histolytica*. Cell. Microbiol. *9*, 1796–1808.

Sánchez, R., Saralegui, A., Olivos-García, A., Scapolla, C., Damonte, G., Sanchez-Lopez, R., Alagón, A., and Stock, R.P. (2005). *Entamoeba histolytica*: intracellular distribution of the sec61alpha subunit of the secretory pathway and down-regulation by antisense peptide nucleic acids. Exp. Parasitol. *109*, 241–251.

Santi-Rocca, J., Weber, C., Guigon, G., Sismeiro, O., Coppée, J.Y., and Guillén, N. (2008). The lysine- and glutamic acid-rich protein KERP1 plays a role in *Entamoeba histolytica* liver abscess pathogenesis. Cell. Microbiol. *10*, 202–217.

Schaenman, J.M., Gilchrist, C.A., Mann, B.J., and Petri, W.A., Jr. (2001). Identification of two *Entamoeba histolytica* sequence-specific URE4 enhancer-binding proteins with homology to the RNA-binding motif RRM. J. Biol. Chem. 276, 1602–1609.

Scherczinger, C.A., Yates, A.A., and Knecht, D.A. (1992). Variables affecting antisense RNA inhibition of gene expression. Ann. N.Y. Acad. Sci. *660*, 45–56.

Sen, A., Chatterjee, S.N., Ali Akbar, M., Nandi, N., and Das, P. (2007). The 29-kilodalton thiol-dependent peroxidase of *Entamoeba histolytica* is a factor involved in pathogenesis and survival of the parasite during oxidative stress. Eukaryot. Cell *6*, 664–673.

Sijen, T., Steiner, F.A., Thijssen, K.L., and Plasterk, R.H. (2007). Secondary siRNAs result from unprimed RNA synthesis and form a distinct class. Science *315*, 244–247.

Solis, C.F., and Guillén, N. (2008). Silencing genes by RNA interference in the protozoan parasite *Entamoeba histolytica*. Methods Mol. Biol. *442*, 113–128.

Song, J.J., Smith, S.K., Hannon, G.J., and Joshua-Tor, L. (2004). Crystal structure of Argonaute and its implications for RISC slicer activity. Science *305*, 1434–1437.

Srivastava, S., Bhattacharya, S., and Paul, J. (2005). Species and strain-specific probes derived from repetitive DNA for distinguishing *Entamoeba histolytica* and *Entamoeba dispar*. Exp. Parasitol. *110*, 303–308.

Stock, R.P., Olvera, A., Sánchez, R., Saralegui, A., Scarfì, S., Sánchez-López, R., Ramos, M.A., Boffa, L.C., Benatti, U., and Alagón, A. (2001). Inhibition of gene expression in *Entamoeba histolytica* with antisense peptide nucleic acid oligomers. Nat. Biotechnol. *19*, 213–215.

Tavares, P., Rigothier, M.C., Khun, H., Roux, P., Huerre, M., and Guillén, N. (2005). Roles of cell adhesion and cytoskeleton activity in *Entamoeba histolytica* pathogenesis: a delicate balance. Infect. Immun. *73*, 1771–1778.

Tillack, M., Nowak, N., Lotter, H., Bracha, R., Mirelman, D., Tannich, E., and Bruchhaus, I. (2006). Increased expression of the major cysteine proteinases by stable episomal transfection underlines the important role of EhCP5 for the pathogenicity of *Entamoeba histolytica*. Mol. Biochem. Parasitol. *149*, 58–64.

Vats, D., Vishwakarma, R.A., Bhattacharya, S., and Bhattacharya, A. (2005). Reduction of cell surface glycosylphosphatidylinositol conjugates in *Entamoeba histolytica* by antisense blocking of *E. histolytica* GlcNAc-phosphatidylinositol deacetylase expression: effect on cell proliferation, endocytosis, and adhesion to target cells. Infect. Immun. 73, 8381–8392.

Vaucheret, H., and Fagard, M. (2001). Transcriptional gene silencing in plants: targets, inducers and regulators. Trends Genet. *17*, 29–35.

Vayssié, L., Vargas, M., Weber, C., and Guillén, N. (2004). Double-stranded RNA mediates homology-dependant gene silencing of γ-tubulin in the human parasite *Entamoeba histolytica*. Mol. Biochem. Parasitol. *138*, 21–28.

Vines, R.R., Purdy, J.E., Ragland, B.D., Samuelson, J., Mann, B.J., and Petri, W.A. (1995). Stable episomal transfection of *Entamoeba histolytica*. Mol. Biochem. Parasitol. *71*, 265–267.

Vines, R.R., Ramakrishnan, G., Rogers, J.B., Lockhart, L.A., Mann, B.J., and Petri, W.A., Jr. (1998). Regulation of adherence and virulence by the *Entamoeba histolytica* lectin cytoplasmic domain, which contains a β2 integrin motif. Mol. Biol. Cell *9*, 2069–2079.

Wang, F., Koyama, N., Nishida, H., Reith, W., and Tsukamoto, T. (2006). The assembly and maintenance of heterochromatin initiated by transgene repeats are independent of the RNA interference pathway in mammalian cells. Mol. Cell Biol. *26*, 4028–4040.

Wassmann, C., Hellberg, A., Tannich, E., and Bruchhaus, I. (1999). Metronidazole resistance in the protozoan parasite *Entamoeba histolytica* is associated with increased expression of iron-containing superoxide dismutase and peroxiredoxin and decreased expression of ferredoxin 1 and flavin reductase. J. Biol. Chem. *274*, 26051–26056.

Weber, C., Blazquez, S., Marion, S., Ausseur, C., Vats, D., Krzeminski, M., Rigothier, M.C., Maroun, R.C., Bhattacharya, A., and Guillén, N. (2008). Bioinformatics and functional analysis of an *Entamoeba histolytica* mannosyltransferase necessary for parasite complement resistance and hepatical infection. PLoS Negl. Trop. Dis. 2, e165.

Wender, N., Villalobo, E., and Mirelman, D. (2007). EhLimA, a novel LIM protein, localizes to the plasma membrane in *Entamoeba histolytica*. Eukaryot. Cell 6, 1646–1655.

Willhoeft, U., Buss, H., and Tannich, E. (2002). The abundant polyadenylated transcript 2 DNA sequence of the pathogenic protozoan parasite *Entamoeba histolytica* represents a nonautonomous non-long-terminal-repeat retrotransposon-like element which is absent in the closely related nonpathogenic species *Entamoeba dispar*. Infect. Immun. *70*, 6798–6804.

Zhang, H., Ehrenkaufer, G.M., Pompey, J.M., Hackney, J.A., and Singh, U. (2008). Small RNAs with 5′polyphosphate termini associate with a PIWI-related protein and regulate gene expression in the single-celled eukaryote *Entamoeba histolytica*. PLoS Pathog. 4, e1000219.

Zhang, X., Zhang, Z., Alexander, D., Bracha, R., Mirelman, D., and Stanley, S.L., Jr. (2004). Expression of amoebapores is required for full expression of *Entamoeba histolytica* virulence in amebic liver abscess but is not necessary for the induction of inflammation or tissue damage in amebic colitis. Infect. Immun. 72, 678–683.

Zhang, Z., Shibahara, K.-I., and Stillman, B. (2000a). PCNA connects DNA replication to epigenetic inheritance in yeast. Nature *408*, 221–225.

Zhang, Z.L., Wang, K.B., Seydel, E., Li, S., Ankri, S., Mirelman, D., and Stanley, S.L. (2000b). *Entamoeba histolytica* cysteine proteinases with interleukin-1 beta converting enzyme (ICE) activity cause intestinal inflamation and tissue damage in amoebiasis. Mol. Microbiol. 37, 542–548.

Giardia: Nuclear and Chromosomal Structure and Replication

8

Rodney D. Adam and Staffan G. Svärd

Abstract

Giardia and the other diplomonads are unique in their possession of two nuclei that are similar in appearance, DNA content, transcription, and time of replication. Trophozoites have a ploidy of four and the ploidy of cysts is eight, which in turn raises the question of how *Giardia* maintains homogeneity between the chromosomes of the same and opposite nuclei. *Giardia* has been assumed to be primitively asexual and with no means of transferring DNA between nuclei. These assumptions make it very difficult to explain the remarkably low level of allelic heterozygosity (< 0.01%) in the genome isolate, WB. However, all these assumptions are now in doubt with the identification of meiotic genes in *Giardia*, evidence for recombination among isolates, and the evidence for exchange of genetic material between nuclei during the process of encystation. These findings raise a new set of questions that will be answered only by further investigation.

Introduction

Giardia trophozoites have long fascinated observers with the two nuclei that are placed symmetrically in the anterior half of the trophozoite and appear to stare back at the observer as a pair of eyes. *Giardia* species are members of the diplomonads (two bodies), which include the *Spironucleus* species, parasites of vertebrates (especially fish), and *Hexamita* species, which can be free-living or parasitic (Leipe *et al.*, 1993; Harper, 1991), as well as *Trepomonas* (Cavalier-Smith and Chao, 1996) and *Octomitus* species. This clade of organisms is unique in its possession of two nuclei that are nearly identical. Although the ciliates, including *Tetrahymena* and *Paramecium* spp., have two nuclei, the ciliate nuclei consist of a transcriptionally inactive, germline-like micronucleus, which contains the complete genome, and a transcriptionally active macronucleus, which contains amplified, fragmented regions of the genome and provides the site for RNA transcription. Thus, the ciliate nuclei have different structure and function and they replicate at different times in the cell cycle. In contrast, the diplomonads have two nuclei with similar function, morphology, structure, time of replication, and both nuclei are transcriptionally active.

The two nuclei replicate at approximately the same time

The initial study of the replication of *Giardia* nuclei using tritiated thymidine suggested that replication occurred at nearly the same time, but was possibly out of phase by about 15 to 30 minutes (Wiesehahn *et al.*, 1984). The slight asynchrony has subsequently been supported by electron and fluorescence microscopy (Solari *et al.*, 2003; Benchimol, 2004b). The daughter nucleus separates from the parent in a dorsal:ventral direction, which maintains a left:right asymmetry between the nuclei.

The nuclei are attached to the anterior flagella and the ventral disc, which appears to be involved in nuclear localization during the process of replication, presumably assuring correct nuclear segregation during division. The nuclear envelope remains intact throughout cell division, unlike the nuclei of higher eukaryotes (Solari *et al.*, 2003;

Benchimol, 2004b; Tumova *et al.*, 2007; Sagolla *et al.*, 2006). The nuclei differ substantially in the number of pore complexes; these pore complexes are markedly reduced in dividing nuclei (Benchimol, 2004a). Rather than being evenly distributed, they are clustered into pore-rich regions. In general, the nuclear pores correlate with transcriptional activity, so this difference is somewhat surprising, since prior data has shown approximately equal levels of transcription in the two nuclei (Kabnick and Peattie, 1990).

Both nuclei are transcriptionally active

Prior studies have used DAPI (Kabnick and Peattie, 1990) or Feulgen (Erlandsen and Rasch, 1994) staining to determine the DNA quantity in the two nuclei. These studies have shown that the nuclei have approximately equal quantities of DNA. In situ hybridization with a tritium-labelled rDNA probe has demonstrated that the two nuclei have equal numbers of rRNA genes and UTP incorporation demonstrated that the two nuclei transcribe equal quantities of rRNA (Kabnick and Peattie, 1990).

Subsequently, a study using fluorescence *in situ* hybridization with chromosome-specific probes for each of the five chromosomes in two Genotype A isolates (Isr and WB) has shown that the two nuclei each contain a complete genome. Although this study did not formally attempt to quantitate the amount of DNA in each nucleus, the amount of hybridization in the two nuclei was usually equal, suggesting equal ploidy numbers for the nuclei (Yu *et al.*, 2002).

Does Giardia *have a nucleolus?*

Nucleoli are distinct compartments within nuclei and can be seen readily by electron microscopy. They have been a universal finding in eukaryotic nuclei and are the centres of ribosomal (r) RNA synthesis and ribosome subunit assembly. However, nucleoli had not been found in *Giardia* nuclei. In addition, fibrillarin, a protein involved in rRNA methylation and ribosome assembly and highly conserved from yeast to mammals, is found in *Giardia* (Narcisi *et al.*, 1998; Ghosh *et al.*, 2001), but monoclonal antibodies to the yeast fibrillarin stained *Giardia* nuclei diffusely, suggesting the absence of discrete nucleoli (Narcisi *et al.*, 1998). Thus, it is of interest that recent reports have identified at least one structure per nucleus consistent with a nucleolus (Saraiya and Wang, 2008; Jimenez-Garcia *et al.*, 2008). One of these studies demonstrated localization of epitope-tagged forms of the nucleolar proteins fibrillarin and rRNA-pseudouridine synthase (CBF5) to the compartment (Saraiya and Wang, 2008). Interestingly, a snoRNA precursor of a microRNA (GlsR17) was detected in only one of the putative nucleoli (Saraiya and Wang, 2008), which suggest functional differences between the two nucleolar regions.

Giardia *has five chromosomes*

The ploidy and DNA content of a cell are not only fundamental biological questions, but also important factors regulating gene expression during cellular differentiation (Birchler *et al.*, 2007). Changes in ploidy affect the copy number of each gene and are often associated with changes in chromatin structure, thereby up- or down-regulating gene expression (Birchler *et al.*, 2007).

In order to understand the chromosomal structure and organization in *Giardia* trophozoites, pulsed field gel electrophoresis (PFGE) separations of chromosomal DNA were performed, revealing 4 or 5 bands ranging in size from approximately 1 to 4 million basepairs (Mb) (Adam *et al.*, 1988). Chromosome-specific probes were used to confirm that for the isolates with four PFGE bands, chromosomes 1 and 2 were co-migrating. These studies confirmed that Genotypes A and B (Groups 1, 2, and 3) organisms have five linkage groups or chromosomes, adding up to a haploid genome size of approximately 12 Mb (Le Blancq and Adam, 1998; Adam *et al.*, 1988). The telomeric sequence TAGGG is found at the end of the chromosomes (Adam *et al.*, 1991; Hou *et al.*, 1995). Up to four size variants of individual chromosomes were documented by hybridization of chromosome-specific probes to PFGE separations of chromosomes (Adam, 1992; Le Blancq *et al.*, 1991; Hou *et al.*, 1995). The central regions of the chromosome homologues were identical, with the size difference being attributed to differences in subtelomeric repetitive (STR) DNA. The size variation found in chromosome homologues was most pronounced in chromosome 1 with sizes varying from 1.1 to 1.9 Mb. Approximately 30% of the variability could be attributed to different copy numbers of the subtelomeric rDNA repeats, while the remaining 70% is attributable to variations in other STR DNA. The rDNA repeats can be found on any of the five chromosomes, but the chromosomal locations vary among isolates, and differ even for otherwise closely related isolates.

Other subtelomeric repeats that have been identified include retroposons and vsp genes. Three different retroposons have been identified in *Giardia*; GilM, GilT and GilD (Arkhipova and Morrison, 2001). All three belong to long interspersed nuclear element (LINE)-like families. GilT and GilM are in tandem arrays at the telomeres as demonstrated by sequence linkage and hybridization studies (Prabhu *et al.*, 2007). GilM and GilT have intact reverse transcriptase open reading frames (ORFs) and the genes have long 3′-UTRs of 2150 and 2644 basepairs in length, respectively. Studies using RT-PCR have shown that the ORFs as well as the long 3′-UTR are expressed (Prabhu *et al.*, 2007). Sequencing of small RNAs from *Giardia* identified small sense and antisense RNAs homologous to the GilT retroposon family (Ullu *et al.*, 2005). These small RNAs may contribute to silencing of GilT expression via a RNA interference mechanism. On the other hand, GilD has multiple deletions, stop codons and frameshifts, so it is a pseudogene. Two STRs consisted of vsp genes (Prabhu *et al.*, 2007). One STR consisted of a single vsp gene that was truncated by the telomeric repeat, and the other had a linear array of two vsp genes in which the second was terminated by the telomeric repeats. Both STRs were probably at one end of chromosome 5, indicating two homologues of chromosome 5. The expression status of the one intact vsp gene in an STR is not known.

The locations of two vsp genes have been correlated with expression status and, in both cases, the vsp gene was chromosome-internal and rearrangement did not occur during antigenic variation. Thus, it is clear that neither a telomeric location nor DNA rearrangement is required for antigenic variation to occur. Whether these subtelomeric genes are capable of being expressed has not yet been determined.

Centromeres have not yet been identified in *Giardia*.

Ploidy of trophozoites and cysts

The haploid genome size of 12 Mb estimated by PFGE has subsequently been confirmed by the genome sequencing project for the WB isolate (Morrison *et al.*, 2007). This value differs markedly from semiquantitative estimates of total DNA content that suggested at least 8 copies of each chromosome in the genome (Adam *et al.*, 1988). As noted above, a ploidy of at least four had been suggested by the identification of up to four different sizes of chromosome homologues. In addition, a ploidy of three to four is also supported by Southern blots of restriction enzyme-digested DNA hybridized to repeat-containing vsp genes that are allelic, but vary in repeat copy number and thus migrate differently (see Fig. 8.1).

The question that remained to be resolved is why there appeared to be a functional ploidy of about four, yet the quantitative data suggested at least double that number. To resolve this issue flow cytometry has been used to determine the DNA content of trophozoites (Bernander *et al.*, 2001). The ploidy of *Giardia* trophozoites in exponential and stationary phase was determined by DNA staining using fluorescent dyes and analysis by flow cytometry. In the dividing trophozoite culture two peaks representing different DNA contents were detected, as expected (Bernander *et al.*, 2001). The lower peak corresponded to the genomic content before DNA replication (G1) and the higher peak to the genomic content after replication (G2 and M). The area between the two peaks corresponds to replicating or S-phase cells. In stationary phase the higher peak was dominant, indicating that the cells arrest in a post-replicative stage (G2). Peaks from exponentially growing trophozoites were calibrated against those of rifampin-treated *Escherichia coli* cells. These cells carry multiple copies of the *E. coli* genome (4.64 Mb) and were used to correlate the intensity of fluorescence measured in the flow cytometer to cellular DNA content. Comparison with *E. coli* intensities showed that the trophozoite DNA content oscillates between 4 and 8 copies of the genome size (12 Mb) during exponential growth. Microscopic analysis showed equal staining of the two nuclei. These results are consistent with the PFGE findings and suggest that each nucleus is diploid in the pre-replicative

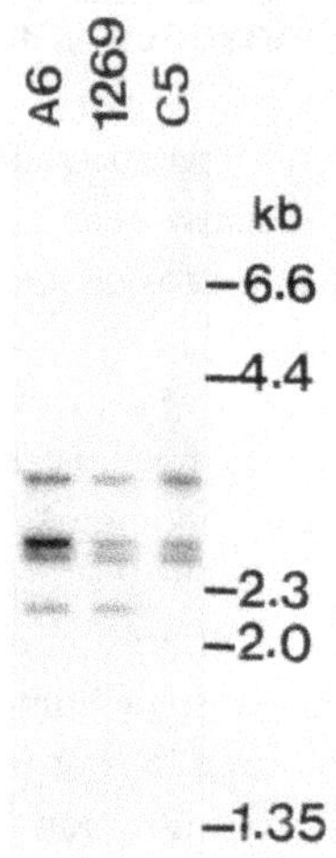

Figure 8.1 Southern blot of restriction digests from WBa6, WB1269, and WBC5 probed with the tandem repeat region of vspC5. Data presented in Yang and Adam (1995) confirmed that the 3 or 4 bands represent different alleles of the vspC5 gene. Thus, these bands represent up to four alleles of vspC5. WB1269 is an antigenic variant derived from WBA6 (Nash *et al.*, 1988) and WBC5 is an antigenic variant derived from WB1269 (Yang *et al.*, 1994). Thus, these three lines are genetically related; however, the smallest vspC5 allele has disappeared from WBC5 without a compensatory increase in another band, raising the possibility that one of the chromosome 5 homologues was deleted in the transition from WB1269 to WBC5. Figure 2C from Yang and Adam (1995), reproduced with permission.

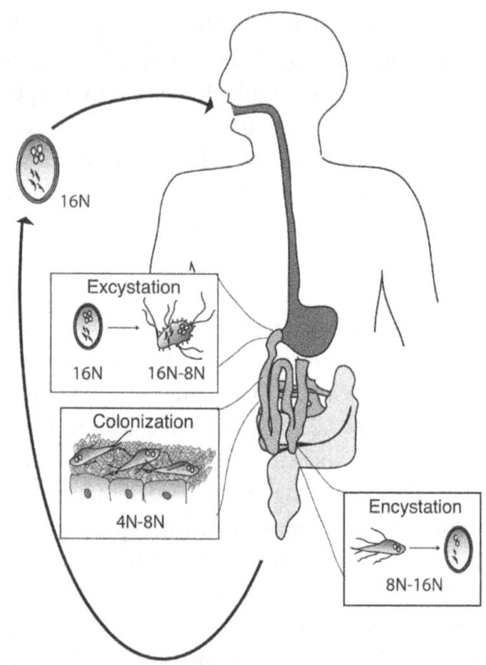

Figure 8.2 Life cycle of *Giardia*. In the vegetative stage the parasite multiplies by binary fission and cycles between cellular ploidies of 4N and 8N. Upon induction of encystation the trophozoite rounds up, the disc disassembles and numerous encystation-specific vesicles (ESVs) containing cyst-wall proteins become visible. Late in encystation the two nuclei divide and the DNA is replicated, generating cysts with a ploidy of 16N. Cysts excyst, releasing the excyzoite with a ploidy of 16N. The excyzoite divides twice without DNA replication, generating four trophozoites with a 4N ploidy.

stage and tetraploid in the post-replicative stage (Fig. 8.2).

While passing through the stomach of the host the cyst-enclosed parasite senses the environmental changes and initiates a process called excystation (Fig. 8.2). Microscopic analysis of recently excysted cells shows that they contain four nuclei and have a ploidy of 16N as they exit the cyst (Bernander *et al.*, 2001). Immediately after excystation they enter cytokinesis and divide. Accordingly, flow cytometry analysis showed that the 8N peak reappeared four hours after excystation (Bernander *et al.*, 2001). Later, the growing trophozoite population goes back to the 4N/8N distribution described above. The newly excysted cell is morphologically different from a trophozoite; it has no adhesive disc, the metabolism is intermediate between trophozoite and cyst (Jarroll *et al.*, 1989) and the transcription pattern is excystation-specific (Hetsko *et al.*, 1998). Importantly, it is the recently excysted cell that initiates infection. Therefore a new name has been proposed for the newly excysted cell: excyzoite. The ploidy of 16N of the excyzoite is potentially important for the success rate of infection because it carries enough DNA to divide two times independent of DNA replication. Thus, four trophozoites can quickly be formed from one excyzoite.

Cells do not enter encystation at the same time, indicating that there is a specific point in the cell cycle for entering encystation (Reiner *et al.*, 1992; Reiner *et al.*, 2008; Bernander *et al.*, 2001)). Early in encystation the pre-replicative peak disappears, indicating that cell division is blocked and that the trophozoites differentiate from the G2 stage. Differentiation from the G2 phase has also been suggested for another cyst-forming eukaryote, *Acanthamoeba* (Jantzen *et al.*, 1988). The arresting point for stationary phase cells is probably beyond the restriction point for encystation as cyst numbers are low when inducing encystation in a stationary phase culture.

Cell cycle arrest

Several agents that arrest the cell cycle of eukaryotic cells have been tested for their effect on *Giardia* trophozoites. Colchicine (G_2+M blocker) had no effect on *Giardia* trophozoites, whereas hydroxyurea (G_2+M blocker) and razoxane (G_1/S or G_2+M blocker) arrested the cell cycle in the G_2+M phase (Hoyne *et al.*, 1989). Until recently, no *in vitro* cultures of *Giardia* had been synchronized. It has now been shown that the drug aphidicolin can be used to arrest trophozoites between the G1 and S phases of the cell cycle. Removal of the drug, followed by addition of fresh medium, generated semi-synchronized *Giardia* trophozoites (Reiner *et al.*, 2008). Growth of released trophozoites for different times generated cells in different stages of the cell cycle. Induction of encystation in synchronized cells at different stages of the cell cycle supported the evidence for a restriction point in early to mid G2 of the cell cycle (Reiner *et al.*, 2008).

Histones

As in other eukaryotes, *Giardia* uses histones as the basic DNA-packaging protein complex for chromosomal organization. *Giardia* has the four core histones (H2a, H2b, H3, and H4) but the linker histone (H1) is missing (Yee *et al.*, 2007). Variants of histone H3 were associated with heterochromatin in centromeric regions and in other chromosomal regions (Dawson *et al.*, 2007). Eukaryotic histones possess long, positively charged N-terminal tails that are targets of several post-translational modifications, such as acetylation, methylation, phosphorylation and ubiquitination, and these tails can also be found in the *Giardia* histones. Enzymes mediating these modifications are a universal feature of eukaryotes and regulate transcription both globally and locally by dynamically remodelling chromatin to allow or restrict access to general and specific transcription factors. Even though the key histone modifications are universal to eukaryotes, the domain architectures of proteins binding post-translationally modified histones vary considerably across eukaryotes and *Giardia* has a reduced repertoire of proteins that can modify and bind histone tails (Iyer *et al.*, 2008; Morrison *et al.*, 2007). Furthermore, recent data show that the genomic DNA of *Giardia* is not methylated (Prucca *et al.*, 2008). This indicates that the histone code might be 'interpreted' differently in *Giardia* compared to yeast and human cells. The non-methylated DNA can also be important during the interaction of the host's immune cells since certain Toll-like receptors recognize non-methylated CpG DNA (Müller *et al.*, 2008).

Is *Giardia* sexual or asexual?

Giardia has long been assumed to be asexual. Sex has never been observed in *Giardia* or in any of the other diplomonads, leading to the proposal that these organisms are primitively asexual. Species for sexual organisms are defined on the basis of reproductive isolation, but different approaches have been used for organisms that are presumed to be asexual. For protists, species have typically been defined on the basis of morphology. Based on the assumption that *Giardia* is asexual, species in this genus have been assigned on the basis of morphological differences seen by light or electron microscopy (see chapter by Andersson *et al.* for details on *Giardia* species). The organisms that are classified as *G. lamblia* are all mammalian parasites and are morphologically identical by light and electron microscopy. Those infecting humans were initially divided into three groups based on differences in DNA restriction fragment length polymorphism (Nash *et al.*, 1985) and patterns of radiolabeled surface proteins (Nash and Keister, 1985). Groups 1 and 2 are quite similar and have subsequently been designated as Genotypes (or Assemblages) A1 and A2. On the other hand, Genotype B (Group 3) is substantially different and has been proposed as a separate species (Nash *et al.*, 1985; Nash and Keister, 1985). An additional five genotypes have been described that are genetically distinct from Genotypes A and B (designated Genotypes C through G) and are found only in non-human mammals. Further comments will focus on only isolates from Genotypes A and B.

Low levels of allelic heterozygosity

Since *Giardia* is a tetraploid organism with two diploid nuclei, sequence divergence is expected to accumulate in the absence of sex. For example, extensive sequence divergence has been observed between the two former haplotypes in asexual bdelloid rotifers (Mark Welch *et al.*, 2008). If *Giardia* species are truly asexual organisms, one might expect a very high degree of allelic sequence divergence, especially since the genome is divided into two different nuclei. However, a surprisingly low level of allelic sequence divergence, less than 0.01%, was reported from analysis of the WB genome (Morrison *et al.*, 2007) In addition, no allelic variations were found in the *Spironucleus salmonicida* genome survey (Andersson *et al.*, 2007).

Not only does the WB isolate have very little allelic heterozygosity within its genome, but Genotype A1 (the genotype of the genome isolate, WB) is a highly homogeneous group of organisms found throughout the world. These organisms are so similar to each other that no polymorphisms have been described that distinguish isolates. The few sequence comparisons between A1 and A2 organisms (because of the paucity of A2 sequence information) indicates a sequence identity of 98–99% (Baruch *et al.*, 1996; Cooper *et al.*, 2007). Some studies have documented heterogeneity among the A2 isolates (Cooper *et al.*, 2007) as well as some degree of allelic heterozygosity.

Our analysis of the recently sequenced GS genome shows that the overall level of allelic sequence heterozygosity was 0.53% in the GS genome. As expected, the allelic sequence heterozygosity is higher in non-coding than in coding regions, 1.25% and 0.3%, respectively. This high frequency of allelic sequence heterozygosity suggests that 'double-peaks' in sequencing chromatograms observed in genotyping studies of clinical samples (Lebbad *et al.*, 2008) could be explained by allelic sequence heterozygosity within a single infecting *Giardia*, in addition to double infections of two different isolates. PCR analysis of different genes from patient samples containing Genotype B isolates often show a high degree of sequence heterozygosity in certain positions (Lebbad *et al.*, 2008), whereas this is never seen in Genotype A1 isolates.

Population genetic evidence that *Giardia* is asexual

Population genetics is very useful for determining whether the reproduction of micro-organisms is predominantly clonal or whether it demonstrates recombination between isolates. The criteria suggestive of clonal reproduction include the lack of segregation or recombination. The lack of segregation is demonstrated by fixed heterozygosity, absence of segregating genotypes, and deviation from Hardy-Weinberg equilibrium. The lack of recombination is demonstrated by over-represented widespread genotypes, absence of recombinant genotypes, linkage disequilibrium, and correlation between independent sets of genetic markers (Tibayrenc *et al.*, 1991). It should be noted that clonal population genetics can still be observed if sex is present but rare enough that it doesn't impact population structure. For example, *Plasmodium falciparum* is known to be sexual, but some population genetic studies suggest primarily clonal reproduction (Tibayrenc *et al.*, 1990). However, it should be noted that more recent sequence-based analyses have confirmed the population genetic impact of recombination and have provided a more precise assessment of the degree of recombination found in *P. falciparum* populations (Mu *et al.*, 2005; Conway *et al.*, 1999).

The initial studies of the population genetics of *Giardia* were performed on axenized isolates using electrophoretic analysis of multiple metabolic enzymes, as reviewed in Tibayrenc *et al.* (1990). The largest study tested 10 different enzymes on 30 axenized isolates. Differences are observed when amino acid substitutions change the charge of the molecule enough to alter its gel migration. These studies suggested that *Giardia* showed a clonal population structure. However, such studies are limited by the requirement for sequence alterations that change the amino acid sequence (and charge). Thus, the majority of nucleotide differences between isolates would probably not be identified. In addition, the lack of heterogeneity among the A1 isolates would give the appearance of clonal replication for those isolates. Significantly, Genotype B isolates generally have been more difficult to axenize, causing these organisms to be underrepresented in some of these studies.

One of the forms of recombination seen in sexual organisms is the independent reassortment of chromosomes. A molecular marker for independent reassortment can be provided by comparing the phylogenies of genes/sequences from different chromosomes for several isolates. If the population genetic structure is primarily clonal, then there should be no chromosome reassortment. Thus, phylogenetic comparisons of different isolates should yield identical results regardless of the chromosomal sources of the loci being compared. However, a recent study of five A2 faecal isolates from a highly endemic region near Lima, Peru, in addition to the A2 reference isolate JH (Nash *et al.*, 1985; Nash and Keister, 1985; Adam *et al.*, 1988), evaluated 6, 1.2, and 9.6 kb loci from chromosomes 3, 4, and 5, respectively (Cooper *et al.*, 2007) (Fig. 8.3). The relative placement of isolates varied according to the chromosomal location of the locus, a result expected with sexual reproduction but not with clonal reproduction.

In addition, there was evidence for crossing over at one site on the 6 kb chromosome 3 locus and two sites on the 9.6 kb chromosome 5 locus. The crossing over sites on chromosome 5 are shown in Fig. 8.4. These data strongly suggest

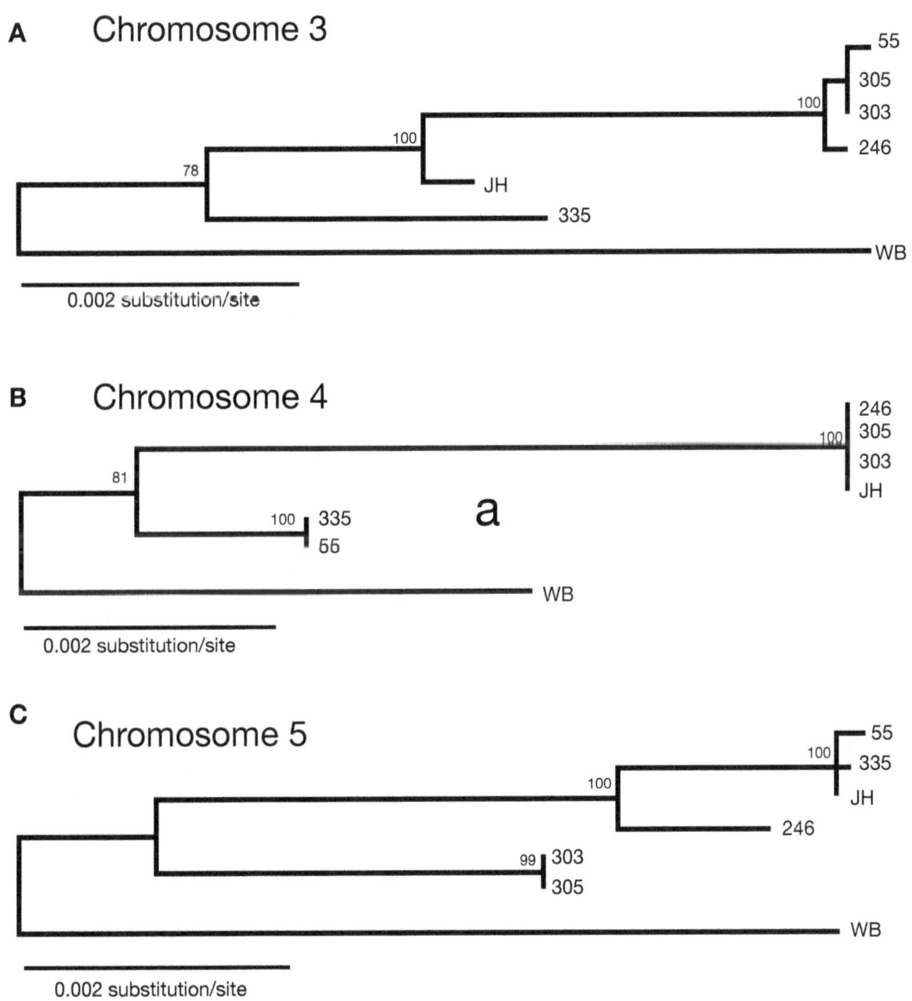

Figure 8.3 Maximum likelihood trees of sequences from loci on chromosomes 3, 4 and 5 showing bootstrap support. WB is the Genotype A1 outgroup. Note that the trees are different for each of the three chromosomes, consistent with independent reassortment of the chromosomes. Isolates 303 and 305 are from the same household and the sequences are identical at all three loci. In contrast, each of the other isolates is assigned to different branches for each chromosome. Figure 2 from Cooper *et al.* (2007), reproduced with permission.

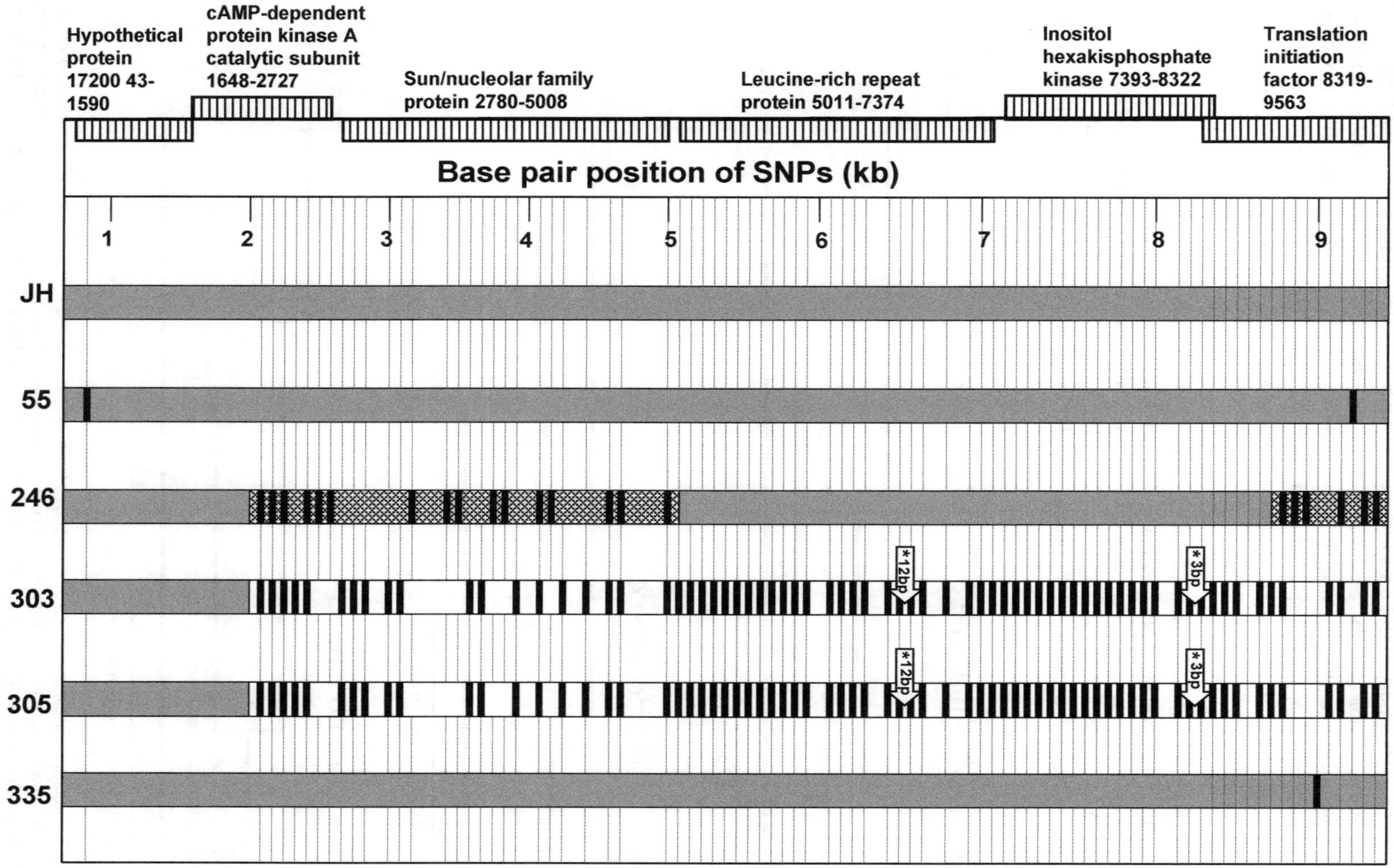

Hypothetical protein 17200 43-1590
cAMP-dependent protein kinase A catalytic subunit 1648-2727
Sun/nucleolar family protein 2780-5008
Leucine-rich repeat protein 5011-7374
Inositol hexakisphosphate kinase 7393-8322
Translation initiation factor 8319-9563
Base pair position of SNPs (kb)
1
2
3
4
5
6
7
8
9
JH
55
246
303
305
335
*12bp
*3bp
*12bp
*3bp

sexual recombination in this population of organisms. It is important to note that these isolates were all from the same subgenotype (A2). So far, there is no evidence for recombination between A1 and A2 isolates in that the A1 isolates are so homogeneous that if recombination with A2 isolates occurred, one result would be identifiable sequence differences. So far, too little Genotype B sequence data exists to rule out recombination between A2 and B organisms, but so far our investigation of isolates from a highly endemic location has not provided evidence for such recombination. This will be an important question to settle, because a lack of recombination between the genotypes would provide further evidence that they belong to separate species.

Meiotic genes are found in *Giardia*

The *Giardia* genome was examined for the presence of genes involved in mitosis and meiosis (Ramesh *et al.*, 2005). This type of analysis is somewhat challenging because many of the proteins involved in DNA repair are utilized in mitosis as well as meiosis. However, there is a group of genes that are involved in meiosis but not mitosis in *Saccharomyces cerevisiae*. In each case, there are null mutations that result in defects of meiosis, but not mitosis (Villeneuve and Hillers, 2001). This analysis resulted in the identification of seven meiosis-specific genes in *S. cerevisiae*, and five of those seven are found in the *Giardia* genome (Table 8.1). The two genes not found in the *Giardia* genome (Msh4 and Msh5) are also absent from *Drosophila* and *Schizosaccharomyces*, so their absence from *Giardia* doesn't necessarily mean that *Giardia* has an incomplete complement of meiotic genes. In a subsequent study, a set of meiosis-specific genes were expressed as GFP fusion proteins from their native promoters in *Giardia* trophozoites and in encysting organisms (Poxleitner *et al.*, 2008). Spo11, Hop1 and Dmc1a localized to the four cyst nuclei, but not to the nuclei of vegetative trophozoites. Mnd1 and Dmc1b localized to the nuclei of both cysts and trophozoites.

The pathogenic yeast *Candida albicans* also has a complement of meiotic genes, but has always been considered to be asexual. In fact, the genus *Candida* refers to the asexual yeasts. *C. albicans* has now been shown to have mating types and has demonstrated recombination between isolates, and is thus a sexual organism. However, recent evidence suggests that recombination in *C. albicans* does not use meiosis. Rather, the nuclei of two diploid organisms fuse, allowing chromosome reassortment and mitotic crossing over, followed by resolution without ever going through meiosis. The 'meiosis-specific' gene Spo11 is specifically involved in this parasexual reproduction in *C. albicans* (Forche *et al.*, 2008). The processes of meiotic sexual reproduction and parasexual reproduction are similar enough that it may be impossible to know which one is occurring without directly observing the process.

All meiosis-related genes identified in WB can be found in GS. Several are well conserved (Spo11: 78% amino acid identity (aaID); Dmc1: 92% aaID; Msh6: 78% aaID) whereas others are not as conserved (Mre11: 62% aaID; Rad50: 60% aaID) and some even show deletions (15 amino acids (aa) in Rad52) or insertions (14 aa in Mlh1) in important regions. Further investigations will show if these differences can explain the observed differences in allelic sequence divergence. Interestingly, the fish commensal *Spironucleus barkhanus*, which is morphologically indistinguishable from the pathogenic *S. salmonicida* (Jørgensen and Sterud, 2006), appears to have a high frequency of allelic sequence variation (unpublished observation), further suggesting that the presence of allelic sequence variation is variable even between closely related diplomonads.

Figure 8.4 (opposite) SNP diagram of a 9.6 kb locus from chromosome 5 for six Genotype A2 *G. lamblia* isolates. ORFs are depicted as vertical striped boxes: top boxes are forward, bottom boxes are reverse strand encoded. ORF names are written above the box and if the ORF is a hypothetical protein, its corresponding *Giardia*DB ORF number is provided. JH is the reference Genotype A2 isolate, while 55, 246, 303/305, and 335 are Genotype A2 field isolates. Isolates 303/305 are from siblings and are identical. Hash marks along each horizontal line represent SNPs in the field isolates, compared to JH. Open arrows with an asterisk depict an insertion/deletion (indel). Fig 1B from Cooper *et al.* (2007), reproduced with permission.

Table 8.1 Meiosis-specific genes found in *Giardia*

Name	Function in yeast
Spo11	Transesterase, creates double-strand break in DNA during meiosis
Hop1	Binds double strand breaks and oligomerizes during meiotic prophase I
Hop2	Forms a complex with Mnd1 for homology-searching during prophase I
Mnd1	Works with Hop2 and is involved in DNA repair
Dmc1	Homologue of Rad51; involved in interhomologue recombination (two forms, a and b, found in *Giardia*)

Exchange of genetic material during encystation

Recent studies have demonstrated that *Giardia* nuclei may fuse during the process of encystation (Poxleitner *et al.*, 2008). The fusion event has been demonstrated by electron microscopy and functionally by following the fate of an episome that replicates in one of the two nuclei. During vegetative growth of WB trophozoites, the transfected episome is never found in both nuclei (Yu *et al.*, 2002). However, during encystation, the episome is occasionally found in three of the four nuclei, a process which most likely occurs during the nuclear fusion. The authors called this process 'diplomixis'. It is important to note that this process could account for resolution of allelic heterozygosity, but cannot explain the population genetic evidence for recombination among isolates. However, it is possible that a mechanistically related process could be involved.

Remaining questions

If sex truly does occur, which stage of replication and where in the host?

Although *Giardia* has been assumed to be asexual, a number of recent findings have cast doubt on that assumption. The genome contains genes that are thought to be specific to meiosis (or at least sexual reproduction), the low level of allelic heterozygosity suggests sex, and the evidence for recombination among isolates of a single genotype (A2) all suggest that meiotic or parasexual reproduction is occurring. The observation that nuclei can fuse during encystation and exchange genetic material also supports the concept of sexual reproduction, although this mechanism obviously cannot explain how recombination can occur between different isolates (outcrossing). The lack of exchange of genetic material during vegetative growth of trophozoites suggests that if outcrossing occurs it is probably occurring during the encystation process.

What is a species? How different must isolates be before they fail to mate?

The evidence for outcrossing also raises the issue of species designation in a way that has not yet been addressed. For sexual organisms, species are defined on the basis of reproductive isolation. To date, there is no convincing evidence of recombination between Genotype A and B isolates. If confirmed by sequences of Genotypes A and B isolates, these genotypes should be designated as separate species.

References

Adam, R.D. (1992). Chromosome-size variation in *Giardia lamblia*: the role of rDNA repeats. Nucleic Acids Res. *20*, 3057–3061.

Adam, R.D., Nash, T.E., and Wellems, T.E. (1988). The *Giardia lamblia* trophozoite contains sets of closely related chromosomes. Nucleic Acids Res. *16*, 4555–4567.

Adam, R.D., Nash, T.E., and Wellems, T.E. (1991). Telomeric location of *Giardia* rDNA genes. Mol. Cell Biol. *11*, 3326–3330.

Andersson, J.O., Sjögren, Å.M., Horner, D.S., Murphy, C.A., Dyal, P.L., Svärd, S.G., Logsdon, J.M., Jr., Ragan, M.A., Hirt, R.P., and Roger, A.J. (2007). A genomic survey of the fish parasite *Spironucleus salmonicida* indicates genomic plasticity among diplomonads and significant lateral gene transfer in eukaryote genome evolution. BMC Genomics *8*, 51.

Arkhipova, I.R., and Morrison, H.G. (2001). Three retrotransposon families in the genome of *Giardia lamblia*: two telomeric, one dead. Proc. Natl. Acad. Sci. U.S.A. *98*, 14497–14502.

Baruch, A.C., Isaac-Renton, J., and Adam, R.D. (1996). The molecular epidemiology of *Giardia lamblia*: a sequence-based approach. J. Infect. Dis. *174*, 233–236.

Benchimol, M. (2004a). *Giardia lamblia*: behavior of the nuclear envelope. Parasitol. Res. *94*, 254–264.

Benchimol, M. (2004b). Participation of the adhesive disc during karyokinesis in *Giardia lamblia*. Biol. Cell *96*, 291–301.

Bernander, R., Palm, J.E., and Svärd, S.G. (2001). Genome ploidy in different stages of the *Giardia lamblia* life cycle. Cell. Microbiol. *3*, 55–62.

Birchler, J.A., Yao, H., and Chudalayandi, S. (2007). Biological consequences of dosage dependent gene regulatory systems. Biochim. Biophys. Acta *1769*, 422–428.

Cavalier-Smith, T., and Chao, E.E. (1996). Molecular phylogeny of the free-living archezoan *Trepomonas agilis* and the nature of the first eukaryote. J. Mol. Evol. *43*, 551–562.

Conway, D.J., Roper, C., Oduola, A.M., Arnot, D.E., Kremsner, P.G., Grobusch, M.P., Curtis, C.F., and Greenwood, B.M. (1999). High recombination rate in natural populations of *Plasmodium falciparum*. Proc. Natl. Acad. Sci. U.S.A. *96*, 4506–4511.

Cooper, M.A., Adam, R.D., Worobey, M., and Sterling, C.R. (2007). Population genetics provides evidence for recombination in *Giardia*. Curr. Biol. *17*, 1984–1988.

Dawson, S.C., Sagolla, M.S., and Cande, W.Z. (2007). The cenH3 histone variant defines centromeres in *Giardia intestinalis*. Chromosoma *116*, 175–184.

Erlandsen, S.L., and Rasch, E.M. (1994). The DNA content of trophozoites and cysts of *Giardia lamblia* by microdensitometric quantitation of Feulgen staining and examination by laser scanning confocal microscopy. J. Histochem. Cytochem. *42*, 1413–1416

Forche, A., Alby, K., Schaefer, D., Johnson, A.D., Berman, J., and Bennett, R.J. (2008). The parasexual cycle in *Candida albicans* provides an alternative pathway to meiosis for the formation of recombinant strains. PLoS Biol. *6*, e110.

Ghosh, S., Ghosh, R., Das, P., and Chattopadhyay, D. (2001). Expression and purification of recombinant *Giardia* fibrillarin and its interaction with small nuclear RNAs. Protein Expr. Purif. *21*, 40–48.

Harper, F.D. (1991). *Hexamita* species present in some avian species in South Wales. Vet. Rec. *128*, 130.

Hetsko, M.L., McCaffery, J.M., Svärd, S.G., Meng, T.C., Que, X., and Gillin, F.D. (1998). Cellular and transcriptional changes during excystation of *Giardia lamblia in vitro*. Exp. Parasitol. *88*, 172–183.

Hou, G., Le Blancq, S.M., E Y., Zhu, H., and Lee, M.G. (1995). Structure of a frequently rearranged rRNA-encoding chromosome in *Giardia lamblia*. Nucleic Acids Res. *23*, 3310–3317.

Hoyne, G.F., Boreham, P.F., Parsons, P.G., Ward, C., and Biggs, B. (1989). The effect of drugs on the cell cycle of *Giardia intestinalis*. Parasitology 99, 333–339.

Iyer, L.M., Anantharaman, V., Wolf, M.Y., and Aravind, L. (2008). Comparative genomics of transcription factors and chromatin proteins in parasitic protists and other eukaryotes. Int. J. Parasitol. *38*, 1–31.

Jantzen, H., Schulze, I., and Stohr, M. (1988). Relationship between the timing of DNA replication and the developmental competence in *Acanthamoeba castellanii*. J. Cell Sci. *91*, 389–399.

Jarroll, E.L., Manning, P., Berrada, A., Hare, D., and Lindmark, D.G. (1989). Biochemistry and metabolism of *Giardia*. J. Protozool. *36*, 190–197.

Jimenez-Garcia, L.F., Zavala, G., Chavez-Munguia, B., Ramos-Godinez, M.P., Lopez-Velazquez, G., Segura-Valdez, M.L., Montanez, C., Hehl, A.B., Arguello-Garcia, R., and Ortega-Pierres, G. (2008). Identification of nucleoli in the early branching protist *Giardia duodenalis*. Int. J. Parasitol. *38*, 1297–1304.

Jørgensen, A., and Sterud, E. (2006). The marine pathogenic genotype of *Spironucleus barkhanus* from farmed salmonids redescribed as *Spironucleus salmonicida* n. sp. J. Eukaryot. Microbiol. *53*, 531–541.

Kabnick, K.S., and Peattie, D.A. (1990). In situ analyses reveal that the two nuclei of *Giardia lamblia* are equivalent. J. Cell Sci. *95*, 353–360.

Le Blancq, S.M., and Adam, R.D. (1998). Structural basis of karyotype heterogeneity in *Giardia lamblia*. Mol. Biochem. Parasitol. *97*, 199–208.

Le Blancq, S.M., Korman, S.H., and Van der Ploeg, L.H. (1991). Frequent rearrangements of rRNA-encoding chromosomes in *Giardia lamblia*. Nucleic Acids Res. *19*, 4405–4412.

Lebbad, M., Ankarklev, J., Tellez, A., Leiva, B., Andersson, J.O., and Svard, S. (2008). Dominance of *Giardia* assemblage B in Leon, Nicaragua. Acta Trop. *106*, 44–53.

Leipe, D.D., Gunderson, J.H., Nerad, T.A., and Sogin, M.L. (1993). Small subunit ribosomal RNA+ of *Hexamita inflata* and the quest for the first branch in the eukaryotic tree. Mol. Biochem. Parasitol. *59*, 41–48.

Mark Welch, D.B., Mark Welch, J.L., and Meselson, M. (2008). Evidence for degenerate tetraploidy in bdelloid rotifers. Proc. Natl. Acad. Sci. U.S.A. *105*, 5145–5149.

Morrison, H.G., McArthur, A.G., Gillin, F.D., Aley, S.B., Adam, R.D., Olsen, G.J., Best, A.A., Cande, W.Z., Chen, F., Cipriano, M.J., Davids, B.J., Dawson, S.C., Elmendorf, H.G., Hehl, A.B., Holder, M.E., Huse, S.M., Kim, U.U., Lasek-Nesselquist, E., Manning, G., Nigam, A., Nixon, J.E., Palm, D., Passamaneck, N.E., Prabhu, A., Reich, C.I., Reiner, D.S., Samuelson, J., Svärd, S.G., and Sogin, M.L. (2007). Genomic minimalism in the early diverging intestinal parasite *Giardia lamblia*. Science *317*, 1921–1926.

Mu, J., Awadalla, P., Duan, J., McGee, K.M., Joy, D.A., McVean, G.A., and Su, X.Z. (2005). Recombination hotspots and population structure in *Plasmodium falciparum*. PLoS Biol. 3, e335.

Müller, T., Hamm, S., and Bauer, S. (2008). TLR9-mediated recognition of DNA. Handb. Exp. Pharmacol., 51–70.

Narcisi, E.M., Glover, C.V., and Fechheimer, M. (1998). Fibrillarin, a conserved pre-ribosomal RNA processing protein of *Giardia*. J. Eukaryot. Microbiol. *45*, 105–111.

Nash, T.E., Aggarwal, A., Adam, R.D., Conrad, J.T., and Merritt, J.W., Jr. (1988). Antigenic variation in *Giardia lamblia*. J. Immunol. *141*, 636–641.

Nash, T.E., and Keister, D.B. (1985). Differences in excretory-secretory products and surface antigens among 19 isolates of *Giardia*. J. Infect. Dis. *152*, 1166–1171.

Nash, T.E., McCutchan, T., Keister, D., Dame, J.B., Conrad, J.D., and Gillin, F.D. (1985). Restriction-endonuclease analysis of DNA from 15 *Giardia* isolates obtained from humans and animals. J. Infect. Dis. *152*, 64–73.

Poxleitner, M.K., Carpenter, M.L., Mancuso, J.J., Wang, C.J., Dawson, S.C., and Cande, W.Z. (2008). Evidence for karyogamy and exchange of genetic material in the binucleate intestinal parasite *Giardia intestinalis*. Science *319*, 1530–1533.

Prabhu, A., Morrison, H.G., Martinez, C.R., III, and Adam, R.D. (2007). Characterisation of the subtelomeric regions of *Giardia lamblia* genome isolate WBC6. Int. J. Parasitol. *37*, 503–513.

Prucca, C.G., Slavin, I., Quiroga, R., Elias, E.V., Rivero, F.D., Saura, A., Carranza, P.G., and Lujan, H.D. (2008). Antigenic variation in *Giardia lamblia* is regulated by RNA interference. Nature *456*, 750–754.

Ramesh, M.A., Malik, S.B., and Logsdon, J.M. Jr. (2005). A phylogenomic inventory of meiotic genes; evidence for sex in *Giardia* and an early eukaryotic origin of meiosis. Curr. Biol. *15*, 185–191.

Reiner, D.S., Ankarklev, J., Troell, K., Palm, D., Bernander, R., Gillin, F.D., Andersson, J.O., and Svärd, S.G. (2008). Synchronisation of *Giardia lamblia*: identification of cell cycle stage-specific genes and a differentiation restriction point. Int. J. Parasitol. *38*, 935–944.

Reiner, D.S., Shinnick, T.M., Ardeshir, F., and Gillin, F.D. (1992). Encystation of *Giardia lamblia* leads to expression of antigens recognized by antibodies against conserved heat shock proteins. Infect. Immun. *60*, 5312–5315.

Sagolla, M.S., Dawson, S.C., Mancuso, J.J., and Cande, W.Z. (2006). Three-dimensional analysis of mitosis and cytokinesis in the binucleate parasite *Giardia intestinalis*. J. Cell Sci. *119*, 4889–4900.

Saraiya, A.A., and Wang, C.C. (2008). snoRNA, a novel precursor of microRNA in *Giardia lamblia*. PLoS Pathog. *4*, e1000224.

Solari, A.J., Rahn, M.I., Saura, A., and Lujan, H.D. (2003). A unique mechanism of nuclear division in *Giardia lamblia* involves components of the ventral disk and the nuclear envelope. Biocell *27*, 329–346.

Tibayrenc, M., Kjellberg, F., Arnaud, J., Oury, B., Brenière, S.F., Dardë, M.L., and Ayala, F.J. (1991). Are eukaryotic microorganisms clonal or sexual? A population genetics vantage. Proc. Natl. Acad. Sci. U.S.A. *88*, 5129–5133.

Tibayrenc, M., Kjellberg, F., and Ayala, F.J. (1990). A clonal theory of parasitic protozoa: the population structures of *Entamoeba, Giardia, Leishmania, Naegleria, Plasmodium, Trichomonas*, and *Trypanosoma* and their medical and taxonomical consequences. Proc. Natl. Acad. Sci. U.S.A. *87*, 2414–2418.

Tumova, P., Hofstetrova, K., Nohynkova, E., Hovorka, O., and Kral, J. (2007). Cytogenetic evidence for diversity of two nuclei within a single diplomonad cell of *Giardia*. Chromosoma *116*, 65–78.

Ullu, E., Lujan, H.D., and Tschudi, C. (2005). Small sense and antisense RNAs derived from a telomeric retroposon family in *Giardia intestinalis*. Eukaryot. Cell *4*, 1155–1157.

Villeneuve, A.M., and Hillers, K.J. (2001). Whence meiosis? Cell *106*, 647–650.

Wiesehahn, G.P., Jarroll, E.L., Lindmark, D.G., Meyer, E.A., and Hallick, L.M. (1984). *Giardia lamblia*: autoradiographic analysis of nuclear replication. Exp. Parasitol. *58*, 94–100.

Yang, Y., and Adam, R.D. (1995). A group of *Giardia lamblia* variant-specific surface protein (VSP) genes with nearly identical 5′ regions. Mol. Biochem. Parasitol. *75*, 69–74.

Yang, Y.M., Ortega, Y., Sterling, C., and Adam, R.D. (1994). *Giardia lamblia* trophozoites contain multiple alleles of a variant-specific surface protein gene with 105-base pair tandem repeats. Mol. Biochem. Parasitol. *68*, 267–276.

Yee, J., Tang, A., Lau, W.L., Ritter, H., Delport, D., Page, M., Adam, R.D., Müller, M., and Wu, G. (2007). Core histone genes of *Giardia intestinalis*: genomic organization, promoter structure, and expression. BMC Mol. Biol. *8*, 26.

Yu, L.Z., Birky, C.W., Jr., and Adam, R.D. (2002). The two nuclei of *Giardia* each have complete copies of the genome and are partitioned equationally at cytokinesis. Eukaryot. Cell *1*, 191–199.

The *Blastocystis* Mitochondrion-like Organelles

9

Anastasios D. Tsaousis, Alexandra Stechmann, Karleigh A. Hamblin, Mark van der Giezen, Vicente Pérez-Brocal and C. Graham Clark

Abstract

The organelles in *Blastocystis* that resemble mitochondria are an enigma as the organism is a strict anaerobe. Recent sequence analysis of the organelle genome and over 12,000 expressed sequence tags (ESTs) has given us many insights into the role these organelles play in the metabolism of the cell. The genome encodes several subunits of NADH dehydrogenase (complex I) but lacks all trace of genes for cytochrome and ATPase subunits (complexes III–V). ESTs confirm the presence of complexes I and II, and indicate that this partial electron transport chain may lead to an alternative oxidase. The ESTs also suggest that many other metabolic pathways characteristic of mitochondria are still present in the *Blastocystis* organelles. However, other findings show that the organelle also has characteristics in common with hydrogenosomes, as a gene encoding [FeFe] hydrogenase is present and the protein has been localized to the organelles. The nuclear genome should clarify many of the remaining questions surrounding these unusual organelles.

Introduction

Blastocystis sp. is a common member of the human eukaryotic intestinal flora. Although it had probably been seen before then, in 1912 it was named by Brumpt (1912) who described it as a yeast. Having gone through many reclassifications over the years *Blastocystis* finally came to rest among the stramenopiles (syn. heterokonts) (Silberman *et al.*, 1996), a diverse group of eukaryotes that includes fresh water and marine, photosynthetic and heterotrophic, free-living, commensal and pathogenic species. Among the best known stramenopiles are the mildews, kelp, diatoms and potato blight. Although morphologically extremely varied, most stramenopiles share characteristic tripartite tubular hairs (mastigonemes) on their cell body and/or flagella. *Blastocystis* lacks both flagella and mastigonemes, which probably contributed to its uncertain classification until molecular data became available.

Although *Blastocystis* is very common in humans, controversy still surrounds its possible role in human disease. Population surveys frequently find no relationship between infection and symptoms but individual case reports often demonstrate a link between eradication of the parasite and resolution of symptoms (Tan *et al.*, 2002). Part of this contradiction may be due to *Blastocystis* being a species complex, with at least nine subtypes having been found in humans alone that differ by up to 7% in their small subunit ribosomal RNA (SSU-rRNA) gene sequence (Stensvold *et al.*, 2007). The nine subtypes group into three clades of three. Population studies that include subtyping data are currently ongoing and will hopefully resolve the question of whether genetic diversity is linked to the organism's pathogenicity.

Within the stramenopiles, *Blastocystis* SSU-rRNA gene sequences cluster with those of a variety of intestinal commensals of vertebrates, of which the most widely known is probably *Opalina*, a commensal of amphibians (Kostka *et al.*, 2004). This clade (the Opalinata (Adl *et al.*, 2005)) consists exclusively of anaerobic organisms. Nevertheless, all the species appear to have mitochondria that are structurally quite typical in appearance in the electron microscope (Delvinquier and Patterson, 1993) (Fig. 9.1). This is exceptional as in most anaerobic eukaryotes mitochondria are replaced by hydrogenosomes or mitosomes (van der Giezen *et al.*, 2005). In the case of *Blastocystis*, another exceptional feature was the demonstration by staining of DNA within the organelles (Matsumoto *et al.*, 1987; Nasirudeen and Tan, 2004). Again, in most other anaerobic eukaryotes no organellar genome is detectable, the exception being the hydrogenosome of the anaerobic ciliate *Nyctotherus ovalis* (Boxma *et al.*, 2005; van der Giezen *et al.*, 2005).

Early biochemical studies allowed some insights into the mitochondrion-like organelles of *Blastocystis*. Zierdt demonstrated that cytochromes were not detectable in axenic cells (Zierdt, 1986). In addition, a variety of typical mitochondrial enzymes could not be found, indicating that no TCA cycle was functioning (Zierdt *et al.*, 1988). In contrast, a transmembrane potential was easily demonstrated, showing that the organelles were metabolically active (Nasirudeen and Tan, 2004).

A combination of the above features has led to significant interest in *Blastocystis* from evolutionary biologists: 1. it is the only common stramenopile parasite of humans; 2. it has secondarily lost many morphological features characteristic of stramenopiles; 3. it is an anaerobe with morphologically normal mitochondrion-like organelles; 4. it is an anaerobe with organellar DNA; 5. its organelles appear to lack the metabolic pathways most widely associated with mitochondria. These features, coupled with the ability to grow the organism axenically, led to *Blastocystis* being selected for genome sequencing. Although the complete genome is not yet available at the time of writing, we have nevertheless been able to generate a very complete view of the *Blastocystis* organelles – their genome and metabolism – and this is the subject of the present chapter.

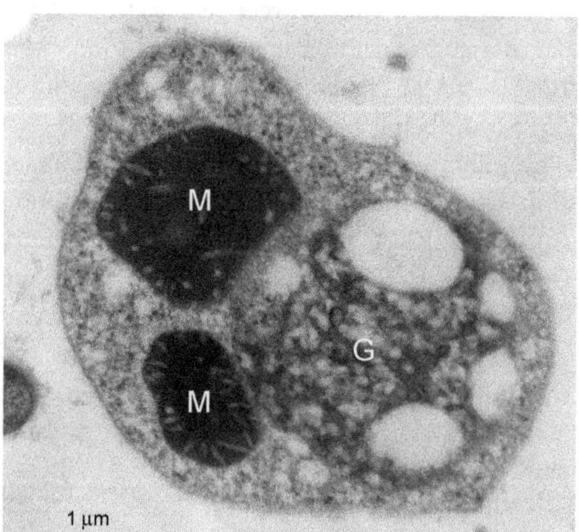

Figure 9.1 Transmission electron microscopy of *Blastocystis* sp. DMP/02–328. This cell from xenic culture is of the granular form (G) and two mitochondrion-like organelles (M) with cristae can be seen clearly. No nucleus is visible in this particular section.

The *Blastocystis* organellar genome

The presence of DNA within the *Blastocystis* organelles was demonstrated by staining with the nucleic acid binding dye DAPI (Matsumoto *et al.*, 1987; Nasirudeen and Tan, 2004), but the structure and coding potential of the DNA was unknown. Gene content of mitochondrial genomes is quite variable across eukaryotes and the apparent absence of cytochromes in *Blastocystis* reduced the range of proteins likely to be encoded in the genome even further. However, all organellar genomes to date encode ribosomal RNAs and this was the starting point for analysis of the *Blastocystis* organelle genome (mtDNA). A few other stramenopile mtDNAs had already been sequenced when the project started and this allowed primers specific for stramenopile mitochondrial rRNA genes to be designed for PCR amplification and sequencing. Initial sequences were extended by anchored PCR and later inverse PCR was used extensively to complete the mtDNA sequences from isolates representing two *Blastocystis* subtypes (Stechmann *et al.*, 2008; Pérez-Brocal and Clark, 2008). Simultaneously, the mtDNA from a third isolate was being sequenced as a consequence of the ongoing genome project (Wawrzyniak *et al.*, 2008). Fortuitously, the three sequences are all from different subtypes (namely ST1 (strain NandII), ST4 (strain DMP/02–328) and ST7 (strain B)) and even more importantly each is in turn a member of one of the three distinct clades. Thus, the features common to all three mtDNA sequences can reasonably be assumed to be present in all nine subtypes found to date in humans.

The size of the *Blastocystis* mtDNA molecule ranges between 27,719 bp and 29,270 bp (Table 9.1). This compares to a range of 31.6 kb – 58.5 kb in the other stramenopile mtDNAs sequenced to date. Pairwise nucleotide identity among the three genomes ranges from 73.5% to 78.1%. The *Blastocystis* mtDNA sequences are A+T-rich (range=78.1–80.1%) and map as circular molecules, although we cannot distinguish between circles and tandem arrays from the data. Despite a size variation of over 1.5 kb between sequences, the gene complement is identical and the three *Blastocystis* genomes are fully syntenic. Phylogenetic analysis of NADH dehydrogenase (*nad*) genes from all available stramenopile mtDNAs shows that ST1 and ST7 are more closely related than either is to ST4 (Pérez-Brocal, unpublished data). This is as expected from the per cent identities between genomes, but also reflects the relationship observed between nuclear SSU-rRNA gene sequences (Stensvold *et al.*, 2007).

As expected from the biochemical data of Zierdt (1986), and in contrast to all other stramenopiles, the *Blastocystis* mtDNAs lack all the genes for the cytochromes (*cox* 1–3 and *cob*) that are part of complexes III and IV of the electron transport chain. In addition, the genes encoding ATPase subunits (complex V) are absent. Finally, the tRNA gene complement is substantially reduced compared to other stramenopiles, despite there being two copies of the elongator tRNA(Met). This implies that several tRNAs must be imported from the cytoplasm to the organelle in order for translation to occur.

The 27 protein-encoding genes that are present are either *nad* genes (complex I of the mitochondrial electron transport chain) or ribosomal protein genes. The latter are quite divergent, so much so that the identification of some genes must be regarded as tentative. Indeed in one report (Pérez-Brocal and Clark, 2008) four genes were classified as unidentified open reading frames due to the low similarity to known genes.

Table 9.1 Comparison of *Blastocystis* mtDNA characteristics with those of other stramenopiles

	Blastocystis sp.	Other
Total length (nt)	27,719–29,270	31,617–58,507
G+C content (%)	19.9–21.9	22.3–38.0
Gene number	45	58–79
Protein coding	27	34–52
Structural RNAs	18	24–34
Gene complement[a]		
nad	10	10–11
rps/rpl	13–16[b]	12–19
atp	-	3–4
cob/cox	-	4
orf	1–4[b]	1–16
Other genes[c]	-	1–2

a Protein coding genes only.

b 3 unidentified ORFs in *Blastocystis* strains DMP/02–328 and NandII (*orf143*, *orf175* and *orf192*) are annotated as ribosomal protein genes (*rps7*, *rpl5* and *rps2*, respectively) in strain B.

c *rpo*, *dpo*, *ymf16/secY/tatC*.

The *nad* genes in contrast are for the most part highly conserved.

No evidence of pseudogenes was found, which indicates that the missing genes are not the result of a recent loss. The genomes are highly compact, with coding regions representing more than the 95% of the total length, and intergenic regions are very short (28–37 nt); indeed in each mtDNA there are several overlapping genes. There are a couple of strain-specific peculiarities; in fact these represent the only significant differences among the three genomes. The gene encoding the putative ribosomal protein S4 is much longer than in most other eukaryotes (range 922–1072 amino acids) and in two of the strains (DMP/02–328 and NandII) there is no start codon. In another gene, identified as ORF118 in strain B (Wawrzyniak *et al.*, 2008) and ORF160 in the others (Pérez-Brocal and Clark, 2008), the sequence from DMP/02–328 has an initiator methionine, but the same region of the gene in B has one in-frame stop codon while that in NandII has two. There is no evidence of alternative codon usage in translation.

EST analyses

In parallel with the mtDNA sequencing, expressed sequence tag (EST) analyses were undertaken (Stechmann *et al.*, 2008). Over 12,000 cDNA sequences were obtained from strain NandII and assembled into 3330 clusters. These clusters were analysed to identify those encoding proteins likely to be involved in the organelle metabolism of *Blastocystis*. By chance, a few ESTs were derived from transcripts of the mtDNA, and were useful in generating that sequence as well as revealing that polycistronic transcription occurs in the organelle.

110 EST clusters encoded putatively organelle-targeted proteins. The amino-terminal sequences were complete for a significant number and most appeared to encode a mitochondrion-targeting signal peptide. These were conserved enough to be identified by predictive programmes modelled on non-stramenopile data (see next section). Among these were proteins typically found in classical mitochondria such as 11 subunits from complex I (complex I of human mitochondria comprises a total of 42 subunits) and all four subunits from complex II of the mitochondrial electron transport chain (ETC). Analysis of the EST data did not reveal any evidence for the presence of complexes III, IV and V in the *Blastocystis* ETC (see Stechmann *et al.* 2008), in agreement with the absence of such genes on the mtDNA (Pérez-Brocal and Clark 2008). These findings suggest a mechanism of proton-pumping similar to that seen in the *Nyctotherus* organelles (Stechmann *et al.* 2008; Boxma *et al.* 2005). Other typically mitochondrial pathways identified from the ESTs are: an incomplete TCA cycle, amino acid metabolism, fatty acid metabolism, pyruvate metabolism, protein import and folding, and iron-sulphur cluster assembly. Interestingly, proteins generally associated with hydrogenosomes were also identified: a 'long type' [FeFe] hydrogenase and a pyruvate: ferredoxin oxidoreductase (PFO). Both proteins appear to have a typical mitochondrion-targeting signal peptide and phylogenetic analyses have shown that they cluster with other eukaryotic homologues (Stechmann *et al.*, 2008).

Of special interest are the proteins found in amino acid metabolism and especially those involved in the glycine decarboxylation complex (GDC) of *Blastocystis.* GDC, along with serine hydroxymethyl transferase (SHMT), is responsible for the interconversion of serine and glycine, which is an essential and ubiquitous step in primary amino acid metabolism. In eukaryotes, GDC is present exclusively in mitochondria and catalyses the oxidative decarboxylation and deamination of glycine. It is comprised of four proteins: H-protein (aminomethyl carrier), P-protein (glycine dehydrogenase), L-protein (dihydrolipoamide dehydrogenase), and T-protein (aminomethyl transferase). Only H- and L- proteins have been identified and characterized in the *Trichomonas* hydrogenosomes, providing the first evidence of an incomplete amino acid metabolism in this type of organelle (Mukherjee *et al*, 2006). In contrast to the *Trichomonas* data, all of the proteins involved in the GDC pathway were identified in the *Blastocystis* ESTs and preliminary bioinformatic analyses suggest functionality similar to the proteins of classical mitochondria (Tsaousis, Stechmann and Roger, unpublished data). All four proteins of the GDC have N-terminal leader peptides supporting the idea that they are localized in the *Blastocystis* organelles.

An essential role of all mitochondria studied to date is the assembly of the iron-sulphur (Fe-S) clusters that function in a variety of proteins (Brzoska *et al.*, 2006). In eukaryotes, the iron-sulphur cluster (ISC) machinery is required for the generation of mitochondrial, cytosolic and nuclear Fe-S proteins. In fact it seems that all anaerobic mitochondrion-derived organelles, including hydrogenosomes and mitosomes, may have retained this machinery (although this still needs to be confirmed in the case of *Entamoeba histolytica*). Among the *Blastocystis* ESTs several proteins of the ISC machinery were identified (Tsaousis, Stechmann and Roger, unpublished data): frataxin – the putative iron donor; IscU – the scaffold for initial cluster assembly; IscS – cysteine desulphurase, the sulphur donor; [2Fe-2S] ferredoxin – involved in electron transfer; and monothiol glutaredoxin – involved in the release of the Fe-S clusters from the scaffold. Most of these proteins have a N-terminal targeting peptide suggesting their potential localization within the organelles. More importantly, the *Blastocystis* IscS and frataxin proteins maintain all the conserved residues that are important for ISC assembly (Tsaousis, Stechmann and Roger,

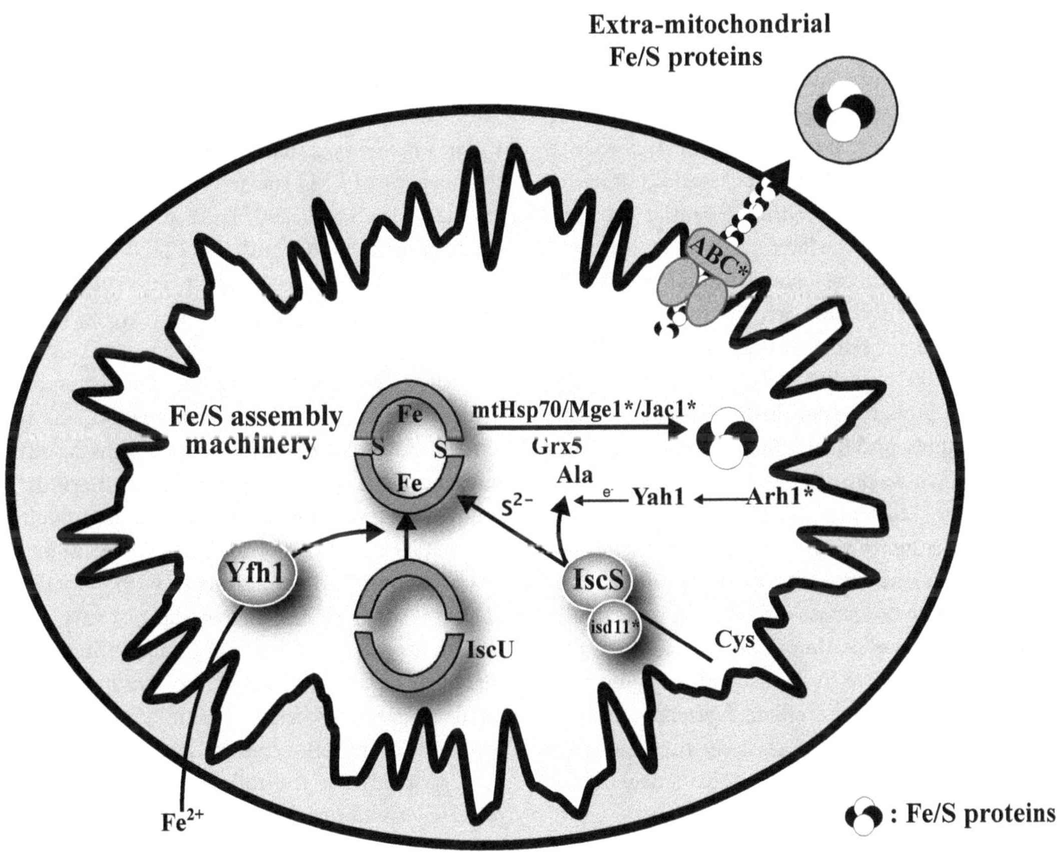

Figure 9.2 A proposed schematic model for the mechanism of ISC biogenesis in *Blastocystis* mitochondrion-like organelles. The imported Fe^{2+} binds to IscU, which serves as a scaffold for synthesis of the iron sulphur cluster (ISC). Metal delivery to IscU is assisted by frataxin (Yfh1), which directly binds to the IscU proteins. ISC synthesis on IscU further requires the cysteine desulphurase (IscS; clustering with Isd11), which mediates the release of sulphur from cysteine, and the electron-transfer chain consisting of NADH, ferredoxin reductase (Arh1) and ferredoxin (Yah1). The chaperone system, which consists of mtHsp70, Jac1 and Mge1, and the glutaredoxin Grx5 are required after ISC assembly on the IscU proteins, e.g. in ISC transfer to the apoproteins. The ABC transporter Atm1 of the inner membrane exports an unknown compound derived from the ISC-assembly machinery to the cytosol. The abbreviations used are derived from the yeast terminology. Asterisks (*) denote components that have not yet been found in the *Blastocystis* EST data.

unpublished data). While details of the chemistry of ISC formation are still under investigation, the current model (Fig. 9.2) proposes that ferrous iron assisted by frataxin associates with conserved cysteine residues of IscU (Gerber *et al.*, 2004). Formation of a complex with IscS leads to the transfer of sulphur from a cysteine persulphide moiety on IscS to a cysteine on IscU and the subsequent formation of the Fe-S cluster (Lill and Mühlenhoff, 2005). Other components that are also required for Fe-S cluster synthesis on the IscU proteins include an electron transfer chain comprising ferredoxin reductase and ferredoxin, proteins that probably function by transferring electrons to sulphur (S^0). These electrons are released from the cysteine, in order to form the sulphide (S^{-2}) present in the cluster (Lange *et al.*, 2000; Li *et al.*, 2001). The ferrous iron in the cluster is introduced by the interaction of frataxin, a protein that is essential for *de novo* ISC assembly on IscU, with the IscU scaffold proteins (Gerber *et al.*, 2003). Other essential proteins that are required in the later stages of Fe-S cluster biosynthesis are the heat shock proteins mtHsp70 (Ssq1 in yeast) and Jac1, which interact with the IscU scaffold protein (Dutkiewicz *et al.*, 2003; Lutz *et al.*, 2001; Voisine *et al.*, 2001; Voisine *et al.*, 2000) and probably transfer Fe-S clusters to the target proteins, and finally the mitochondrial glutaredoxin, whose precise role in ISC biogenesis is still unclear (Lill *et al.*, 2006).

Another interesting feature of the *Blastocystis* organelles is the presence of the multienzyme pyruvate dehydrogenase (PDH) complex. All anaerobic organelles described to date lack the PDH complex, the only exception being the hydrogenosome of the ciliate *Nyctotherus ovalis* (Boxma *et al.* 2005). In classical mitochondria the PDH complex carries out oxidative decarboxylation of pyruvate to form acetyl-CoA, which is subsequently used in the TCA cycle. The complex consists of four subunits: the alpha and beta subunits of the PDH E1 component, which is the pyruvate dehydrogenase itself, catalysing the oxidative decarboxylation of pyruvate, and E2 and E3 components, which are required for the subsequent production of acetyl-CoA. From the EST analyses, two of the four subunits were identified (PDH E1α and PDH E3). An EST with similarities to the E2 subunit of both the PDH complex and the α-ketoglutarate dehydrogenase complex has been reported previously (Stechmann *et al.* 2008). These data suggest the presence in *Blastocystis* of a typical mitochondrial pathway for pyruvate decarboxylation as well as the one typically found in anaerobic organelles, which involves the anaerobic enzymes PFO or PNO (see above and below). Interestingly, in *Trichomonas* hydrogenosomes only the PFO homologue has been identified, whereas the anaerobic *Nyctotherus* organelles seem to possess PDH only. The presence of both systems in the *Blastocystis* organelles raises questions about their functionality, and whether this is a unique adaptation of the organelles to different environmental conditions. A switch from aerobic metabolism to anaerobic fermentation has been reported for the green alga *Chlamydomonas* and for *Euglena* (Mus *et al.* 2007; Hoffmeister *et al.* 2004). In *Chlamydomonas,* it has been demonstrated that the amount of PFO transcribed increases during anoxia (Mus *et al.* 2007). As for the presence of both PFO/PNO and the PDH complex in *Blastocystis,* which is a strict anaerobe, more biochemical testing needs to be carried out to establish their functional significance.

All organelles need to be able to import proteins that have been synthesized in the cytosol but are functional in the organelles. The mitochondrial protein import machinery is a unique eukaryote invention with no homologues in prokaryotes and seems to be present in all anaerobic organelles, although with variations (Dolezal *et al.*, 2006). Thus it was not surprising that an incomplete mitochondrial protein import machinery was identified in the *Blastocystis* EST data. So far, only a small number of proteins involved in protein import have been identified, probably for the following reasons: 1. the relatively small number of EST clusters analysed and 2. the likelihood that some proteins are very divergent from their homologues in other eukaryotes and thus not easily recognized (see Waller *et al.* 2009). Components of the mitochondrial import machinery that were identified from the *Blastocystis* EST data are units of the translocase of the inner membrane (TIM): Tim50 (part of the TIM23 complex; regulates channel opening), Tim17 (part of the translocation channel of the TIM23 complex),

Tim21 (regulates module docking of the TIM23 complex), and Tim9 (one of the tiny TIMS that are responsible for the transfer of substrates to either the S-adenosylmethionine carrier or TIM22 complex). Other proteins related to the protein import system that were identified in the EST data are heat shock protein mtHsp70 (binding of unfolded preproteins), metalloprotease MP1 (probably a processing peptidase), the oxidase assembly-like protein Oxa1 (required for assembly of proteins in the inner mitochondrial membrane), and the translocase of the outer membrane component Tom70 (promotes substrate binding) (Stechmann *et al.*, 2008). The identification of an EST coding for Tom70 was surprising, as this protein was thought to be present only in animals and fungi (Chan *et al*, 2006). Currently, it is not clear whether this Tom70 was acquired by lateral gene transfer from either the animal or fungal lineage or whether it was present in the ancestral eukaryote and subsequently lost in those eukaryotes that lack Tom70. The completion of the *Blastocystis* genome and the identification of additional components of the mitochondrial import machinery will hopefully give us further insights into the function of this important apparatus in the *Blastocystis* organelles.

Classical mitochondria contain up to 54 mitochondrial carrier family proteins (MCFs), which mediate exchange of diverse substrates between the mitochondrial matrix and the host cytosol (Kunji, 2004). Insertion of a host encoded ATP/ADP carrier (AAC) protein into the ancestral mitochondrial endosymbiont inner membrane is considered a key step in the transition from symbiont to organelle (John and Whatley, 1975), as it allows the host to harvest organelle-produced ATP. The *Blastocystis* EST survey identified several clusters that encode proteins with similarity to MCF proteins. Preliminary results based on sequence similarity and bioinformatic analyses of these clusters suggest that the *Blastocystis* genome encodes at least six MCF proteins: ATP/ADP carrier (AAC), phosphate carrier (PiC), carnitine/acylcarnitine carrier (ORC), aspartate/glutamate carrier (AGC), manganese trafficking factor for mitochondrial SOD2 (MTM1), and S-adenosylmethionine carrier (SAM) (Stechmann *et al.*, 2008). Further work will be needed to confirm the function of these transporters and to determine the metabolic interactions between the cytosol and the *Blastocystis* organelles.

Metabolic modelling

As already mentioned, the EST project generated 12,767 sequences, which were assembled into 3330 individual clusters (Stechmann *et al.*, 2008). Because the main goal of the project was to identify proteins putatively targeted to the *Blastocystis* organelles, the use of different prediction tools was employed. Publicly available software tools for identifying mitochondrial proteins are typically designed and tested against animal, fungal and plant genomes. Thus, identification of mitochondrial targeted proteins in other groups of organisms using these tools could be difficult and the results are often unreliable (Gaston *et al.*, 2009). To resolve this issue, Stechmann *et al.*, developed a simple bioinformatics tool CBOrg (Comparative Blast for Organelles), for analysis of various data from the Protist EST Program (PEP) (Stechmann *et al.*, 2008; Gill *et al.*, 2007; Gaston *et al.*, 2009). Based on the human and yeast mitochondrial databases' mitoproteome and Mitop2 (Cotter *et al.*, 2004; Elstner *et al.*, 2008) subtractive databases were created for *Homo sapiens* and *Saccharomyces cerevisiae* that comprised only non-mitochondrial proteins. The *Blastocystis* clusters were then used in comparative BLAST searches against the four databases, human/yeast subtractive and human/yeast mitochondrial. All putative mitochondrial proteins detected for *Blastocystis* were then subjected to another round of BLAST searches against the non-redundant Genbank database.

In parallel, all *Blastocystis* ESTs were analysed using the EGassembler server, an automated online tool that clusters and assembles EST data. The results were used as a gene set for the KEGG Automatic Annotation Server with the bidirectional best-hit method to assign orthologues. These KEGG orthology assignments were used to automatically generate KEGG metabolic pathways. Pathways were screened for sub-cellular location (mitochondrial) and results were finally compared to the output from the CBOrg analysis (Fig. 9.3). All putative mitochondrial proteins with a complete N-terminal sequence were further analysed for the presence of mitochondrial

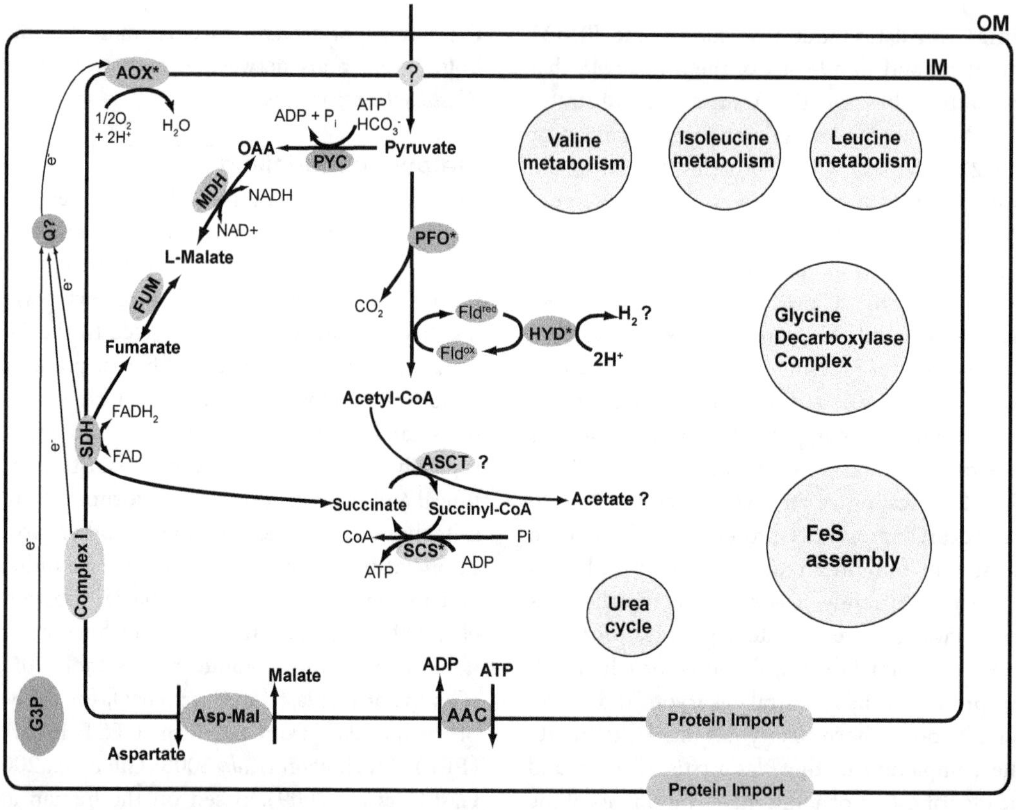

Figure 9.3 *Blastocystis* mitochondrion-like organelle: proposed metabolic map (modified from Stechmann *et al.*, 2008). The main pathway represents the conversion of pyruvate to acetyl-CoA via the pyruvate: ferredoxin oxidoreductase (PFO). PFO reduces flavodoxin (Fld) and the [FeFe] hydrogenase (HYD) oxidizes flavodoxin with the simultaneous production of molecular hydrogen; the question mark indicates that hydrogen production has not been demonstrated for *Blastocystis*. Acetyl-CoA is converted by the acetate: succinate CoA transferase (ASCT) to acetate. Acetate production has not been tested for *Blastocystis*, but an acetyl-CoA hydrolase like the one known to function as ASCT in *Trichomonas* hydrogenosomes is present in *Blastocystis* EST data. ATP production in the *Blastocystis* MLOs is probably via the succinyl-CoA synthetase (SCS). Complexes III and IV of the electron transport chain are proposed to be missing and electrons from complex I and complex II (succinate dehydrogenase; SDH) are passed via the quinone pool to the alternative oxidase (AOX). An incomplete citric acid cycle is present that could function in the reverse direction as part of the branched pathway of malate dismutation where malate is reduced to succinate (see also Stechmann *et al.* 2008). Abbreviations: AAC, ADP/ATP translocator; AOX: alternative oxidase; Asp-Mal, aspartate malate shuttle; FUM, fumarate hydratase; G3P, glycerol-3-phosphate shuttle; IM, Inner Membrane; MDH, malate dehydrogenase; OAA, oxaloacetate; OM, Outer Membrane; PYC, pyruvate carboxylase; Q, quinone; SCS, succinyl-CoA synthetase; SDH, succinate dehydrogenase. Asterisks (*) denote components that have been experimentally localized in the *Blastocystis* organelles.

targeting sequences, short N-terminal extensions that are enriched in hydrophobic and positively charged amino acids and are capable of forming amphiphilic alpha helices, using MitoProt, TargetP, SignalP and iPSORT (Claros and Vincens 1996; Emanuelsson *et al.* 2007; Bannai *et al.* 2002).

Combining these subcellular predictions along with cellular localization and biochemical characterization of putative organellar proteins will ultimately improve our understanding of the organellar metabolism in this unique anaerobe. For example, antibodies against three of the predicted organellar proteins have been used in immunofluorescence microscopy to confirm their subcellular location: [FeFe] hydrogenase (Stechmann *et al.* 2008), succinyl-CoA synthetase (Hamblin *et al.* 2008), and alternative oxidase (AOX) (Hamblin, Elliott, Moore and van der Giezen, unpublished) (see below).

Twenty years before the EST data were published, Zierdt *et al.* (1988) biochemically characterized the *Blastocystis* organelles by trying to detect activities of some key mitochondrial proteins. Their enzyme assays show no activity for pyruvate dehydrogenase complex, citric acid cycle enzymes and cytochrome *c* oxidase. Yet, they proposed that the organelles were functional and based on microscopic evidence suggested that they may be involved in lipid synthesis and storage. Following up on those results, Lantsman *et al.* (2008) have studied the activity of several hydrogenosomal, mitochondrial, cytosolic and lysosomal enzymes in whole cell extracts and subcellular fractions from *Blastocystis* and established a metabolic model of the organelles based on their results. In agreement with the proposed metabolism based on the EST survey, the results from Lantsman *et al.* (2008) suggest that pyruvate is metabolized to acetate. This is accomplished by the acetate: succinate CoA transferase (ASCT), which transfers the CoA moiety from acetyl-CoA to succinate yielding acetate and succinyl-CoA. Hamblin *et al.* (2008) clearly showed that the *Blastocystis* SCS uses ADP as a substrate in preference to GDP to produce ATP (see also below). ATP production by coupling energy from the thioester bond of CoA is suggested in the Lantsman metabolic model as well.

However, in contrast to the EST data, Lantsman *et al.* (2008) report that pyruvate: NADP oxidoreductase (PNO) is responsible for the conversion of pyruvate to acetyl-CoA rather than PFO, based on substrate and cofactor requirements. PNO was first characterized in *Euglena gracilis* and is made up of a NADPH-cytochrome P450 reductase gene fused to the C-terminus of a PFO gene (Inui *et al.* 1987; Rotte *et al.* 2001). Lantsman *et al.* (2008) identified an EST clone in the *Blastocystis* cDNA library with similarity to the *Euglena* PNO to support their theory. That EST library clone (BHL00002376), however, shows similarity to the N-terminal PFO domain of the *Euglena* PNO gene and thus does not necessarily encode a PNO protein. A PFO sequence was identified in the EST survey (Stechmann *et al.* 2008) that has a poly-A tail at the C-terminus, ruling out the possibility that this might in fact be a PNO. While biochemical evidence suggests that PNO is responsible for the conversion of pyruvate to acetyl-CoA, sequence data clearly show a PFO to be present. Further, Lantsman *et al.* (2008) show that the PNO localizes to a large granular fraction obtained after differential centrifugation of *Blastocystis* cell homogenates, which contains the organelles, while transmission electron microscopy analyses using immunogold staining also showed that PFO localizes to the *Blastocystis* organelles (Wawrzyniak *et al.* 2008). *Blastocystis* is thus the first organism for which it has been shown that the two anaerobic enzymes PNO and PFO are present in its organelles.

The role and function of the PDH complex in *Blastocystis* is still uncertain, as previous studies did not detect any activity of this enzyme (Zierdt *et al.*, 1988) (see also next section). Another interesting result from the study by Lantsman *et al.* (2008) is that they show the presence of three citric acid cycle enzymes, which were not identified by the EST survey (see Stechmann *et al.* 2008). However, if the results from the two studies were combined one would obtain a near complete citric acid cycle typical for aerobic mitochondria (citrate synthase being the only enzyme missing).

It is not certain whether the discrepancies between the results from Lantsman *et al.* (2008), Wawrzyniak *et al.* (2008) and the EST survey (Stechmann *et al.* 2008) are due to artefacts or whether they are real. However, it is noteworthy that the biochemical characterization in these studies was carried out using different *Blastocystis* subtypes (subtype 7: Wawrzyniak *et al.* 2008; Lantsman *et al.* 2008 vs subtype 1: Stechmann *et al.* 2008). As mentioned before, there is a remarkable genetic diversity between different *Blastocystis* subtypes and such a diversity could explain the differences seen among the recent studies of the *Blastocystis* organelles.

The organelle

Morphologically, the organelles in *Blastocystis* resemble mitochondria in many other organisms. They are small, elongated globular structures with cristae and 2–4 are present in each cell (Zierdt, 1991). As *Blastocystis* is reportedly a strict anaerobe (Zierdt *et al.*, 1967) with morphologically normal mitochondria, establishing the localization of the atypical enzymes [FeFe] hydrogenase and PFO was important. These two enzymes are often encountered in anaerobic eukaryotes, but their cellular distribution varies depending on the level of metabolic compartmentalisation. Organisms containing mitosomes, for example *Giardia intestinalis* (Tovar *et al.*, 2003) and *Entamoeba histolytica* (Tovar *et al.*, 1999), do not compartmentalise these two oxygen-sensitive enzymes. On the other hand, in anaerobic eukaryotes with hydrogenosomes [FeFe] hydrogenase and PFO are compartmentalised into these organelles (Bui and Johnson, 1996; Hrdý and Müller, 1995; Marvin-Sikkema *et al.*, 1993; Yarlett *et al.*, 1986). Therefore, their location in *Blastocystis* would help in the classification of its unusual organelles. Strictly speaking, to be able to label an organelle as a 'hydrogenosome' requires further physiological characterization of the organelle. Hydrogenosomes *sensu* Müller (1993) are defined as organelles that produce molecular hydrogen (by using either malate or pyruvate as a substrate that is decarboxylated into acetate with the production of carbon dioxide and ATP). However, detecting production of molecular hydrogen is not trivial, as shown in the case of *G. intestinalis* (Lloyd *et al.*, 2002).

Recombinant *Blastocystis* [FeFe] hydrogenase protein was used to raise polyclonal antisera which, when tested on total cell lysate, identified a single band of approximately 84 kDa (Stechmann, Tsaousis and Roger, unpublished). Based on the EST sequence, the theoretical molecular weight of the [FeFe] hydrogenase is 83.3 kDa, suggesting that the antiserum is indeed recognizing *Blastocystis* [FeFe] hydrogenase protein. This antiserum was used in immunofluorescence confocal microscopy. It was shown previously that active dyes such as Janus Green, Rhodamine 123 (Zierdt *et al.*, 1988) and MitoLight (Nasirudeen and Tan, 2004) accumulated in *Blastocystis* organelles. A similar dye, MitoTracker, was therefore used as a second label to identify the organelles. In addition, because *Blastocystis* organellar DNA can be detected using the nucleic acid binding dye DAPI (Nasirudeen and Tan, 2004), this dye was also included in the [FeFe] hydrogenase localization experiments. Co-localization experiments using immunofluorescence confocal microscopy showed that MitoTracker, DAPI and anti-[FeFe] hydrogenase antiserum all label the same organelle in *Blastocystis* (Stechmann *et al.*, 2008). Similarly, Wawrzyniak *et al.* (2008) demonstrate the localization of hydrogenase in the *Blastocystis* organelles by transmission electron microscopy. These data are highly suggestive of a hydrogenosomal nature for the *Blastocystis* organelles. In contrast, Lantsman *et al.* (2008) unsuccessfully attempted to detect the production of molecular hydrogen in *Blastocystis* using methyl viologen under anoxic conditions according to the method of Lindmark and Müller. However, unlike the measurements by Lloyd *et al.* (2002) on *Giardia* lysates using mass spectrometry, the spectrophotometric assay using methyl viologen does not detect hydrogen production directly. Rather, this method detects the presence of an active hydrogenase enzyme. The failure of Lantsman *et al.* (2008) to detect [FeFe] hydrogenase activity in their whole cell *Blastocystis* lysate contrasts with the finding of [FeFe] hydrogenase transcripts amongst the ESTs and [FeFe] hydrogenase protein in the organelles. Primary sequence analysis suggests that typical [FeFe] hydrogenase protein domains are present. However, the *Blastocystis* [FeFe] hydrogenase does have an extension at the C-terminus, a flavodoxin domain of unknown function (Stechmann *et al.*, 2008). Further biochemical work and alternative methodologies will be needed to elucidate the activity of the *Blastocystis* [FeFe] hydrogenase.

Apart from establishing that the [FeFe] hydrogenase protein is located in the *Blastocystis* organelles, and hence that they might be hydrogenosomes, it was also important to establish whether the organelles are involved in ATP production. Our EST analyses suggest that SCS, a citric acid cycle enzyme, is also located in the *Blastocystis* organelles. This prediction is based on the presence of recognisable mitochondrial targeting sequences at the amino-termini of both the alpha and beta subunits of this enzyme. As with [FeFe] hydrogenase, SCS is found in hydrogenosomes but not, to date, in mitosomes (Dacks *et al.*, 2006). In addition, SCS is the only citric acid cycle enzyme that produces ATP or GTP by substrate level phosphorylation and therefore can be of energetic significance for *Blastocystis*. For these reasons, a biochemical characterization of this enzyme was conducted. In agreement with the presence of putative mitochondrial targeting signals, heterologous antisera used to detect the beta-SCS subunit colocalised with MitoTracker (Hamblin *et al.*, 2008). As SCS in model organisms has been shown to perform different roles, the *Blastocystis* SCS was characterized enzymatically. SCS can produce either GTP or ATP and the primary sequence suggested the *Blastocystis* enzyme would be GTP specific (Fraser *et al.*, 2000). However, kinetic studies using recombinant SCS indicated that the *Blastocystis* enzyme is ATP specific. Together, the enzymatic and localization data clearly indicate that *Blastocystis* organelles have maintained a role in energy provision.

Early biochemical analyses established that *Blastocystis* did not contain haem iron and therefore had no cytochromes (Zierdt, 1986). Sequencing of the mitochondrial genome in several *Blastocystis* subtypes and analysis of the ESTs did not detect the presence of the classical mitochondrial genes encoding cytochromes *cox* 1–3 and *cob* that are part of complexes III and IV of the electron transport chain. The *Blastocystis* ESTs survey has identified all subunits that comprise complex II of the electron transport chain: the flavoprotein (succinate dehydrogenase (sdh) subunit A), the iron-sulphur subunit (sdhB) and the transmembrane/quinone binding subunits (sdhC and sdhD). Interestingly, the membrane subunits of complex II (sdhC and sdhD) generally do contain haem-coordinated iron (Lemire and Oyedotun, 2002; Oyedotun and Lemire, 2004). Sequence analysis and molecular docking simulations of *Blastocystis* sdhC and sdhD proteins suggest that iron-coordinating histidines are present (Standley and van der Giezen, unpublished data) bringing into question the supposed absence of cytochromes. Nonetheless, since complexes III and IV are absent, the nature of the terminal electron acceptor in this short electron transport chain is unknown. Surprisingly, recent biochemical studies suggest the absence of detectable succinate dehydrogenase activity in *Blastocystis* (Lantsman *et al.*, 2008), raising questions regarding the potential function of complex II. In addition, the oxidation of fatty acids and some amino acids is linked to the mitochondrial electron transport chain via electron transfer flavoprotein (ETF) and ETF-ubiquinone oxidoreductase (Ruzicka and Beinert, 1977). ESTs for both proteins have been identified in *Blastocystis* indicating that the electron flow is more significant than in, for example, the *T. vaginalis* hydrogenosome. Moreover, the glycerol-3-phosphate dehydrogenase that feeds into the mitochondrial electron transport chain has been identified in the EST data as well (Stechmann *et al.*, 2008). Interestingly, an unusual terminal electron acceptor has been discovered among the *Blastocystis* ESTs: an alternative oxidase (AOX). AOX is found attached to the mitochondrial inner membrane in animals, plants, several fungi and some protists (McDonald and Vanlerberghe, 2006). AOX accepts electrons from the quinone pool, which are used to reduce oxygen to water. AOX does not pump protons across the membrane and by directly reducing oxygen it short-circuits the electron transport chain before complexes III and IV (Berthold and Stenmark, 2003). A monoclonal antibody raised against *Sauromatum guttatum* AOX (Voodoo Lily) recognizes a band of the correct size in total *Blastocystis* protein extract (Hamblin, Elliott, Moore and van der Giezen, unpublished). Immunofluorescence confocal microscopy shows clear colocalisation of this protein with MitoTracker, indicating that AOX is localized to the *Blastocystis* organelles. Overexpression of the *Blastocystis* AOX in haem-deficient *Escherichia coli* cells indicated that the *Blastocystis* enzyme uses oxygen, accepts electrons from quinone, and is sensitive to typical AOX inhibitors such as octyl gallate (Hamblin, Elliott, Moore and van der Giezen, unpublished). Although the presence of an enzyme that uses molecular oxygen as its substrate seems at odds given the strict anaerobic nature of *Blastocystis*, it might possibly serve as an oxygen scavenger (Gomes *et al.*, 2001).

Resources

A variety of subtypes (1, 3 and 4) are represented currently among the *Blastocystis* isolates available from the American Type Culture Collection (ATCC), most of which were deposited by Zierdt. The cDNA library used for EST sequencing was from *Blastocystis* sp. Nand II (ATCC 50177), which belongs to subtype 1 (Stensvold *et al.* 2007). The *Blastocystis* mitochondrial genomes sequenced to date come from isolates NandII,

DMP/02–328 (subtype 4) and B (subype 7), and the latter subtype is being used to generate the first nuclear genome sequence (Genoscope). It is to be hoped that the genome isolate will also be deposited in the ATCC collection.

Sequencing of the *Blastocystis* ESTs was part of the now completed Protist EST program (PEP), which set out to investigate the diversity of eukaryote genomes by sequencing ESTs from a number of unicellular eukaryotes. All EST data generated through the PEP program, including those of *Blastocystis*, were initially deposited in the TBestDB database. Here, ESTs were assembled into clusters and annotated using AutoFACT (implemented in the TBestDB database), a rapid annotation procedure, and TBLASTX (also implemented in TBestDB) with the non-redundant database from Genbank (O'Brien *et al.* 2007). All ESTs reported by Stechmann *et al.* (2008) have also been deposited in Genbank as are the individual proteins where full length gene sequences were obtained.

Future directions

The immediate future holds the promise of the entire *Blastocystis* genome (Wawrzyniak *et al.*, 2008). Indeed it may be released before this chapter is published. Although the genome data will be from a distinct subtype from that which gave us the EST data, it seems unlikely that substantial differences will exist between the nuclear genomes of different subtypes, at least as far as gene content is concerned. The EST data do not provide us with the complete gene complement of this organism. Indeed, the absence of ESTs encoding 'missing' enzymes from otherwise complete pathways makes this quite clear (Stechmann *et al.*, 2008). Nevertheless, such differences as are detected will need to be experimentally 'cross-referenced' between the subtypes. Despite the divergence in sequence that will probably be found, the EST data should prove helpful in annotation of the genome sequence. Only a few nuclear gene sequences are available, but it is already clear that *Blastocystis* genes can contain multiple short introns (Hamblin *et al.* 2008).

By chance, the three mtDNAs that have been sequenced to date do not include the most common subtype found in humans (ST3), which is responsible for over 50% of all human infections (Stensvold *et al.*, 2009). It seems logical that this subtype will be investigated in more detail in the future. As sequence costs continue to fall and with a reference genome already available, it also seems logical to predict that additional *Blastocystis* genomes will eventually be sequenced. Whether these will include examples from ST3 and other human-infective subtypes or will focus on the more distantly related species from reptiles and insects remains to be seen. To a certain extent this will probably depend on the questions being asked by the investigators involved. The more distantly related *Blastocystis* species clearly have the potential to answer some questions about the order of events during the loss of genes from the mitochondrial genome, as do analyses of homologous organelles in other Opalinata.

At present the number of research groups working on *Blastocystis* molecular and cellular biology can be counted on one hand. As a result it is not surprising that no methods exist for the construction of transgenic organisms. Since these depend on some understanding of gene regulation in the organism, the lack of available gene flanking sequences alone has precluded such work to date. The imminent arrival of the genome sequence will dramatically change this. Promoter/terminator motifs predicted by bioinformatic analyses will immediately become available and although these may prove to be subtype-specific, *Blastocystis* may well become amenable to transfection, opening up a whole new area of research for this intriguing organism.

Acknowledgements

ADT and AS are supported by a grant (MOP 62809) to Andrew J. Roger from the Canadian Institutes for Health Research and Tula Foundation postdoctoral funding awarded to AJR from the Centre for Comparative Genomics and Evolutionary Bioinformatics at Dalhousie University. Work performed in London and Exeter has been supported by grant 078566/A/05/Z from the Wellcome Trust awarded to MvdG and CGC. KAH was supported by a Queen Mary, University of London, PhD-studentship.

References

Adl, S.M., Simpson, A.G., Farmer, M.A., Andersen, R.A., Anderson, O.R., Barta, J.R., Bowser, S.S., Brugerolle,

G., Fensome, R.A., Fredericq, S., James, T.Y., Karpov, S., Kugrens, P., Krug, J., Lane, C.E., Lewis, L.A., Lodge, J., Lynn, D.H., Mann, D.G., McCourt, R.M., Mendoza, L., Moestrup, O., Mozley-Standridge, S.E., Nerad, T.A., Shearer, C.A., Smirnov, A.V., Spiegel, F.W., and Taylor, M.F. (2005). The new higher level classification of eukaryotes with emphasis on the taxonomy of protists. J. Eukaryot. Microbiol. *52*, 399–451.

Bannai, H., Tamada, Y., Maruyama, O., Nakai, K., and Miyano, S. (2002). Extensive feature detection of N-terminal protein sorting signals. Bioinformatics *18*, 298–305.

Berthold, D.A., and Stenmark, P. (2003). Membrane-bound diiron carboxylate proteins. Annu. Rev. Plant. Biol. *54*, 497–517.

Boxma, B., de Graaf, R.M., van der Staay, G.W., van Alen, T.A., Ricard, G., Gabaldón, T., van Hoek, A.H., Moon-van der Staay, S.Y., Koopman, W.J., van Hellemond, J.J., Tielens, A.G., Friedrich, T., Veenhuis, M., Huynen, M.A., and Hackstein, J.H. (2005). An anaerobic mitochondrion that produces hydrogen. Nature *434*, 74–79.

Brumpt, E. (1912). *Blastocystis hominis* n. sp. et formes voisines. Bull. Soc. Pathol. Exot. *5*, 725–730.

Brzóska, K., Meczyńska, S., and Kruszewski, M. (2006). Iron-sulfur cluster proteins: electron transfer and beyond. Acta Biochim. Pol. *53*, 685–691.

Bui, E.T.N., and Johnson, P.J. (1996). Identification and characterization of [Fe]-hydrogenases in the hydrogenosome of *Trichomonas vaginalis*. Mol. Biochem. Parasitol. *76*, 305–310.

Chan, N.C., Likić, V.A., Waller, R.F., Mulhern, T.D., and Lithgow, T. (2006). The C-terminal TPR domain of Tom70 defines a family of mitochondrial protein import receptors found only in animals and fungi. J. Mol. Biol. *358*, 1010–1022.

Claros, M.G., and Vincens, P. (1996). Computational method to predict mitochondrially imported proteins and their targeting sequences. Eur. J. Biochem. *241*, 779–786.

Cotter, D., Guda, P., Fahy, E., and Subramaniam, S. (2004). MitoProteome: mitochondrial protein sequence database and annotation system. Nucleic Acids Res. *32*, D463–467.

Dacks, J.B., Dyal, P.L., Embley, T.M., and van der Giezen, M. (2006). Hydrogenosomal succinyl-CoA synthetase from the rumen-dwelling fungus *Neocallimastix patriciarum*; an energy-producing enzyme of mitochondrial origin. Gene 373, 75–82.

Delvinquier, B.L.J., and Patterson, D.J. (1993). The Opalines. In Parasitic Protozoa, vol. 3, J.P. Kreier and J.R. Baker eds. (San Diego, USA: Academic Press), pp. 247–325.

Dolezal, P., Likic, V., Tachezy, J., and Lithgow, T. (2006). Evolution of the molecular machines for protein import into mitochondria. Science *313*, 314–318.

Dutkiewicz, R., Schilke, B., Knieszner, H., Walter, W., Craig, E.A., and Marszalek, J. (2003). Ssq1, a mitochondrial Hsp70 involved in iron-sulfur (Fe/S) center biogenesis. Similarities to and differences from its bacterial counterpart. J. Biol. Chem. *278*, 29719–29727.

Elstner, M., Andreoli, C., Ahting, U., Tetko, I., Klopstock, T., Meitinger, T., and Prokisch, H. (2008). MitoP2: an integrative tool for the analysis of the mitochondrial proteome. Mol. Biotechnol. *40*, 306–315.

Emanuelsson, O., Brunak, S., von Heijne, G., and Nielsen, H. (2007). Locating proteins in the cell using TargetP, SignalP and related tools. Nat. Protoc. *2*, 953–971.

Fraser, M.E., James, M.N., Bridger, W.A., and Wolodko, W.T. (2000). Phosphorylated and dephosphorylated structures of pig heart, GTP-specific succinyl-CoA synthetase. J. Mol. Biol. *299*, 1325–1339.

Gaston D., Tsaousis, A.D., and Roger, A.J. (2009). Predicting proteomes of mitochondria and related organelles from genomic and expressed sequence tag data. Methods Enzymol. *457*, 21–41.

Gerber, J., Mühlenhoff, U., and Lill, R. (2003). An interaction between frataxin and Isu1/Nfs1 that is crucial for Fe/S cluster synthesis on Isu1. EMBO Rep. *4*, 906–911.

Gerber, J., Neumann, K., Prohl, C., Muhlenhoff, U., and Lill, R. (2004). The yeast scaffold proteins Isu1p and Isu2p are required inside mitochondria for maturation of cytosolic Fe/S proteins. Mol. Cell Biol. *24*, 4848–4857.

van der Giezen, M., Tovar, J., and Clark, C.G. (2005). Mitochondrion-derived organelles in protists and fungi. Int. Rev. Cytol. *244*, 175–225.

Gill, E.E., Diaz-Trivino, S., Barbera, M.J., Silberman, J.D., Stechmann, A., Gaston, D., Tamas, I., and Roger, A.J. (2007). Novel mitochondrion-related organelles in the anaerobic amoeba *Mastigamoeba balamuthi*. Mol. Microbiol. *66*, 1306–1320.

Gomes, C.M., Le Gall, J., Xavier, A.V., and Teixeira, M. (2001). Could a diiron-containing four-helix-bundle protein have been a primitive oxygen reductase? Chembiochem. *2*, 583–587.

Hamblin, K., Standley, D.M., Rogers, M.B., Stechmann, A., Roger, A.J., Maytum, R., and van der Giezen, M. (2008). Localization and nucleotide specificity of *Blastocystis* succinyl-CoA synthetase. Mol. Microbiol. *68*, 1395–1405.

Hoffmeister, M., van der Klei, A., Rotte, C., van Grinsven, K.W., van Hellemond, J.J., Henze, K., Tielens, A.G., and Martin, W. (2004). *Euglena gracilis* rhodoquinone: ubiquinone ratio and mitochondrial proteome differ under aerobic and anaerobic conditions. J. Biol. Chem. *279*, 22422–22429.

Hrdý, I., and Müller, M. (1995). Primary structure and eubacterial relationships of the pyruvate: ferredoxin oxidoreductase of the amitochondriate eukaryote *Trichomonas vaginalis*. J. Mol. Evol. *41*, 388–396.

Inui, H., Ono, K., Miyatake, K., Nakano, Y., and Kitaoka, S. (1987). Purification and characterization of pyruvate:NADP+ oxidoreductase in *Euglena gracilis*. J. Biol. Chem. *262*, 9130–9135.

John, P., and Whatley, F.R. (1975). *Paracoccus denitrificans* and the evolutionary origin of the mitochondrion. Nature *254*, 495–498.

Kostka, M., Hampl, V., Cepicka, I., and Flegr, J. (2004). Phylogenetic position of *Protoopalina intestinalis* based on SSU rRNA gene sequence. Mol. Phylogenet. Evol. *33*, 220–224.

Kunji, E.R. (2004). The role and structure of mitochondrial carriers. FEBS Lett. *564*, 239–244.

Lange, H., Kaut, A., Kispal, G., and Lill, R. (2000). A mitochondrial ferredoxin is essential for biogenesis of cellular iron-sulfur proteins. Proc. Natl. Acad. Sci. U.S.A. *97*, 1050–1055.

Lantsman, Y., Tan, K.S., Morada, M., and Yarlett, N. (2008). Biochemical characterization of a mitochondrial-like organelle from *Blastocystis* sp. subtype 7. Microbiology *154*, 2757–2766.

Lemire, B.D., and Oyedotun, K.S. (2002). The *Saccharomyces cerevisiae* mitochondrial succinate:ubiquinone oxidoreductase. Biochim. Biophys. Acta *1553*, 102–116.

Li, J., Saxena, S., Pain, D., and Dancis, A. (2001). Adrenodoxin reductase homolog (Arh1p) of yeast mitochondria required for iron homeostasis. J. Biol. Chem. *276*, 1503–1509.

Lill, R., Dutkiewicz, R., Elsasser, H.P., Hausmann, A., Netz, D.J., Pierik, A.J., Stehling, O., Urzica, E., and Mühlenhoff, U. (2006). Mechanisms of iron-sulfur protein maturation in mitochondria, cytosol and nucleus of eukaryotes. Biochim. Biophys. Acta *1763*, 652–667.

Lill, R., and Mühlenhoff, U. (2005). Iron-sulfur-protein biogenesis in eukaryotes. Trends Biochem. Sci. *30*, 133–141.

Lloyd, D., Ralphs, J.R., and Harris, J.C. (2002). *Giardia intestinalis*, a eukaryote without hydrogenosomes, produces hydrogen. Microbiology *148*, 727–733.

Lutz, T., Westermann, B., Neupert, W., and Herrmann, J.M. (2001). The mitochondrial proteins Ssq1 and Jac1 are required for the assembly of iron sulfur clusters in mitochondria. J. Mol. Biol. *307*, 815–825.

Marvin-Sikkema, F.D., Pedro Gomes, T.M., Grivet, J., Gottschal, J.C., and Prins, R.A. (1993). Characterization of hydrogenosomes and their role in glucose metabolism of *Neocallimastix* sp. L2. Arch. Microbiol. *160*, 388–396.

Matsumoto, Y., Yamada, M., and Yoshida, Y. (1987). Light-microscopical appearance and ultrastructure of *Blastocystis hominis*, an intestinal parasite of man. Zentralbl. Bakteriol. Mikrobiol. Hyg. [A] *264*, 379–385.

McDonald, A.E., and Vanlerberghe, G.C. (2006). Origins, evolutionary history and taxonomic distribution of alternative oxidase and plastoquinol terminal oxidase. Comp. Biochem. Physiol. D *1*, 357–364.

Mukherjee, M., Brown, M.T., McArthur, A.G., and Johnson, P.J. (2006). Proteins of the glycine decarboxylase complex in the hydrogenosome of *Trichomonas vaginalis*. Eukaryot. Cell 5, 2062–2071.

Müller, M. (1993). The hydrogenosome. J. Gen. Microbiol. *139*, 2879–2889.

Mus, F., Dubini, A., Seibert, M., Posewitz, M.C., and Grossman, A.R. (2007). Anaerobic acclimation in *Chlamydomonas reinhardtii*: anoxic gene expression, hydrogenase induction, and metabolic pathways. J. Biol. Chem. *282*, 25475–25486.

Nasirudeen, A.M., and Tan, K.S. (2004). Isolation and characterization of the mitochondrion-like organelle from *Blastocystis hominis*. J. Microbiol. Methods *58*, 101–109.

O'Brien, E.A., Koski, L.B., Zhang, Y., Yang, L., Wang, E., Gray, M.W., Burger, G., and Lang, B.F. (2007). TBestDB: a taxonomically broad database of expressed sequence tags (ESTs). Nucleic Acids Res. *35*, D445–D451.

Oyedotun, K.S., and Lemire, B.D. (2004). The quaternary structure of the *Saccharomyces cerevisiae* succinate dehydrogenase. Homology modeling, cofactor docking, and molecular dynamics simulation studies. J. Biol. Chem. *279*, 9424–9431.

Pérez-Brocal, V., and Clark, C.G. (2008). Analysis of two genomes from the mitochondrion-like organelle of the intestinal parasite *Blastocystis*: complete sequences, gene content and genome organization. Mol. Biol. Evol. *25*, 2475–2482.

Rotte, C., Stejskal, F., Zhu, G., Keithly, J.S., and Martin, W. (2001). Pyruvate: NADP+ oxidoreductase from the mitochondrion of *Euglena gracilis* and from the apicomplexan *Cryptosporidium parvum*: a biochemical relic linking pyruvate metabolism in mitochondriate and amitochondriate protists. Mol. Biol. Evol. *18*, 710–720.

Ruzicka, F.J., and Beinert, H. (1977). A new iron-sulfur flavoprotein of the respiratory chain. A component of the fatty acid beta oxidation pathway. J. Biol. Chem. *252*, 8440–8445.

Silberman, J.D., Sogin, M.L., Leipe, D.D., and Clark, C.G. (1996). Human parasite finds taxonomic home. Nature *380*, 398.

Stechmann, A., Hamblin, K., Pérez-Brocal, V., Gaston, D., Richmond, G.S., van der Giezen, M., Clark, C.G., and Roger, A.J. (2008) Organelles in *Blastocystis* that blur the distinction between mitochondria and hydrogenosomes. Curr. Biol. *18*, 580–585.

Stensvold, C.R., Alfellani, M.A., Nørskov-Lauristen, S., Prip, K., Victory, E.L., Maddox, C., Nielsen, H.V., and Clark, C.G. (2009). Subtype distribution of *Blastocystis* isolates from synanthropic and zoo animals and identification of a new *Blastocystis* sp. subtype. Int. J. Parasitol. 39, 473–479.

Stensvold, C.R., Suresh, G.K., Tan, K.S., Thompson, R.C.A., Traub, R.J., Viscogliosi, E., Yoshikawa, H., and Clark, C.G. (2007). Terminology for *Blastocystis* subtypes – a consensus. Trends Parasitol. 23, 93–96.

Tan, K.S., Singh, M., and Yap, E.H. (2002). Recent advances in *Blastocystis hominis* research: hot spots in *terra incognita*. Int. J. Parasitol. 32, 789–804.

Tovar, J., Fischer, A., and Clark, C.G. (1999). The mitosome, a novel organelle related to mitochondria in the amitochondrial parasite *Entamoeba histolytica*. Mol. Microbiol. 32, 1013–1021.

Tovar, J., León-Avila, G., Sánchez, L., Sutak, R., Tachezy, J., van der Giezen, M., Hernández, M., Müller, M., and Lucocq, J.M. (2003). Mitochondrial remnant organelles of *Giardia* function in iron-sulphur protein maturation. Nature 426, 172–176.

Voisine, C., Cheng, Y.C., Ohlson, M., Schilke, B., Hoff, K., Beinert, H., Marszalek, J., and Craig, E.A. (2001). Jac1, a mitochondrial J-type chaperone, is involved in the biogenesis of Fe/S clusters in *Saccharomyces cerevisiae*. Proc. Natl. Acad. Sci. U.S.A. *98*, 1483–1488.

Voisine, C., Schilke, B., Ohlson, M., Beinert, H., Marszalek, J., and Craig, E.A. (2000). Role of the mitochondrial Hsp70s, Ssc1 and Ssq1, in the maturation of Yfh1. Mol. Cell Biol. *20*, 3677–3684.

Waller, R.F., Jabbour, C., Chan, N.C., Celik, N., Likic, V.A., Mulhern, T.D., and Lithgow, T. (2009). Evidence of a reduced and modified mitochondrial protein import apparatus in microsporidian mitosomes. Eukaryot. Cell *8*, 19–26.

Wawrzyniak, I., Roussel, M., Diogon, M., Couloux, A., Texier, C., Tan, K.S., Vivares, C.P., Delbac, F., Wincker, P., and El Alaoui, H. (2008). Complete circular DNA in the mitochondria-like organelles of *Blastocystis hominis*. Int. J. Parasitol. *38*, 1377–1382.

Yarlett, N., Orpin, C.G., Munn, E.A., and Greenwood, C. (1986). Hydrogenosomes in the rumen fungus *Neocallimastix patriciarum*. Biochem. J. *236*, 729–739.

Zierdt, C.H. (1986). Cytochrome-free mitochondria of an anaerobic protozoan: *Blastocystis hominis*. J. Protozool. 33, 67–69.

Zierdt, C.H. (1991). *Blastocystis hominis*: past and future. Clin. Microbiol. Rev. *4*, 61–79.

Zierdt, C.H., Donnolley, C.T., Muller, J., and Constantopoulos, G. (1988). Biochemical and ultrastructural study of *Blastocystis hominis*. J. Clin. Microbiol. *26*, 965–970.

Zierdt, C.H., Rude, W.S., and Bull, B.S. (1967). Protozoon characteristics of *Blastocystis hominis*. Am. J. Clin. Pathol. *48*, 495–501.

Index

A

Actin
- *Entamoeba histolytica* 82–83, 86–87
- *Giardia* 128
- *Trichomonas* 144–145

Actin-binding proteins
- *Entamoeba histolytica* 84–85, 92–95
- *Giardia* 128–130
- *Trichomonas* 144

Actin-related proteins
- *Entamoeba histolytica* 83, 87–90
- *Giardia* 128–130
- *Trichomonas* 144–146

Amoebiasis 1

B

Blastocystis
- morphology 206
- subtypes 205

Blastocystis mitochondrion-like organelle
- carrier proteins 211
- genome 206–208
- iron sulphur clusters 208 210
- metabolism 208, 210–215
- protein import 210–211

C

Comparative genomics
- *Entamoeba* spp. 10–14
- meiosis 63–65

D

Diplomonads 23
- codon usage 29–30
- life cycle 25–26
- morphology 24–26
- phylogeny 24–25

E

Entamoeba histolytica
- cytoskeleton *see Entamoeba histolytica* cytoskeleton
- genetic manipulation *see Entamoeba histolytica* genetic manipulation
- genome *see Entamoeba histolytica* genome
- microarrays *see Entamoeba histolytica* microarrays
- motility 106–108
- phagocytosis 108–111
- phospholipids 104–106
- prevalence 1
- protein kinases 8
- proteomics 158, 169–170
- small GTPases *see Entamoeba histolytica* small GTPases
- tRNA genes 4
- web resources 14–20

Entamoeba histolytica cytoskeleton
- actin 82–83, 86–87
- actin-binding proteins 84–85, 92–95
- actin nucleators 90–92
- actin-related proteins 83, 87–90
- myosin 85, 95–96

Entamoeba histolytica genetic manipulation
- antisense 177–179
- dominant-negative 179–181
- epigenetic gene silencing 183–187
- over-expression 176–177
- RNAi 181–183

Entamoeba histolytica genome
- characteristics 2, 3, 8
- HAPPY mapping 9

Entamoeba histolytica microarrays 157–169
- amoebic colitis 158–161
- amoebic liver abscess 159, 161–162
- DNA methylation 168
- encystation 159, 162–165
- stress response 167–168
- transcriptional pathways 168–169
- virulence 164–166

Entamoeba histolytica small GTPases 96–99
- effectors 103–104
- regulators 99–103

Entamoeba spp.
- life cycle 1
- comparative genomics 10–14
- genomes cytoskeleton *see Entamoeba* spp. genomes
- transposable elements 4–7

Entamoeba spp. genomes
- assembly 2–4

G+C content 3. 13
introns 3, 13

G

Giardia
assemblages 27
attachment 135–136
cell cycle 31
cell division 136–137
chromosomes 29, 194–519, 197
classification 27–28
cytoskeleton *see Giardia* cytoskeleton
encystation 137–138
genome cytoskeleton *see Giardia* genome
genotyping 28
high-cysteine membrane proteins (HCMps) 37–38
histones 197
history 23
life-cycle 26–27, 121
microtubule organizing centre 126–128
mitosis 136–137
morphology 26–27
motility 135
mRNA 34–5
nuclear replication 193–194
nuclear structure 194
ploidy 29, 195–197
protein kinases 38
sexuality 197–202
small RNAs 35–6
sterile transcripts 33
variant-specific surface proteins (VSPs) 36–37
web resources 24, 121
Giardia cytoskeleton 122–138
actin 128
actin-binding proteins 128–130
actin-related proteins 128–130
coiled-coil proteins 134
flagella 123–124
giardin 38–40, 132–134
median body 124–125
microtubule-associated proteins 126–127
motor proteins 130–132
tubulins 125–127
ventral disc 124
Giardia genome
database 24
G+C content 29–30
introns 34–35
Giardiasis 27, 121

L

Lateral gene transfer 31–32

P

Parabasalids 45
Protein kinases
Entamoeba histolytica 8
Giardia 38
Trichomonas 52, 54, 56–57, 139

S

Spironucleus salmonicida genome survey 24
Stramenopiles 205

T

Trichomonads 45–46
Trichomonas
Trichomonas vaginalis
autophagy 70–71
classification 45
coat proteins 68–9
hydrogenosome 45–46, 53
meiosis 61–63
membrane trafficking 65–71
metabolism 51–52
microtubule organising centre 143–144
mitosis 148
pathogenesis 52–53
prevalence 46
protein kinases 52, 54, 56–57, 139
small GTPases 66–68
small RNAs 56, 58–60
transcription 54–56
transposable elements 48–51
tRNA genes 51
web resources 71–74
Trichomonas vaginalis cytoskeleton 138–148
actin 144–145
actin-related proteins 144–146
coiled-coil proteins 146–148
costa 140–141
flagella 139–140
microtubule-associated proteins 141–143
motor proteins 146–147
pelta and axostyle 139–140
tubulins 141–142
Trichomonas vaginalis genome
assembly 48–49
characteristics 49, 138–139
G+C content 56, 58
history 47–48
introns 51
repeated sequences 48–51
Trichomoniasis 46–47
Tubulin post-translational modification
Giardia 125–127
Trichomonas 143